Wind Turbines

Wind Turbines

Topical Collection Editors

Frede Blaabjerg
Elizaveta Liivik

MDPI • Basel • Beijing • Wuhan • Barcelona • Belgrade

MDPI

Topical Collection Editors
Frede Blaabjer
Aalborg University
Denmark

Elizaveta Liivik
Tallinn University of Technology
Estonia

Editorial Office
MDPI
St. Alban-Anlage 66
Basel, Switzerland

This is a reprint of articles from the Topical Collection published online in the open access journal *Energies* (ISSN 1996-1073) from 2016 to 2017 (available at: https://www.mdpi.com/books)

For citation purposes, cite each article independently as indicated on the article page online and as indicated below:

LastName, A.A.; LastName, B.B.; LastName, C.C. Article Title. *Journal Name* **Year**, *Article Number*, Page Range.

ISBN 978-3-03897-360-7 (Pbk)
ISBN 978-3-03897-361-4 (PDF)

Cover image courtesy of Frede Blaabjerg.

Contents

Topic (1): Wind Prediction and Aerodynamics

Topic (2): Reliability and Fault Diagnosis

Topic (3): Off Shore Wind Farms

Topic (4): Energy System Integration Including Smart Grid

About the Topical Collection Editors

Frede Blaabjerg (S'86–M'88–SM'97–F'03) was with ABB-Scandia, Randers, Denmark, from 1987 to 1988. From 1988 to 1992, he got the PhD degree in Electrical Engineering at Aalborg. He became an assistant professor in 1992, an associate professor in 1996, and a full professor of Power Electronics and Drives in 1998. From 2017 he became a Villum investigator. He is honoris causa at the University Politehnica Timisoara (UPT), Romania and Tallinn Technical University (TTU) in Estonia. His current research interests include power electronics and its applications, such as in wind turbines, PV systems, reliability, harmonics, and adjustable speed drives. He has published more than 500 journal papers in the field of power electronics as well as its applications. He is the co-author of two monographs and the editor of seven books on power electronics and its applications. He has received 26 IEEE Prize Paper Awards, the IEEE PELS Distinguished Service Award in 2009, the EPE-PEMC Council Award in 2010, the IEEE William E. Newell Power Electronics Award 2014, and the Villum Kann Rasmussen Research Award 2014. He was the editor-in-chief of the IEEE Transactions on Power Electronics, from 2006 to 2012. He has been a distinguished lecturer for the IEEE Power Electronics Society from 2005 to 2007 and for the IEEE Industry Applications Society from 2010 to 2011, as well as from 2017 to 2018. In 2018, he became the president elect of the IEEE Power Electronics Society. He was nominated in 2014, 2015, 2016, and 2017 by Thomson Reuters as one of the 250 most cited researchers in engineering in the world.

Elizaveta Liivik (SM'18) received Dipl.-Eng, M.Sc. and Ph.D. degrees in Electrical Engineering from the Department of Electrical Drives and Power Electronics, Tallinn University of Technology, Tallinn, Estonia, in 1998, 2000, and 2015, respectively. She is currently a Researcher at the Department of Electrical Engineering, Tallinn University of Technology. From 2002 to 2007, she was a lecturer in the Department of Electrical Drives and Power Electronics, Tallinn University of Technology. She is the currently Guest Postdoctoral Researcher in Aalborg University from 2017. Her main research interests include impedance-source power electronic converters, renewable energy and distributed generation, as well as control and reliability issues of power electronic converters in active distribution networks. She has authored or co-authored more than 40 research papers and 1 book.

Preface to "Wind Turbines"

Wind energy occupies a leading place among renewable energy sources for electric power system generation. The cumulative installations of wind turbines are continuously growing, and last year the cumulative capacity increased by nearly 11%, to around 539 GW [Wind Energy Systems. *Proceedings of the IEEE, 2017*], and it is expected to transcend 760 GW by 2020 [Renewables 2018 Global Status Report—REN21]. Today, the global installed renewable generation capacity has passed 2000 GW, where hydro-power counts for around 1000 GW. Moreover, in 2017, the offshore wind power sector had its best year yet and its total capacity increased by 30%. This constant increase in the rapid pace of wind power is due to significant technological achievements in recent years, and it has made it possible to lower the cost of energy dramatically from that coming from wind turbines. The size of wind turbines continues to increase, and several manufacturers have announced plans to produce wind turbines of 10 MW and larger [Wind Energy Systems. Proceedings of the IEEE, 2017]. All of the above is achieved through technological advancements, including advanced wind speed prediction; advanced control methods; power electronics and its control; advanced manufacturing techniques; new materials, but also fault detection and diagnostic methods, which can provide a high level of reliability; availability; maintainability; and safety for the wind turbines. The competition between the different renewable energy sources is intense—photovoltaic power is constantly pushing the limits of lowering the cost of energy—and thereby also challenges wind power technology to come up with better and more cost-effective solutions.

Wind power is very multi-disciplinary in terms of subjects, ranging from atmospheric physics, aerodynamics, material science and technology, foundations, the wind power conversion technology itself, to how to integrate it into the overall energy system. New and better solutions might appear in the different specialized disciplines, as well across the disciplines.

The Contributions in This Book

This Special Issue on wind power has collected some of the most promising advancements presented in selected papers from *Energies* for the last two years into a book, where the focus of the selected papers has been on fault prediction and reliability, energy system integration, wind power generation forecasting methods, and off shore wind farms.

In particular, the following topics have been chosen by grouping together the papers:
(1) Wind prediction and aerodynamics (six papers);
(2) Reliability and fault diagnosis (three papers);
(3) Off shore wind farms (three papers);
(4) Energy system integration including smart grid (four papers)

Topic (1): Wind Prediction and Aerodynamics

In the first paper, a procedure has been developed to analyze and predict the wind speed by using standard meteorological variables. The authors are using traditional statistical techniques, like the ARIMA model, and are then using a multivariate artificial neural network technique, as follows: thereby the NARX model is proposed. The paper describes the wind speed predictions given by both models, including analyzing and comparing them qualitatively and quantitatively with a number of measured data.

Spatio-temporal (multi-channel) linear models are explored in the second paper, where the neighboring measurements around the target location are used and investigate the short-term wind speed forecasting problem. Clear definitions of the problems of the multi-channel

ARMA models (also called MARMA) are presented, and efficient multi-channel prediction coefficient estimation techniques are proposed. The important result is that the proposed multi-channel linear model can predict the Δ hour wind speed value in milliseconds, by using an ordinary desktop computer, which is suitable for very short term (in seconds) wind speed forecasting.

Predicting the wind energy using a hybrid model has been proposed in the third paper. The authors used EEMD technology to decompose the original wind power generation series. Then, a principal component (PCA) was applied to select the most important modelling inputs; five significant variables were selected from nine available inputs. Thus, the proposed method has been demonstrated to be a credible and promising algorithm for wind power generation prediction.

In the fourth paper, a vertical axis wind turbine (VAWT) was demonstrated to have some potential for being a reliable means of wind energy extraction compared with the conventional horizontal axis wind turbine (HAWT) system. The authors used a previously validated large-eddy simulation framework, in which an actuator linear model was employed to parameterize the blade forces on the flow, thereby being able to simulate the atmospheric boundary layer flow for stand-alone VAWTs placed on a flat terrain.

A methodology for wind energy has been presented in the fifth paper, which allows for assessing the statistical annual wind energy yield (AEY) using a high spatial resolution (50 m × 50 m) grid in an area with a mosaic-like land cover pattern, as well as complex topography. It is further based and validated on a long-term (1979–2010) near-surface wind speed time series measured at 58 stations of the German Weather Service (DWD).

In the sixth paper, the authors focus on the analysis of an innovative, extensible blade technology that aims to utilize wind energy in areas with low-class wind resources. A computational model and method is developed based on the blade element momentum (BEM) theory, which determines the aerodynamic load and the output power of the blade at different wind conditions.

Topic (2): Reliability and Fault Diagnosis

In the seventh paper, an interesting problem is discussed regarding the fault detection and diagnosis (FDD) of the wind turbine blades. The idea is to use macro-fiber composites to detect cracks in the blades in a structural health monitoring (SHM) system. This approach, based on non-destructive testing (NDT), automatically identifies and locates failure by using an acoustic emission source coming from a fiber's breakage in a wind turbine blade section, by applying a novel signal processing method.

In the eighth paper, the principal component analysis (PCA) method is used as a way to condense and extract information from a number of collected signals from the turbine. The objective is focused on the development of a wind turbine fault detection strategy, which combines a data driven baseline model with a reference pattern obtained from a healthy structure. This is all based on PCA and as well as hypothesis testing.

The authors of the ninth paper investigated a new fault diagnosis scheme, which is composed of multiple extreme learning machines (ELM) in a hierarchical structure, where a forwarding list of ELM layers is concatenated, and each of them is processed independently for its corresponding role. The framework is successfully applied to recognize the fault patterns coming from the wind turbine generator system.

Topic (3): Off Shore Wind Farms

In the tenth paper, the goal is to optimize the maintenance management of wind farms through the estimation of the fault probability of each wind turbine. In order to evaluate it qualitatively, a fault tree analysis (FTA) method of wind turbines (WT) is applied using a binary decision diagram (BDD). The approach is based on the fault probabilities of each component of the WT, which depends on the statistical function of the probability of occurrence over time. The fault probability of the WT has been set using the Boolean expression, which was obtained by the BDD.

The application of optimal coordinated control was investigated in the eleventh for a finite-sized wind farm using large eddy simulations, extending the work done by Goit and Meyers into a regime where the entrance effects are important in order to increase the total energy extraction in wind farms. The individual wind turbines are considered as flow actuators, and their energy extractions are dynamically regulated in time, so they are optimally influenced by the wind flow field.

In the twelfth paper, new research is presented indicating that logistics make up to 18% of the levelized cost of energy (LCoE) for offshore wind power plants. This case study's findings, which conservatively show this number to be 18% of the LCoE, are based on the definition of logistics throughout the whole offshore wind farm (OWF) life-cycle. It uses the idea from the conceptualization and planning of the farm, through the construction, operations/service, and, finally, the de-commissioning/abandonment of the complete OWF site. This case study is timely and highly relevant from different perspectives of society, such as policy, governance, academic, and practitioner.

Topic (4): Energy System Integration Including Smart Grid

In the thirteenth paper, an enhanced hybrid approach to forecast the electricity market price (EMP) signals is proposed, which is composed of an innovative combination of wavelet transform (WT), differential evolutionary particle swarm optimization (DEEPSO), and the adaptive neuro-fuzzy inference system (ANFIS) used in different electricity markets. The geographical case is the wind power in Portugal, which, in the short-term only consider the historical data.

In the fourteenth paper, the authors use the EFI's (Norwegian Electric Power Research Institute) multi-area power market simulator (EMPS) model to simulate the Nordic energy market, and shows that increasing the wind power capacity in Mid-Norway can reduce the energy balance deficit. The deficit becomes almost nil during high a consumption/price period (i.e., in winter), although the deficit remains important at a yearly time scale.

A combined heat and power dispatch model considering both the dynamic thermal performance (PDTP) of the pipelines and the buildings' thermal inertia (BTI) is discussed in the fifteenth paper (abbreviated as the CPB-CHPD model), emphasizing the importance of a coordinated operation between the electric power and the district heating systems, in order to break the strong coupling without impacting the end users' heat supply quality.

In the sixteenth paper, a Demand Side Management (DSM) controller is designed, where five different heuristic algorithms—the genetic algorithm (GA), the binary particle swarm optimization algorithm (BPSO), the wind-driven optimization algorithm (WDO), the bacterial foraging optimization algorithm (BFOA), and the proposed hybrid genetic wind-driven algorithm (GWD)—are evaluated. These algorithms were used for scheduling the residential loads between peak hours (PHs) and off-peak hours (OPHs) in a real-time pricing (RTP) environment, and by maximizing the user comfort (UC) and minimizing both the electricity cost and the peak to average ratio (PAR). They were tested in the following two ways: scheduling the load of a single home and scheduling the load of multiple homes.

Frede Blaabjerg and Elizaveta Liivik
Topical Collection Editors

energies

MDPI

Article

Wind Speed Prediction Using a Univariate ARIMA Model and a Multivariate NARX Model

Erasmo Cadenas [1], Wilfrido Rivera [2,†], Rafael Campos-Amezcua [2,*,†] and Christopher Heard [3,†]

[1] Facultad de Ingenieria Mecanica, Universidad Michoacana de San Nicolas de Hidalgo, Santiago Tapia No. 403, Col. Centro, CP 58000 Morelia, Michoacan, Mexico; ecadenas@umich.mx

[2] Instituto de Energias Renovables, Universidad Nacional Autonoma de Mexico, Apartado postal 34, CP 62580 Temixco, Morelos, Mexico; wrgf@ier.unam.mx

[3] Division de Ciencias de la Comunicacion y Diseno, Departamento de Teoria y Procesos del Diseno, Diseno Ambiental, Universidad Autonoma Metropolitana Unidad Cuajimalpa, Torre III, 5to. piso, Av. Vasco de Quiroga 4871, Col. Santa Fe Cuajimalpa, Del. Cuajimalpa, Mexico D.F. 11850, Mexico; cheard@correo.cua.uam.mx

* Correspondence: rca@ier.unam.mx; Tel.: +52-777-362-0090 (ext. 38010)

† These authors contributed equally to this work.

Academic Editor: Guido Carpinelli
Received: 17 June 2015; Accepted: 22 January 2016; Published: 17 February 2016

Abstract: Two on step ahead wind speed forecasting models were compared. A univariate model was developed using a linear autoregressive integrated moving average (ARIMA). This method's performance is well studied for a large number of prediction problems. The other is a multivariate model developed using a nonlinear autoregressive exogenous artificial neural network (NARX). This uses the variables: barometric pressure, air temperature, wind direction and solar radiation or relative humidity, as well as delayed wind speed. Both models were developed from two databases from two sites: an hourly average measurements database from La Mata, Oaxaca, Mexico, and a ten minute average measurements database from Metepec, Hidalgo, Mexico. The main objective was to compare the impact of the various meteorological variables on the performance of the multivariate model of wind speed prediction with respect to the high performance univariate linear model. The NARX model gave better results with improvements on the ARIMA model of between 5.5% and 10.6% for the hourly database and of between 2.3% and 12.8% for the ten minute database for mean absolute error and mean squared error, respectively.

Keywords: wind speed prediction; NARX; ARIMA; multivariate analysis

1. Introduction

At the end of 2014, the worldwide installed wind energy generating capacity was 369,597 MW; Europe having 134,007 MW, of which Germany and Spain stood out with 39,165 and 22,987 MW, respectively. During 2015, 42% of electric power in Denmark was generated from wind [1]. In the Asia-Pacific region, China had a reported capacity of 114,609 MW of a total of 141,964 MW. In North America, the reported U.S. installed capacity was 65,879 MW with the Mexican and Canadian installed capacities being 9694 and 2551 MW, respectively. In Latin America, Brazil was the leader, with 5939 MW of 8526 MW total [2].

Onshore wind based power generation has reached the technological maturity of being competitive with the lowest cost power generation options in many places. For example in Mexico in 2012 installed capacity increased by 76% with respect to the total installed wind energy generation capacity at the end of 2011 due to increasing exploitation of the intense resource in the state of Oaxaca.

In Oaxaca in the corridor from La Venta to La Mata passing through La Ventosa, the annual average wind speed is over 9 m/s at 30 m above ground level with a dominant wind direction of North-Northwest/North-Northeast 70% of the time [3]. These highly favorable intense wind conditions in Oaxaca represent an appreciable source of inexpensive renewable energy in addition to Mexico's large fossil fuel reserves, which makes its exploitation a priority.

In Mexico, the National Center for Energy Control (CENACE) is responsible for the dispatch control of energy for the National Electric System. CENACE uses an information system to prepare pre-dispatch strategies. This system takes into account: availability, derating, restrictions and other factors that affect the dispatch capacity of generating units, as well as the electricity demand forecast. These models are produced by CENACE. An hourly operation plan is essential for each unit [4]. Energy producers have a responsibility to provide forecasts of wind and net energy production to CENACE a day ahead.

Recently, a considerable number of wind speed prediction models have been developed using a range of methods, some simple and others combining various techniques. Cadenas and Rivera [5] have reported short-term wind speed forecasting in a region of Oaxaca using an artificial neural network (ANN) with a representative hourly time series for the site. The model showed good accuracy for energy supply prediction.

Salcedo-Sanz *et al.* [6] presented a hybrid model between a fifth generation mesoscale model (MM5) and a neural network for short-term wind speed prediction at specific points.

Cadenas *et al.* [7] analyzed and forecasted wind velocity in Chetumal, Quintana Roo, Mexico, with a single exponential smoothing method. The method was found to be good for wind forecasting when the field data had alpha values close to one.

Li and Shi [8] compared three artificial neural networks for wind speed forecasting. These were: adaptive linear element, back propagation and radial basis function. None of these outperformed the others on all of the metrics evaluated.

A new short-term hybrid method based on wavelet and classical time series analysis to predict wind speed and power was proposed by Liu *et al.* [9]. The mean relative error in multi-step forecasting using this method was smaller than that from classical time series and back-propagation network methods.

A wind speed forecasting model for three regions of Mexico was developed using a hybrid autoregressive integrated moving average technique (ARIMA-ANN) by Cadenas and Rivera [10]. Initially, the ARIMA models were used to generate wind speed forecasts for the time series. The resulting errors were used to build the ANN to account for the non-linear behavior that the ARIMA technique could not model. This reduced the errors. The results showed that the hybrid model produced higher accuracy wind speed predictions than those of the separate ARIMA and ANN models for all three sites.

Kavasseri and Seetharaman [11] used the fractional-ARIMA models to predict wind speed and power production one or two days ahead for North Dakota. Forecasting errors in wind speed and power were compared to the persistence model. Significant improvements were obtained.

Li *et al.* [12] presented a robust two-step method for accurate wind speed forecasting based on a Bayesian combination algorithm and three neural network models: an adaptive linear element network (ADALINE), back propagation (BP) and a radial basis function (RBF). The results were that the neural networks were not consistent for one hour ahead wind speed. However, the Bayesian combination method could always give adaptive, reliable and comparatively accurate forecasts.

Liu *et al.* [13] evaluated the effectiveness of autoregressive moving average-generalized autoregressive conditional heteroscedasticity (ARMA-GARCH) for modeling mean wind speed and its volatility. The results showed that ARMA-GARCH could capture the trend changes of these parameters. In this study, it was found that none of the models were consistently better than the others over the whole range of heights considered. The authors recommended that for a given dataset, all of the models should be evaluated to find the most appropriate (Given that the range of heights considered was from 10 to 80 m and that the swept diameter of wind turbines is of the order of 90 m centered at a height of 70 m for 3-MW units, the heights covered by the study needed to be greater).

Guo *et al.* [14] developed an empirical mode decomposition (EMD) based feed-forward neural network (FFNN) learning model, which resulted in improved accuracy over each of the two methods individually for predicting daily and monthly mean wind speeds.

Liu *et al.* [15] have proposed hybrid ARIMA-ANN and ARIMA-Kalman methods for hourly wind speed forecasting. The authors concluded that both methods gave good results and can be applied to dynamic wind speed forecasting for wind power systems.

Kalman filtering was optimized for application to very short-term wind forecasting and applied to wind energy for a site at Varese Ligure in Italy by Cassola and Burlando [16]. A numerical meteorological model BOLAM (Bologna Limited Area Model) was used, and the results with the application of Kalman filtering showed a considerable reduction in error.

On the basis of parameter selection and data decomposition, two combined strategies and four modified models based on the first-order and second-order adaptive coefficient (FAC and SAC) were proposed by Zhang *et al.* [17] for wind speed forecasting in four different sites in China. It was shown that the approaches derived from the combined strategies obtained higher prediction accuracy than the individual FAC and SAC models at the four sampled sites.

A hybrid model based on the EMD and ANNs named EMD-ANN for wind speed prediction was proposed by Liu *et al.* [18]. The results were compared to an ANN model and an autoregressive integrated moving average model. These showed that the performance of the model was very good compared to the individual methods.

Two prediction methods were studied by Peng *et al.* [19] for short-term wind power forecasting on a wind farm. Three key factors were used in the models: temperature, wind speed and direction. One was an artificial neural network and the other a hybrid model based on physical and statistical methods. The hybrid model produced higher accuracy results than the individual ANN model.

Chen and Yu [20] developed a hybrid model that integrated a support vector regression (SVR)-based state-space model with an unscented Kalman filter (UKF). This was to predict short-term wind speed sequences. The results gave much better performance for both one step and multi-step ahead wind speed forecasts than support vector machine, autoregressive and ANNs.

Hocaoglu *et al.* [21] developed a model for the artificial prediction of wind speed data, from atmospheric pressure measurements using the hidden Markov models (HMMs) technique. The model accuracy was evaluated from Weibull distribution parameters. The relevance of the technique is in its use of an additional meteorological variable (atmospheric pressure).

Hocaoglu *et al.* [22] used the Mycielski algorithm for wind speed forecasting. The algorithm performs a prediction using the total exact history of the data samples. The basic idea of the algorithm was to search for the longest suffix string at the end of the data sequence that had been repeated at least once in the history of the sequence. It was concluded that the model was robust for different behaviors of wind speed patterns. Experimental results also showed that the model not only provided very consistent time variations in agreement with the actual measured data, but also provides accurate distribution model parameters for estimating the wind power potential of a region.

Of all of the models reviewed, only two use additional meteorological parameters other than wind speed, such as pressure and temperature.

In this work, wind speed forecasting for La Mata, Oaxaca, and Metepec, Hidalgo, was carried out using univariate and multivariate models. To achieve higher accuracy forecasts, wind speed models using non-linear auto-regressive exogenous (NARX) modeling were developed. This technique uses additional exogenous variables (*i.e.*, other than wind speed) to generate more accurate forecasts with respect to ARIMA models solely based on wind speed time series. The meteorological variables used in this study were: wind speed and direction, solar radiation, temperature and pressure. In the generation of the NARX model, only solar radiation or relative humidity was used due to the results from a correlation study.

2. Experimental Data

Two weather databases were used as sources of information to allow the wind speed prediction models to be tested under a range of conditions. The time series of the variables used in this analysis are shown in Figures 1–6.

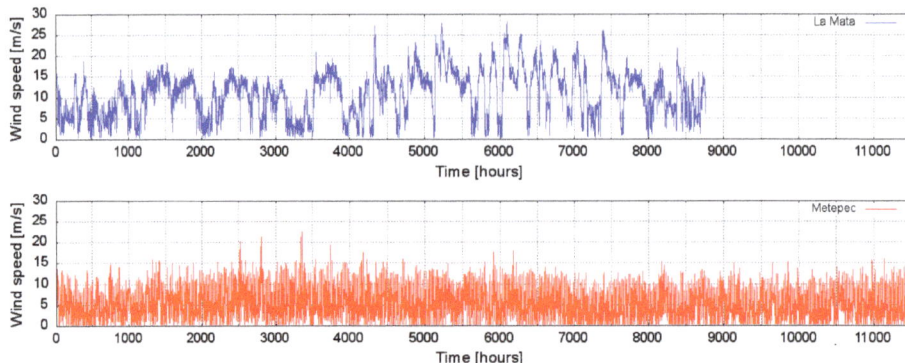

Figure 1. Wind speed time series from La Mata, Oaxaca (hourly averages), and Metepec, Hidalgo (ten-minute averages).

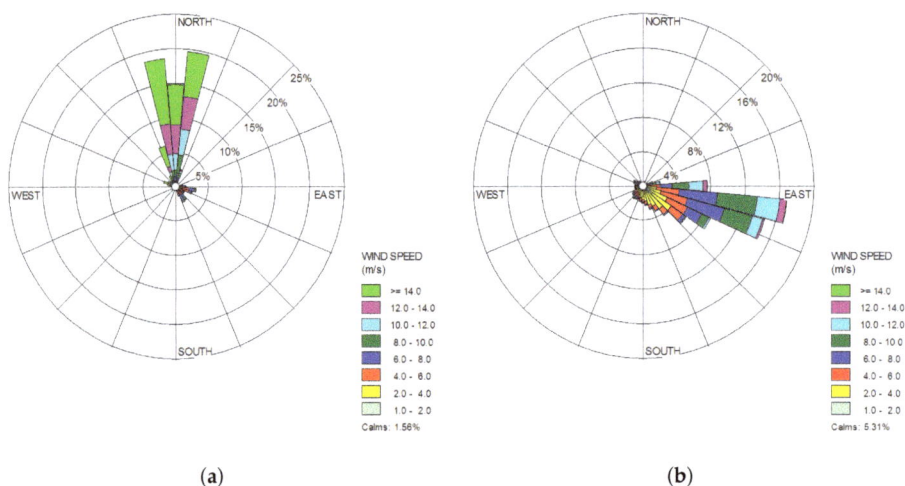

(a)   (b)

Figure 2. Wind rose of the studied sites. (a) La Mata, Oaxaca; (b) Metepec, Hidalgo.

One data set was from observations in the town of La Mata in the state of Oaxaca, Mexico. This was provided by the Mexican Federal Electricity Commission (Comision Federal de Electricidad (CFE)) and has 8759 data points corresponding to one year of hourly averaged measurements taken at 40 m above ground level from 1 May 2006 to 30 April 2007. The measurements were: wind speed, WS (m/s); wind direction, WD (°); barometric pressure, P (mbar), air temperature, T (°C), and solar radiation, SR (W/m^2).

The other data set was from observations in the town of Metepec in the state of Hidalgo, Mexico. This was provided also by the CFE and has 68,550 data points which correspond to just over a year and three months of ten minutely averaged measurements. The measurements were made at a height of 50 m above ground level from 22 November 2007 to 12 March 2009. The measurements were: wind speed, WS (m/s); wind direction, WD (°); barometric pressure, P (mbar), air temperature, T (°C), and relative humidity, RH (%).

Figure 1 shows the measured wind speed for both stations. It can be appreciated that there are no tendencies for neither periodic nor seasonal wind speed variations in the time series. The average wind speeds are 10.9 and 5.2 m/s for La Mata and Metepec, respectively.

Figure 2 shows the wind rose for both sites, where 0°, 90°, 180° and 270° denote North, East, South and West directions, respectively. In the case of La Mata, Oaxaca, the dominant wind direction is from the South (S). It should be noted that the wind direction is in the range from 335° to 15° for 59.4%. Periods of calm (<1 m/s) represent 1.56% of the total sample. In the case of Metepec, Hidalgo, the dominant wind direction is from West–Northwest (W–NW). It should be noted that 53.4% of the time the wind directions is in the range from 85° to 135°. Periods of calm (<1 m/s) represent 5.31% of the total sample.

Figures 3 and 4 show hourly average air temperature and barometric pressure, respectively, for both sites. The solar radiation series shown in Figure 5 for the La Mata site of course shows a daily cycle. The relative humidity for the Metepec site is shown in Figure 6.

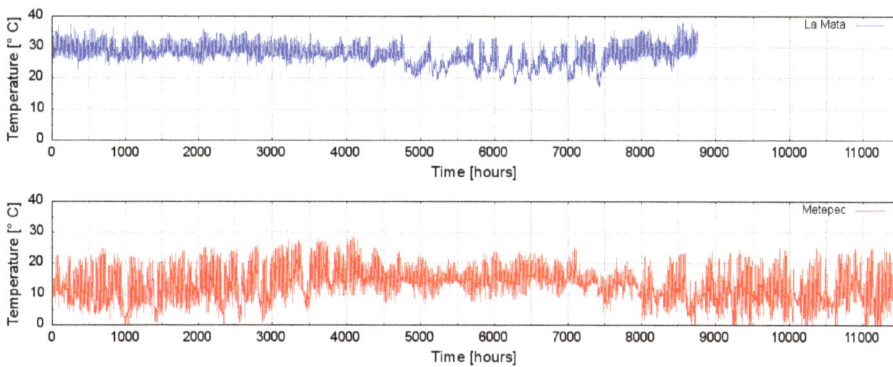

Figure 3. Air temperature time series from La Mata, Oaxaca (hourly averages) and Metepec, Hidalgo (ten-minute averages).

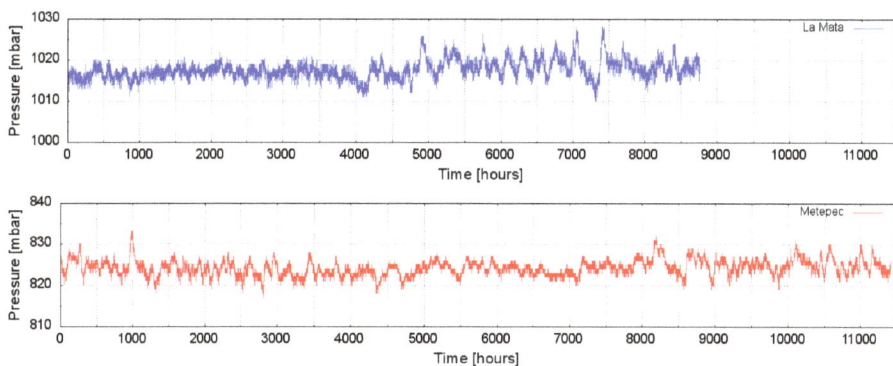

Figure 4. Barometric pressure time series from La Mata, Oaxaca (hourly averages), and Metepec, Hidalgo (ten-minute averages).

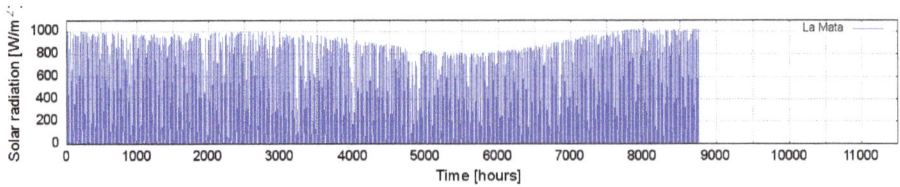

Figure 5. Solar radiation time series from La Mata, Oaxaca (hourly averages).

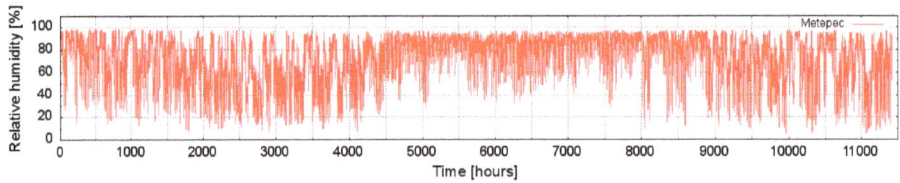

Figure 6. Relative humidity time series from Metepec, Hidalgo (ten-minute averages).

Tables 1 and 2 give the basic statistical characteristics (Central tendency and dispersion) for each of the measured variables for La Mata and Metepec respectively. The calculation of the mean wind direction and standard deviation requires special attention because the wind direction is a circular function resulting in a discontinuity (0°–360°), so that the arithmetic mean cannot be used. Therefore, the mean wind direction was calculated using the arctangent function of the averages of the sine and cosine of the wind directions data.

Table 1. Descriptive statistics of the involved variables of La Mata. WS, wind speed; WD, wind direction; T, temperature; P, pressure; SR, solar radiation.

Variable	Minimum	Maximum	Mean	Mode	Standard Deviation
WS (m/s)	0.4	28.3	10.9	13.6	5.5
WD (°)	0	360	4.8	9.5	36.5
T (°C)	17.3	37.9	27.6	27.2	3.4
P (mbar)	1010	1028	1017.6	1017	2.5
SR (W/m²)	0	1026	249.6	0	332.7
RH (%)	–	–	–	–	–

Table 2. Descriptive statistics of the involved variables of Metepec. RH, relative humidity.

Variable	Minimum	Maximum	Mean	Mode	Standard Deviation
WS (m/s)	0.4	19.3	5.2	0.4	3.2
WD (°)	6.6	355.9	113.3	103.3	50.5
T (°C)	−2.4	28.3	13	11.7	5.3
P (mbar)	817.2	833	824.2	824	2
SR (W/m²)	–	–	–	–	–
RH (%)	5.7	98.4	71.2	92.8	22.3

3. Time Series Models

A time series model (y_t) reproduces the patterns of the prior movements of a variable over time and uses this information to predict its future movements. It is possible in this way to construct a simplified model of the time series that represents its randomness, so that it is useful for prediction [23]. The present study uses univariate and multivariate techniques for wind speed prediction. The univariate method employs an autoregressive integrated moving average (ARIMA) model with only the wind speed as a variable. The multivariate method uses a non-linear autoregressive exogenous (NARX) model using wind direction, air temperature, barometric pressure, solar radiation and relative humidity, in addition to wind speed.

Multivariate analysis allows simultaneous consideration of diverse datasets allowing optimal decisions to be made considering all of the information.

3.1. Autoregressive Integrated Moving Average Models

ARIMA models have been used in a great number of time series prediction problems, because they are robust, as well as easy to understand and implement. However, difficulties exist with atypical values influencing the estimation of future values. A further disadvantage of stochastic models is generally their high order.

In the early 1970s, ARIMA models were popularized by Box and Jenkins [24], their names being associated with general ARIMA models applied to time series analysis and forecasting.

There are many ARIMA models. The non-seasonal general model is known as ARIMA(*p*,*d*,*q*), where:

AR:*p* = order of the autoregression of the model;

I:*d* = degree of differencing to make the model stationary;

MA:*q* = order of the moving average aspect of the model.

The linear expression to define the above notation is:

$$y_t = \sum_{i=1}^{p} \phi_i y_{t-i} + \sum_{j=1}^{p} \theta_j e_{t-i} + \epsilon_t \tag{1}$$

where ϕ_i for the purpose of stabilizing the variance, $_i$ is the *i*-th autoregressive parameter, θ_j is the *j*-th moving average parameter and ϵ_t is the error term at time *t*.

ARIMA models are used in a wide range of applications from engineering to economics. In cases such as the prediction of power demand, wind speed and stock market value behavior, that is things that can be represented as a time series with sufficient measurements, these can be modeled by this technique.

The Box-Jenkins method was followed to model the time series from La Mata and Metepec. This is basically a three-step iterative process: model identification, parameter estimation and diagnostic checking [24]:

1. Identification. Identification methods are approximate procedures applied to a dataset to find the kind of model worth further investigation. This involves determining suitable values for parameters *p* and *q* and determining the degree of differencing, *d*, to obtain stationarity. At this stage, graphs of the original and differenced time series together with their estimated autocorrelation and partial autocorrelation functions are useful tools.
2. Estimation. Having an initial model specification, its parameters are estimated from the maximum likelihood or conditional least squares methods. These are used iteratively starting from values estimated during the identification stage.
3. Diagnostic Checking. Having identified the model and estimated its parameters, diagnostic checks are used to reveal its inadequacies and indicate suitable improvements. Residuals and their autocorrelations are inspected. If the model is a good fit to the data, then the residuals would correspond to white noise and have very little autocorrelation.

Proposed ARIMA Models

As described above, in the Identification step, the data were differenced to obtain a stationary or trend-free series (Figure 7). A transformation of the original series was obtained to stabilize the mean and variance and to identify potential models from the autocorrelation function (ACF) and the partial autocorrelation function (PACF). At this stage of data preparation, it was determined whether or not it

7

should be transformed to stabilize the variance. In the following Estimation step, the parameters in potential models were calculated, and suitable criteria were used to select the best model (Figure 8). Finally, ACF and PACF were used to test the residuals as the Diagnostic Checking stage. Normality and "*t*" tests were applied to the residuals to find their closeness to white noise (Figures 9 and 10).

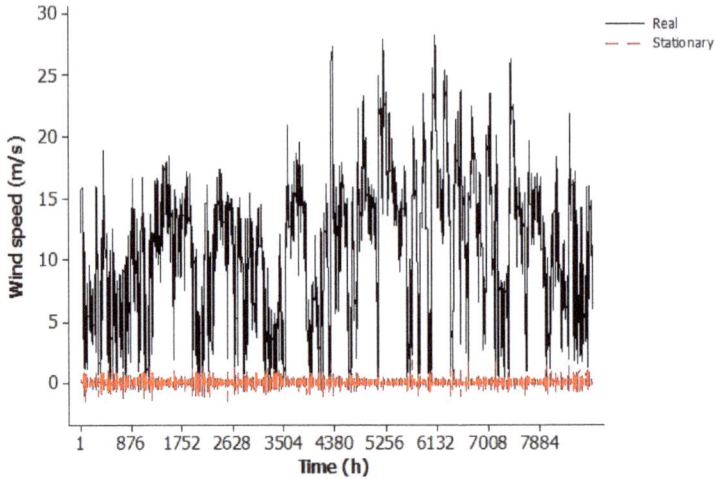

Figure 7. Real and stationary series from La Mata.

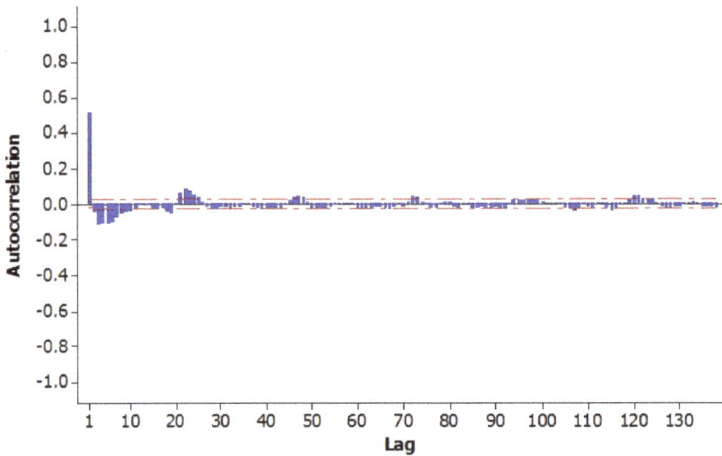

Figure 8. Autocorrelation function of the stationary series.

For La Mata and Metepec, the ARIMA models were ARIMA(1,1,0) and ARIMA(1,1,1), respectively. Table 3 shows the details of each of the two models. In the case of La Mata, the coefficient corresponding to the first term of equation: 1.1 y_{t-1} indicates the importance of the wind speed one hour earlier. This shows that the wind speed is persistent in this region. In the case of Metepec, the second term of the model apparently has a higher relevance: 0.6258 y_{t-2}; however, the third term, the error coefficient 0.660 e_{t-1}, which only involves a delay, has a similar order of magnitude. These two terms are thus of similar importance.

Table 3. ARIMA models for the two studied sites.

Site	ARIMA(p,d,q)	Model
La Mata	ARIMA$(1,1,0)$	$y_t = 1.1y_{t-1} - 0.1y_{t-2}$
Metepec	ARIMA$(1,1,1)$	$y_t = 0.3742y_{t-1} + 0.6258y_{t-2} + 0.6601e_{t-1} + e_t$

Histograms of the residuals between the models and the measured time series for La Mata and Metepec are shown in Figures 9 and 10, respectively. The histograms show the form of the residuals' distribution. It can be seen qualitatively that both cases are sharp normal distributions. The Metepec database has a lower scatter and higher frequencies because of the larger dataset. Both histograms are symmetrical, and this is borne out by the measures of central tendency, such as the mean, median and mode coinciding. This shows that there are no other patterns present in the data.

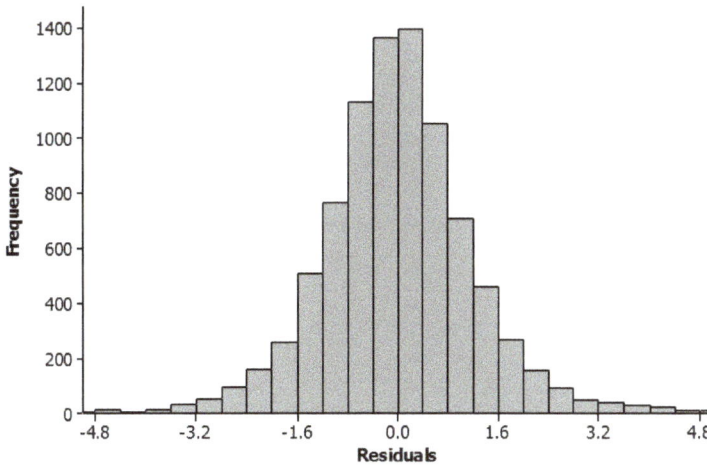

Figure 9. Error histogram for the ARIMA model in La Mata.

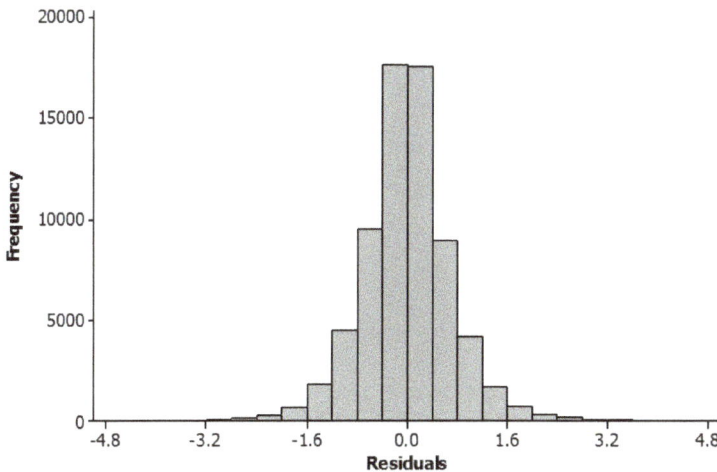

Figure 10. Error histogram for the ARIMA model in Metepec.

3.2. Nonlinear Autoregressive with Exogenous Inputs Models

The NARX model is a type of dynamically-driven recurrent ANN. Recurrent networks have one or more feedback loops, which can be either local or global. Global loops reduce the computational memory requirements. There are two basic uses for recurrent networks:

1. associative memory;
2. input-output mapping networks.

Two applications of input-output are signal modeling and prediction in the form of time series. The most obvious advantage of the NARX models is that the same structure makes up different models and it thus has a reasonable computation cost. Thus, a NARX network can gain degrees of freedom when it includes a time period forecast as an input for subsequent periods compared to a feedforward network. This allows summary information of exogenous variables to be included, as well as a lesser number of residuals, which reduces the number of parameters that have to be estimated.

NARX networks have a more effective learning process compared to other types of neural networks (the learning gradient descent is better). These networks converge, and generalization is improved compared to other types of networks [25].

Figure 11 shows the simplest architecture for a NARX model. The model in this case has only one input, which represents the value of the exogenous variables, which in turn provides feed-forward to a q number of delayed memory neurons. It has only one output $y(t+1)$, which represents the value of the predicted variable one step ahead. In other words, the output is one time unit ahead of the input. In turn, the output provides feedback to the network through a number of q delayed memory neurons.

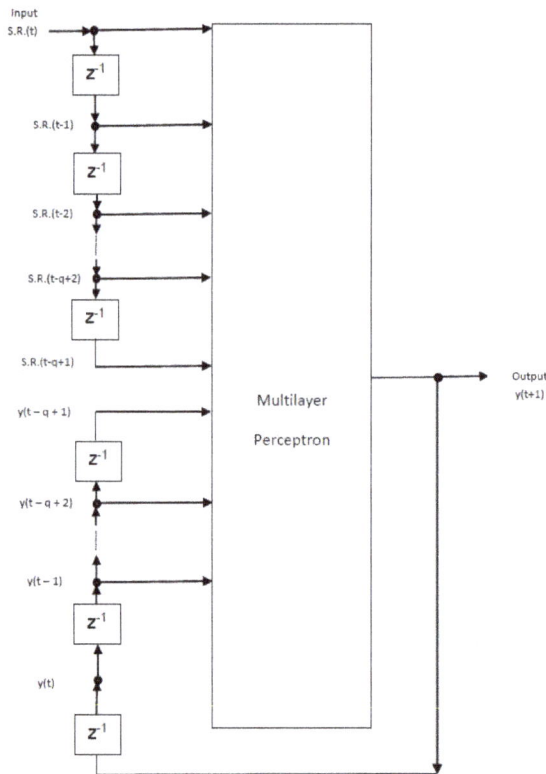

Figure 11. Architecture of the NARX.

These two lines make up the input neural layer of a multilayer perceptron [26]. The following expression describes the model's dynamic behavior:

$$y(n+1) = F\left(y(n), \cdots, y(n-q+1), u(n), \cdots, u(n-q+1)\right) \tag{2}$$

where F is a nonlinear function of its arguments.

3.2.1. Learning Algorithm for the NARX Network

The neurons in the NARX models are sigmoid, and the performance function used in the training of the ANN is the mean squared error (MSE). For the NARX network, it is replaced as follows:

$$MSE = \frac{1}{n}\sum_{i=1}^{n}(e_i)^2 = \frac{1}{n}\sum_{i=1}^{n}(t_i - y_i)^2 \tag{3}$$

$$MSW = \frac{1}{n}\sum_{j=1}^{n}(W_i)^2 \tag{4}$$

$$MSE_{reg} = \gamma \cdot MSE + (1-\gamma) \cdot MSW \tag{5}$$

where: t_i = target, γ = performance ratio and MSW = mean squared weight.

This performance function results in smaller weights and biases in the network and, thus, makes the response smoother and less likely to over fit. The training function that updates the weights and bias values uses Levenberg–Marquardt optimization, which was modified to include the regularization technique [27].

3.2.2. Proposed NARX Models

The NARX model that had the best forecast performance used five input variables with two delays per variable. These were wind speed and direction, air temperature, barometric pressure and solar radiation (for La Mata) or relative humidity (for Metepec). There were ten hidden neurons. The final configuration is shown in Figure 12.

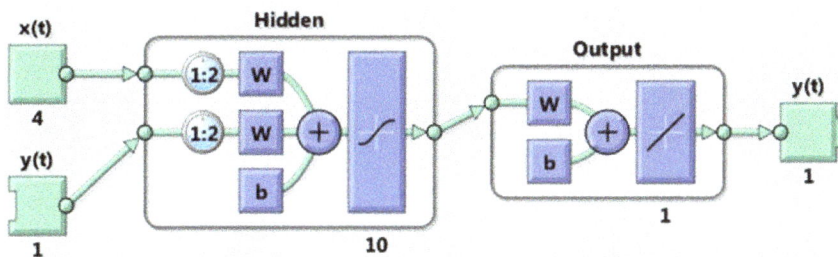

Figure 12. NARX generated in the MATLAB software.

3.3. Statistical Error Measures

The models' performance was evaluated via statistical error measurements. These were mean absolute error (MAE), mean squared error (MSE) and mean absolute percentage error (MAPE), described by the following expressions:

$$MAE = \frac{1}{n}\sum_{t=1}^{n}|e_t| \tag{6}$$

$$MSE = \frac{1}{n} \sum_{t=1}^{n} e_t^2 \tag{7}$$

Hindcasts are a way of evaluating the difference between model results and measurements. The MAE is a measure of the average of the absolute error whose advantage is that it is easier for non-specialists to understand. MSE is similar, but the values are all positive due to the squaring; this makes it easier to use in an optimization technique [28]. To achieve a higher degree of certainty when comparing models, the MAPE was calculated by means of the percentage error (PE_t) and the mean percentage error (MPE) using the following expressions:

$$PE_t = \left(\frac{Y_t - F_t}{Y_t} \right) \cdot 100 \tag{8}$$

$$MPE = \frac{1}{n} \sum_{t=1}^{n} PE_t \tag{9}$$

$$MAPE = \frac{1}{n} \sum_{t=1}^{n} |PE_t| \tag{10}$$

The MAPE makes the comparison of results between from the two models easier because it is percentage based. This gives an indication of the size of the prediction errors in comparison to the measured values in the series.

4. Wind Speed Forecasting Results

Two different sites were selected to demonstrate the effectiveness of the methods proposed for wind speed forecasting and to verify the influence of other atmospheric variables besides wind speed on their accuracy. Each dataset was divided into three parts: 70% for training, 15% for validation and 15% for testing. Table 4 shows the details of these datasets.

Table 4. Datasets for the forecasting analysis.

Site	Training Set (70%)	Validation Set (15%)	Testing Set (15%)	Total Data (100%)
La Mata	6131	1314	1314	8759
Metepec	47,994	10,278	10,278	68,550

To see how the prediction models fit to the real data, Figures 13 and 14 show the last 50 h of data allowing a qualitative comparison. These correspond to 50 data points for La Mata and 300 for Metepec. In these figures, the solid curve is the measured data, the long dashed curve is the ARIMA modeling and the shorter dashed curve is the NARX modeling. From these figures, the importance of the meteorological values of the previous time step (previous hour for La Mata and previous 10 min for Metepec) is obvious for one step ahead wind speed prediction, indicating the persistence of the wind speed on the short-term for these sites.

The mean absolute error and the mean squared error were used to quantitatively evaluate which model best predicts the wind speed behavior. Table 5 shows these measures of forecasting error. The improvement of the NARX model over the univariate ARIMA model based on the MAE and MSE results was calculated using the following equations:

$$P_{MAE} = \left| \frac{MAE_{ARIMA} - MAE_{NARX}}{MAE_{ARIMA}} \right| \cdot 100 \tag{11}$$

$$P_{MSE} = \left| \frac{MSE_{ARIMA} - MSE_{NARX}}{MSE_{ARIMA}} \right| \cdot 100 \tag{12}$$

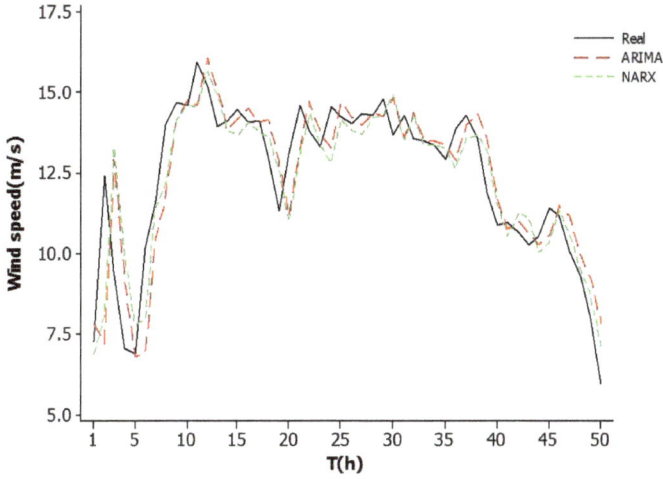

Figure 13. Qualitative comparison of wind speed forecasting for La Mata.

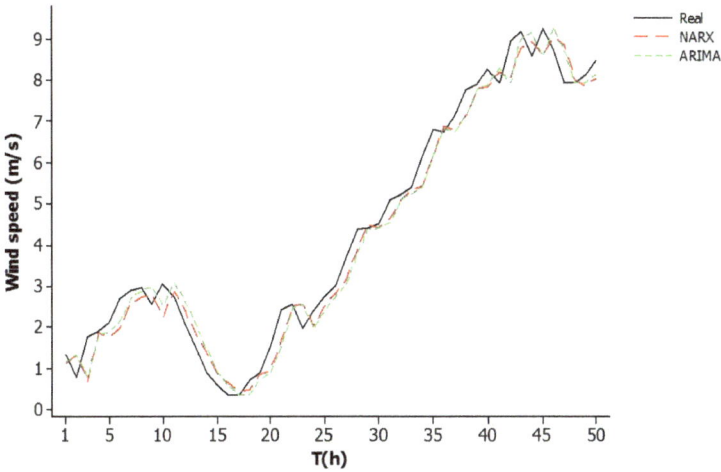

Figure 14. Qualitative comparison of wind speed forecasting for Metepec.

Table 5. Statistical errors generated by the ARIMA and NARX models.

Model	La Mata		Metepec	
	MAE	**MSE**	**MAE**	**MSE**
ARIMA	0.91	1.51	0.44	0.39
NARX	0.86	1.35	0.43	0.34
Improved Percentages	5.5%	10.6%	2.3%	12.8%

From Figures 13 and 14, it can be seen that both models are effective, but it is not obvious which is the best. The error measures in Table 5 confirm the satisfactory performance of both models; however, it is clear that the NARX model is significantly better than the ARIMA model. The MAE of the NARX model for La Mata is 5.5% better than the ARIMA model and 2.3% better for Metepec. The MSE percentage improvements are better than for MAE with values of 10.6% and 12.8% for La Mata and Metepec, respectively.

The additional information provided by more meteorological variables and the non-linear nature of the multivariate NARX model explain the superiority of this model.

5. Conclusions

A procedure to analyze and predict wind speed using standard meteorological variables was developed. Firstly, using traditional statistical techniques, such as the ARIMA model, and, secondly, by using a multivariate artificial neural network technique: the NARX model. Wind speed predictions given by both models were analyzed and compared qualitatively and quantitatively with measured data.

The results obtained show reasonable one step ahead wind speed prediction can be made with the univariate ARIMA model. However, by using a multivariate NARX model, more accurate results were obtained. The inclusion of additional meteorological variables is thus recommended in wind speed forecasting models if they are available.

As well as being a multivariate model, the NARX neural network is a class of discrete-time non-linear techniques that can represent a variety of non-linear dynamic systems, as in the case of wind speed time series.

Acknowledgments: We are thankful to Comision Federal de Electricidad (CFE) for allowing the use of the databases of the measurements of the variables used in this study.

Author Contributions: Erasmo Cadenas: Proposed the original idea for the paper, compiled the data bases, carried out the calculations for wind speed and wrote the original manuscript. Replied to referees' comments. Wilfrido Rivera: Proposed the original paper and contributed to the results analysis and discussion. Replied to referees' comments. Rafael Campos-Amezcua: Contributed to the compilation of the data bases, results analysis and discussion, the writing of the manuscript and replies to the referees' comments. Christopher Heard: Contributed to the writing of the manuscript and the improvement of the grammar and English style, the bibliographic review and replies to the referees' comments.

Conflicts of Interest: The authors declare no conflict of interest.

References

1. Denmark Broke World Record for Wind Power in 2015. Available online: http://www.theguardian.com/environment/2016/jan/18/denmark-broke-world-record-for-wind-power-in-2015 (accessed on 16 February 2016).
2. *Global Wind Report. Annual Market Update, 2014*; Global Wind Energy Council: Brussels, Belgium, 2015.
3. Cadenas, R.; Saldivar, G. Wind power plant La Venta II. Available online: http://www.revista.unam.mx/vol.8/num12/art90/int90.htm (accessed on 16 February 2016).
4. *Reglas de Despacho y Operacion del Sistema Electrico Nacional*; Centro Nacional de Control de Energía: Mexico City, Mexico, 2001. (In Spanish)
5. Cadenas, E.; Rivera, W. Short term wind speed forecasting in La Venta, Oaxaca, México, using artificial neural networks. *Renew. Energy* **2009**, *34*, 274–278.
6. Salcedo-Sanz, S.; Perez-Bellidoa, A.; Ortiz-Garcia, E.; Portilla-Figueras, A.; Prieto, L.; Paredes, D. Hybridizing the fifth generation mesoscale model with artificial neural networks for short-term wind speed prediction. *Renew. Energy* **2009**, *34*, 1451–1457.
7. Cadenas, E.; Jaramillo, O.; Rivera, W. Analysis and forecasting of wind velocity in Chetumal, Quintana Roo, using the single exponential smoothing method. *Renew. Energy* **2010**, *35*, 925–930.
8. Li, G.; Shi, J. On comparing three artificial neural networks for wind speed forecasting. *Appl. Energy* **2010**, *87*, 2313–2320.
9. Liu, H.; Tian, H.; Chen, C.; Li, Y. A hybrid statistical method to predict wind speed and wind power. *Renew. Energy* **2010**, *35*, 1857–1861.
10. Cadenas, E.; Rivera, W. Wind speed forecasting in three different regions of Mexico, using a hybrid ARIMA-ANN model. *Renew. Energy* **2010**, *35*, 2732–2738.

11. Kavasseri, R.; Seetharaman, K. Day-ahead wind speed forecasting using f-ARIMA models. *Renew. Energy* **2009**, *34*, 1388–1393.
12. Li, G.; Shi, J.; Zhou, J. Bayesian adaptive combination of short-term wind speed forecasts from neural network models. *Renew. Energy* **2011**, *36*, 352–359.
13. Liu, H.; Erdem, E.; Shi, J. Comprehensive evaluation of ARMA-GARCH(-M) approaches for modeling the mean and volatility of wind speed. *Appl. Energy* **2011**, *88*, 724–732.
14. Guo, Z.; Zhao, W.; Lu, H.; Wang, J. Multi–step forecasting for wind speed using a modified EMD-based artificial neural network model. *Renew. Energy* **2012**, *37*, 241–249.
15. Liu, H.; Tian, H.; Li, Y. Comparison of two new ARIMA-ANN and ARIMA-Kalman hybrid methods for wind speed prediction. *Appl. Energy* **2012**, *98*, 415–424.
16. Cassola, F.; Burlando, M. Wind speed and wind energy forecast through Kalman filtering of numerical weather prediction model output. *Appl. Energy* **2012**, *99*, 154–166.
17. Zhang, W.; Wu, J.; Wang, J.; Zhao, W.; Shen, L. Performance analysis of four modified approaches for wind speed forecasting. *Appl. Energy* **2012**, *99*, 324–333.
18. Liu, H.; Chen, C.; Tian, H.; Li, Y. A hybrid model for wind speed prediction using empirical mode decomposition and artificial neural networks. *Renew. Energy* **2012**, *48*, 545–556.
19. Peng, H.; Liu, F.; Yang, X. A hybrid strategy of short term wind power prediction. *Renew. Energy* **2013**, *50*, 590–595.
20. Chen, K.; Yu, J. Short-term wind speed prediction using an unscented Kalman filter based state-space support vector regression approach. *Appl. Energy* **2014**, *113*, 690–705.
21. Hocaoglu, F.; Gerek, O.; Kurban, M. A novel wind speed modeling approach using atmospheric pressure observations and hidden Markov models. *J. Wind Eng. Ind. Aerodyn.* **2010**, *98*, 472–481.
22. Hocaoglu, F.; Fidan, M.; Gerek, O. Approach for wind speed prediction. *Energy Convers. Manag.* **2009**, *50*, 1436–1443.
23. Burton, T.; Jenkins, N.; Sharpe, D.; Bossanyi, E. *Wind Energy Handbook*, 2nd ed.; John Wiley & Sons, Inc.: Hoboken, NJ, USA, 2011.
24. Box, G.; Jenkins, G. *Time Series Analysis: Forecasting and Control*; Holden Day: San Francisco, CA, USA, 1970.
25. Gao, Y.; Er, M. NARMAX time series model prediction: Feedforward and recurrent fuzzy neural network approaches. *Fuzzy Sets Syst.* **2005**, *150*, 331–350.
26. Gardner, M.; Dorling, S. Artificial neural networks (the multilayer perceptron)—A review of applications in the atmospheric sciences. *Atmos. Environ.* **1998**, *32*, 2627–2636.
27. Diaconescu, E. Use of NARX neural networks to predict chaotic time series. *WSEAS Trans. Comput. Res.* **2008**, *3*, 182–191.
28. Makridakis, H.; Wheelwright, S. *Forecasting Methods and Applications*, 3rd ed.; John Wiley & Sons, Inc.: Hoboken, NJ, USA, 1998.

energies

MDPI

Article

Improved Spatio-Temporal Linear Models for Very Short-Term Wind Speed Forecasting

Tansu Filik

Department of Electrical and Electronics Engineering, Anadolu University, Eskisehir 26555, Turkey; tansufilik@anadolu.edu.tr; Tel.: +90-222-3350580-6460

Academic Editor: Simon J. Watson
Received: 5 January 2016; Accepted: 25 February 2016; Published: 7 March 2016

Abstract: In this paper, the spatio-temporal (multi-channel) linear models, which use temporal and the neighbouring wind speed measurements around the target location, for the best short-term wind speed forecasting are investigated. Multi-channel autoregressive moving average (MARMA) models are formulated in matrix form and efficient linear prediction coefficient estimation techniques are first used and revised. It is shown in detail how to apply these MARMA models to the spatially distributed wind speed measurements. The proposed MARMA models are tested using real wind speed measurements which are collected from the five stations around Canakkale region of Turkey. According to the test results, considerable improvements are observed over the well known persistence, autoregressive (AR) and multi-channel/vector autoregressive (VAR) models. It is also shown that the model can predict wind speed very fast (in milliseconds) which is suitable for the immediate short-term forecasting.

Keywords: wind energy; wind speed; very short-term; forecasting; prediction; spatio-temporal; multi-channel; autoregressive moving average model

1. Introduction

Electricity consumption of the developing countries increases annually [1,2]. However, the authorities are aiming to reduce the greenhouse gas emission and also the electricity consumption by increasing the amount of renewable energy and improving the energy efficiency respectively [3]. Since wind energy is sustainable, emission-free and cost-effective, it is very attractive and a good candidate to achieve the above ambitious aims. In order to use these energy sources reliably in the future's optimum economic power system operations, it is critically important to accurately forecast wind power generation [4–6]. Since wind power is a function of the cube of wind speed, accurate wind power output prediction depends on wind speed prediction [7].

Wind speed prediction problem is widely investigated in literature and various methods are presented [5,7–13]. The available methods are generally separated as physical and statistical methods. However, for very short-term wind speed forecasting, physical model-based methods such as numerical whether prediction (NWP) have high computational complexity and lower accuracy [7,8]. Therefore, some hybrid of physical (NWP) and statistical methods are proposed in literature as in [8,9]. Computationally efficient but accurate and reliable statistical methods for very short-term wind speed forecasting are required especially for the electricity market-wind forecasting control [14]. The statistical methods can be classified as point and probabilistic forecasting approaches [8]. In point forecasting approach, future wind speed is given as a single value. However, in probabilistic forecasting case, the future wind speed value is modelled as random variable and its probability density function (pdf) is given as a result.

Recently, spatial correlation models, which also known as "spatio-temporal" methods, are appeared as a new trend in short-term wind speed forecasting [5]. These methods use measurements from neighbourhood of target location (wind farm) for more accurate wind speed forecasting with a modest processing overhead [15–21]. Since wind is a horizontal movement in atmosphere, its spatial correlation carries important information for such spatial models. However, the spatial correlation of low level wind directly depends on the complexity of the terrain. In [15], space-time forecasting model is proposed which promises more accurate results according to conventional time series models. However, this model which is called as calibrated probabilistic forecasting is designed only for the selected region. This region specific forecasting model is improved in [16], so it does not require any prior geographic information for the target region. In [17] a graph-learning based spatio-temporal analysis techniques are used to characterize probabilistic models for short-term forecasting. In [5], a methodology is proposed for optimum probabilistic forecasting of geographically dispersed information. In [19], multichannel adaptive filtering technique is applied for short-term prediction which promise lower complexity, improved robustness and ability to track seasonal variations. Most of the above methods are based on the statistical analysis and interpretation of the location specific multi-channel data collected in years.

On the other hand, the conventional linear time series models are easy to implement and requires no preliminary analysis for model development. Hence these models are widely preferred for short-term wind speed forecasting [12,22–24]. However, the multi-channel (spatio-temporal) linear methods, which uses the measurements from neighbourhood of target location, have not been addressed sufficiently for short term wind speed forecasting. The vector autoregressive (VAR) method is applied to geographically dispersed (multi-channel) wind speed data in [25]. There are also some other hybrid artificial neural network (ANN) based methods [26–28].

The multi-channel autoregressive moving average (ARMA) models are commonly used for blind identification of single input multi output (SIMO) systems in communications, source localization and medical imaging [29–31]. These multi-channel blind linear system models can also be applied for multi-channel wind speed prediction problem for more accurate results on target location [32,33].

In this paper, the multi-channel linear prediction models for short-term wind speed forecasting using neighbouring wind speed measurements around the target location which is sketched in Figure 1 are investigated and reviewed. These multi-channel linear prediction models can also be called as multi-channel ARMA or MARMA. The problem formulation, compact matrix forms and efficient multi-channel coefficient estimation approaches are presented and tested using hourly averaged real wind speed/direction values. These values are collected from the five synchronized measurements station of the Turkish State Meteorological Service. These stations are selected around the Canakkale Canel of the Turkey, namely Bozcaada (BOZ), Ipsala (IPS), Gonen (GON), Bandirma (BAN), and Sile (SIL). The root mean square error (RMSE) and mean absolute error (MAE) are used as the performance measurements of the prediction models. It is shown that MARMA model's prediction performance is better than uni-variable AR and multi-variable vector AR (VAR). It is also observed that the performance's of the MARMA increases when the forecast lead time is increased compared to other methods.

The paper is structured as follows: (1) Multi-channel linear prediction models and their compact matrix forms for short-term wind speed forecasting is presented and reviewed. (2) Computationally efficient and accurate linear solution techniques with a new linear channel selection approach for multichannel coefficients are proposed and discussed. (3) MARMA forecasting models are tested using the real wind speed data which are collected from three different locations from the Canakkale region of Turkey. The RMSE and MAE performances are compared for various cases. The section organization of the paper is as follows. In Section 2, problem formulation of the multi-channel linear prediction models and the coefficient estimation techniques are presented. In Section 3, the selected region where the real multi-channel wind data collected is introduced and prediction performances of the models in Section 2 are tested and compared with other methods. We conclude the results in Section 4.

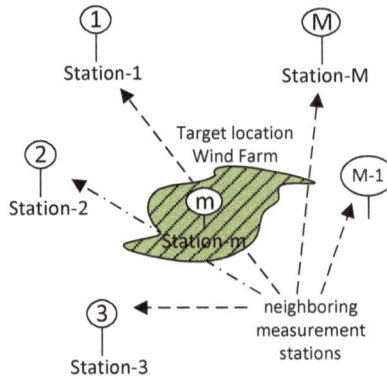

Figure 1. Multi-channel wind data measurement stations around the target location.

2. Multi-Channel Wind Data and Linear Prediction Models

2.1. Multi-Channel Wind Data

We consider M spatially distinct (geographically separated) measurement stations with known positions as shown in Figure 1. At each mth station (channel), discrete measurements are assumed to be collected as:

$$y_m[n] = \frac{1}{\Delta t} \int_{t_n - \Delta t}^{t_n} y_m(t)dt \tag{1}$$

where $y_m[n]$ is averaged wind speed values at discrete time index n respectively. Δt is the averaging time duration and can be chosen as a minute or a hour.

The problem is to forecast short-term wind speed value at m^{th} station, using M spatially distributed (multi-channel) averaged wind data measurements as in Figure 1. Since wind directions are spread to all directions, wind measurement stations should surround the target location for the best result.

2.2. Multi-Channel Linear Prediction Models

In this part, multi-channel ARMA model which is used for blind identification of SIMO systems in [31] are modified and implemented for the multichannel wind speed prediction model. AR model is applied to multi-channel real wind speed data which is called as vector autoregressive model (VAR) in [25]. The VAR predictor's Δ hour ahead output for the mth channel (target location) is given as:

$$y_m[n+\Delta] = \sum_{i=1}^{M} \sum_{p=1}^{P} a_{i,p}^m y_i[n-p] + w_m[n], \quad m = 1, ..., M \tag{2}$$

where M is the number of channels, P is the number of coefficients and $w_m[n]$ is the additive noise (model error) terms at each channel and assumed as temporally and spatially white random process with variance σ_w^2.

Two different multi-channel ARMA models are proposed for short-term wind speed prediction. First model is called as MARMA-1 and Δ hour forecast lead time output at m^{th} location is defines as:

$$y_m[n + \Delta] = \sum_{i=1}^{M} \sum_{p=1}^{P} a_{i,p}^m y_i[n - p] + \sum_{q=1}^{Q} b_q^m s[n - q] + w_m[n], \ m = 1, ..., M \tag{3}$$

where $s[n]$ is the common input signal which is white noise random process with constant power spectrum and statistically independent from the additive channel noises $w_m[n]$ with variance σ_s^2. Second model is called as MARMA-2 and Δ hour forecast lead time output at m^{th} channel which differently using multi-channel spatially and temporally white noise inputs, $s_k[n]$, as follows:

$$y_m[n + \Delta] = \sum_{i=1}^{M} \sum_{p=1}^{P} a_{i,p}^m y_i[n - p] + \sum_{k=1}^{M} \sum_{q=1}^{Q} b_{k,q}^m s_k[n - q] + w_m[n], \ m = 1, ..., M \tag{4}$$

It is possible to put M channel wind data for the above linear prediction models in matrix form as:

$$\mathbf{y}[n + \Delta] = \mathbf{A}\mathbf{x}[n] + \mathbf{w}[n] \tag{5}$$

where $\mathbf{y}[n + \Delta] = [y_1[n + \Delta] \dots y_M[n + \Delta]]^T$ is a $M \times 1$ vector and this vector (also known as snapshot) includes M channel wind values from different locations at the same time. $\mathbf{x}[n]$ is the input data for the multichannel linear prediction models and defined for MARMA-1 in Equation (6) and MARMA-2 in Equation (7) respectively as:

$$\mathbf{x}[n] = [\underline{\mathbf{y}}_1^T[n - 1] \dots \underline{\mathbf{y}}_M^T[n - 1] \ \underline{\mathbf{s}}^T[n - 1]]^T \tag{6}$$

$$\mathbf{x}[n] = [\underline{\mathbf{y}}_1^T[n - 1] \dots \underline{\mathbf{y}}_M^T[n - 1] \ \underline{\mathbf{s}}_1^T[n - 1] \dots \underline{\mathbf{s}}_M^T[n - 1]]^T \tag{7}$$

where $\underline{\mathbf{y}}_m[n - 1] = [y_m[n - 1] \dots y_m[n - P]]^T$ for $m = 1 \dots M$ and $\underline{\mathbf{s}}[n - 1] = [s[n - 1] \dots s[n - Q]]^T$. Similarly multi-channel white noise process in Equation (7) is defined as $\underline{\mathbf{s}}_m[n - 1] = [s_m[n - 1] \dots s_m[n - Q]]^T$ for $m = 1 \dots M$. $\mathbf{x}[n]$ in Equation (6) is a $(MP + Q) \times 1$ vector and $\mathbf{x}[n]$ in in Equation (7) is a $(M(P + Q)) \times 1$ vector. $\mathbf{w}[n] = [w_1[n] \dots w_M[n]]^T$ is $M \times 1$ additive channel noise vector. Finally the multi-channel prediction filter coefficient matrix ($\mathbf{A}$) for MARMA-1 is defined as:

$$\mathbf{A} = \begin{bmatrix} \mathbf{a}_1^1 & \cdots & \mathbf{a}_M^1 & \mathbf{b}^1 \\ \mathbf{a}_1^2 & \cdots & \mathbf{a}_M^2 & \mathbf{b}^2 \\ \vdots & & \vdots & \vdots \\ \mathbf{a}_1^M & \cdots & \mathbf{a}_M^M & \mathbf{b}^M \end{bmatrix} \tag{8}$$

where $\mathbf{a}_i^m = [a_{i,1}^m \dots a_{i,P}^m]$ and $\mathbf{b}^m = [b_1^m \dots b_Q^m]$ for $i = 1, \dots, M, \ m = 1, \dots, M$. $\mathbf{A}$ is a $M \times (MP + Q)$ matrix and it includes all the unknown coefficients in Equation (3). Similarly for MARMA-2, $\mathbf{A}$ matrix is defined as:

$$\mathbf{A} = \begin{bmatrix} \mathbf{a}_1^1 & \cdots & \mathbf{a}_M^1 & \mathbf{b}_1^1 & \cdots & \mathbf{b}_M^1 \\ \mathbf{a}_1^2 & \cdots & \mathbf{a}_M^2 & \mathbf{b}_1^2 & \cdots & \mathbf{b}_M^2 \\ \vdots & & \vdots & \vdots & & \vdots \\ \mathbf{a}_1^M & \cdots & \mathbf{a}_M^M & \mathbf{b}_1^M & \cdots & \mathbf{b}_M^M \end{bmatrix} \tag{9}$$

where $\mathbf{a}_i^m = [a_{i,1}^m \dots a_{i,P}^m]$ and $\mathbf{b}_k^m = [b_{k,1}^m \dots b_{k,Q}^m]$ for $i = 1, \dots, M, \ k = 1, \dots, M, \ m = 1, \dots, M$. In this case, $\mathbf{A}$ is a $M \times M(P + Q)$ matrix and it includes all the unknown coefficients in Equation (4).

It is required to efficiently solve linear prediction model coefficients in Equations (8) and (9). The matrix form of the multi-channel linear prediction models, which is given in Equation (5), is similar to well known array signal model in array theory [34]. Array signal processing area deals with the space-time signals which are collected by an array of sensors. It is possible to solve these coefficients using the subspace methods in [30]. Another computationally efficient way of solving these coefficients is given in [31].

In the next section, computationally efficient and accurate linear solution technique with a new linear channel selection approach for multichannel coefficient estimation is presented.

2.3. Multi-Channel Linear Prediction Coefficient Estimation

In order to find multi-channel linear prediction coefficients for more accurate results, N snapshot measurements are collected and the data in Equation (5) is extended as:

$$
\begin{bmatrix} \mathbf{y}[n+\Delta] \\ \mathbf{y}[n+\Delta-1] \\ \cdots \\ \mathbf{y}[n+\Delta-N] \end{bmatrix} = \begin{bmatrix} \mathbf{A} & & \\ & \mathbf{A} & \\ & & \ddots & \\ & & & \mathbf{A} \end{bmatrix} \begin{bmatrix} \mathbf{x}[n] \\ \mathbf{x}[n-1] \\ \cdots \\ \mathbf{x}[n-N] \end{bmatrix} + \begin{bmatrix} \mathbf{w}[n] \\ \mathbf{w}[n-1] \\ \cdots \\ \mathbf{w}[n-N] \end{bmatrix} \tag{10}
$$

which can be rewritten as:

$$
\mathbf{Y}[n+\Delta] = \bar{\mathbf{A}}\mathbf{X}[n] + \mathbf{W}[n] \tag{11}
$$

where $\mathbf{Y}[n+\Delta]$ is extended multi-channel linear prediction output vector with size $MN \times 1$ and $\mathbf{X}[n]$ is extended prediction input vector. $\mathbf{W}[n]$ is extended model error vector with size $MN \times 1$ and $\bar{\mathbf{A}}$ is the extended coefficient matrix. In order to solve MARMA-1 and MARMA-2 coefficients in Equations (8) and (9) respectively, it is possible to apply a selection matrix for the specified m^{th} target location as:

$$
\mathbf{S}_m\mathbf{Y}[n+\Delta] = \mathbf{S}_m(\bar{\mathbf{A}}\mathbf{X}[n] + \mathbf{W}[n]) \tag{12}
$$

where the selection matrix for the mth location is defined as:

$$
\mathbf{S}_m = [\mathbf{e}_1 \ \mathbf{e}_2 \ \cdots \ \mathbf{e}_M]^T. \tag{13}
$$

$\mathbf{e}_k$ is a $1 \times MN$ row vector as:

$$
\mathbf{e}_k = [\ \underbrace{0, \ \cdots \ ,0}_{M(k-1)+m-1} ,1, \ \underbrace{0, \ \cdots \ ,0}_{M(N-k+1)-m}\ ] \tag{14}
$$

If we multiply N multi-channel data in Equation (10) with m^{th} selection matrix $\mathbf{S}_m$ as in Equation (12) we get linear set of equations for the mth location as:

$$
\mathbf{Y}_m = \mathbf{H}\bar{\mathbf{a}}_m^T + \mathbf{w}_m \tag{15}
$$

where $\mathbf{Y}_m = [y_m[n+\Delta] \ldots y_m[n+\Delta-N]]^T$ is a $N \times 1$, m^{th} channel output data vector. $\mathbf{H}$ matrix is equivalent to the $\mathbf{X}[n]$ in Equation (11) and it is the measurement data matrix which consist from the previous multichannel wind data and white noise signal. $\bar{\mathbf{a}}_m$ is the mth row of the $\mathbf{A}$ matrix in Equations (8) or (9) which is the prediction coefficients of MARMA-1 and MARMA-2 respectively for the mth target location. This model is the well known linear model in classical estimation theory [35] and it is possible to apply linear least squares (LS) techniques to find the optimum prediction coefficients. In this case, it is required to minimize the following cost function:

$$
\min_{\bar{\mathbf{a}}_m} J(\bar{\mathbf{a}}_m) = (\mathbf{Y}_m - \mathbf{H}\bar{\mathbf{a}}_m^T)^T(\mathbf{Y}_m - \mathbf{H}\bar{\mathbf{a}}_m^T) \tag{16}
$$

where $()^T$ is for transpose operation and the optimum LS solution for the unknown prediction coefficients is:

$$
\hat{\bar{\mathbf{a}}}_m^T = (\mathbf{H}^T\mathbf{H})^{-1}\mathbf{H}^T\mathbf{Y}_m \tag{17}
$$

There are some computationally efficient ways to solve the above matrix pseudoinverse solutions as in [36,37].

3. Data and Test Results

3.1. Data Set

The accuracy of the proposed multi-channel linear prediction models are tested with hourly averaged wind speed and direction data which were collected from five stations around the Canakkale region of Turkey. Data is available in [38]. The three years hourly averaged wind speed and direction values between the years 2008 and 2010 are used. These five stations (Bozcaada, Ipsala, Gonen, Bandirma and Sile) belong to the Turkish State Meteorological Service and the locations are shown in Figure 2.

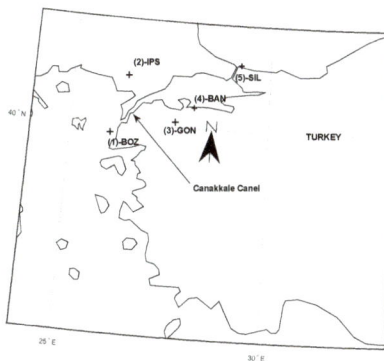

Figure 2. The measurement stations, Bozcaada (BOZ), Ipsala (IPS), Gonen (GON), Bandirma (BAN), and Sile (SIL) of Turkey where **N** indicates the North.

All wind measurements are taken from 10 meters height above ground. The region is known as having one of the highest wind energy potential in Turkey. These stations are selected arbitrary from the available measurement locations in that region. The topographic map of the region is shown in Figure 3. The topography is indicated by different colors; green colors indicated low altitude and white colors indicate hight altitude. As shown in Figure, these measurement stations are not close each other and the canal. BOZ is located at the highest point of an island. IPS is located in a valley. BAN is close to GON but it is separated from the canal. SIL is approximately 250 km far away from GON which is completely separated from the canal and other stations.

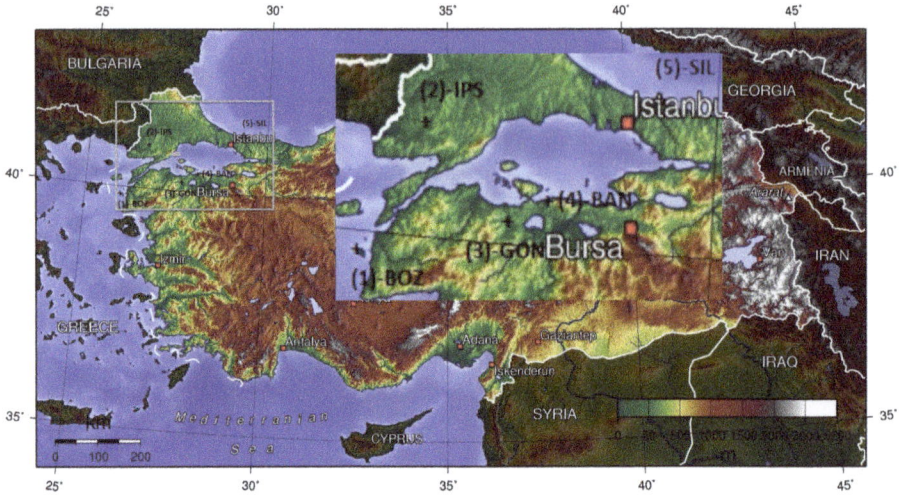

Figure 3. The topographic map of Turkey—A portion is zoomed for visual purposes. ("Turkey topo" by Captain Blood—Licensed under CC BY-SA 3.0)

Figure 4 shows the Auto and Cross-Correlation Coefficients of the stations with the target location GON for different time delays. All the correlations demonstrate a decline with time delay, except for maximum at diurnal periods (multiples of 24 h). It can be seen from Figure 4 that cross-correlation coefficient values of BOZ and IPS are higher than the other two (BAN and SIL) stations for short time delays, $1 \leq \Delta \leq 4$. SIL station has the lowest correlation values as expected. Since these stations are selected arbitrarily from the available stations their spatial dependencies are limited as shown. So it is not possible to apply a region specific space time method such as [15].

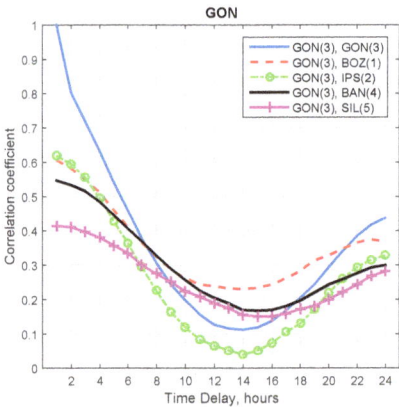

Figure 4. Autocorrelation and cross-correlation coefficients of wind measurements at Bozcaada (BOZ), Ipsala (IPS), Bandirma (BAN), and Sile (SIL) with Gonen (GON) in October–November 2008.

Figure 5 shows the frequency of the wind directions at the measurement stations as polar histograms. These polar plots show that prevailing wind directions at the stations are similar and along the Canakkale Canal from North East (NE) to South West (SW) and vice versa due to the large-scale circulation in that region.

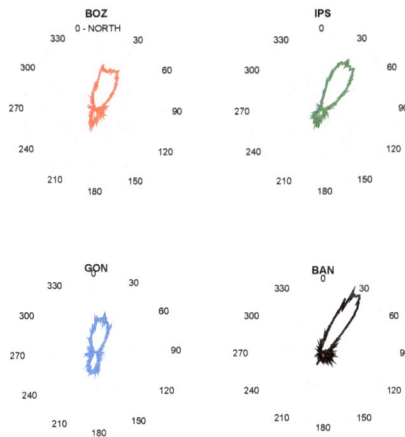

Figure 5. The frequency of the wind directions at the measurement stations for three year hourly averaged data where North is zero degree.

Some of the basic statistics (annual maximum, mean and variance values) of the used multi-channel data set are summarized in Table 1.

Table 1. Some basic statistics of the used multi-channel data set.

Year	(1)-BOZ (m/s)			(2)-IPS (m/s)			(3)-GON (m/s)			(4)-BAN (m/s)			(5)-SIL (m/s)		
	max	mean	var	max	mean	var	max	mean	var	max	mean	var	max	mean	var
2008	27.7	5.7	13.4	16.5	2.8	3.3	12.4	2.1	2.5	17.7	3.7	6.9	12.7	2.1	2.6
2009	22.8	5.6	12.9	14.7	2.7	3.1	11.5	1.9	2.1	17.3	3.7	7.1	11.0	2.2	1.9
2010	39.7	6.1	20.8	22.5	3.2	6.0	17.6	2.2	3.9	37.7	4.04	11.1	15.5	2.3	2.2

3.2. Test Results

In this section, real wind speed forecasting performances of the proposed multi-channel models are compared with the persistence, AR, VAR models. In order to compare and show the performances of the forecasting models, RMSE and MAE are calculated as,

$$RMSE_m(\Delta) = \sqrt{\frac{1}{K}\sum_k(\hat{y}_m[k+\Delta] - y_m[k+\Delta])^2}$$

$$MAE_m(\Delta) = \frac{1}{K}\sum_k|\hat{y}_m[k+\Delta] - y_m[k+\Delta]| \tag{18}$$

where $\hat{y}_m$ is the predicted value and y_m is the actual value. Δ is for forecast lead time in hours and m indicates the index of target location. K is the number of total predictions to calculate the RMSE and MAE in Equation (18). In this study, K is selected to cover the whole data between the years 2009 and 2010. In the following calculations of RMSE and MAE results total $K = 17280$ prediction values are used as in Equation (18) respectively. The persistence forecasting method in [25,39] is used as a benchmark to compare all the results. In persistence forecasting, the Δ ahead future value is taken as the current value. The prevailing wind directions are along the NE to SW and vice versa as shown in Figure 5. GON station is in the midst of the prevailing wind directions according to other stations. Therefore in the following case study, third station ($m = 3$) is selected as the target station which is also surrounded by other stations. However, it is also possible to select the other stations as the target station.

3.2.1. Model Order

The linear prediction model orders of P and Q in Equations (2), (3) and (4) can be selected using the information criteria in [40] or the minimum description length in [41]. Figures 6 and 7 show the RMSE and MAE performances of the AR, VAR and MARMA models according to the model order for the 3rd station, GONEN ($m = 3$) and for the forecast lead time $\Delta = 2$, respectively. It is observed that the AR has minimum error for $P = 2$ and VAR and MARMA-1 gives minimum error when $P = 1$. On the other hand, MARMA-2's RMSE and MAE values are reducing when the filter order increased. Therefore, MARMA-2 model gives the best performance when $P = 4$ compared with other models.

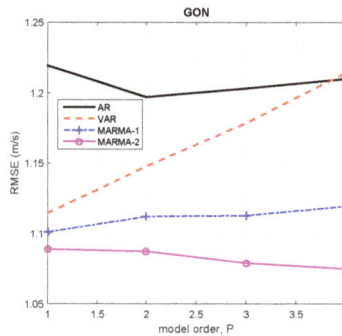

Figure 6. Root mean square error (RMSE) performances of the autoregressive (AR), vector autoregressive (VAR) and multi-channel autoregressive moving average (MARMA) models with respect to model order P when m = 3 (GONEN) and $\Delta = 2$ h.

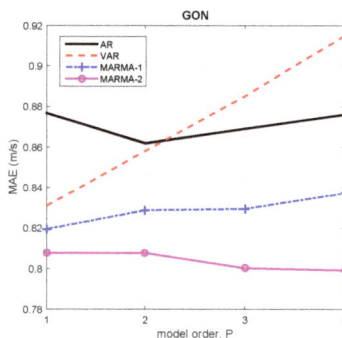

Figure 7. MAE performances of the AR, VAR and MARMA models with respect to model order P when m = 3 (GONEN) and $\Delta = 2$ h.

The similar confirmation is repeated for the Q parameter's of the MARMA models. Figures 8 and 9 show the RMSE and MAE performances of the MARMA-1 and MARMA-2 models respectively according to Q. It is observed that the increasing the model order Q slightly reduces the performances. For the best performance, Q is selected as 1 for MARMA models.

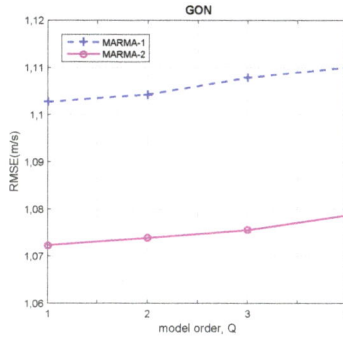

Figure 8. RMSE performances of the MARMA-1 and MARMA-2 models with respect to model order Q when m = 3 (GONEN) and Δ = 2 h.

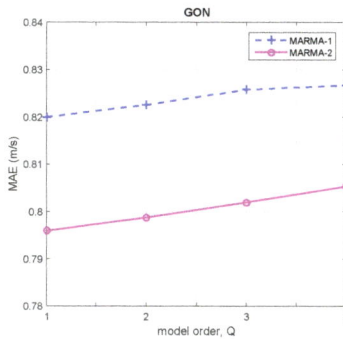

Figure 9. Mean absolute error (MAE) performances of the MARMA-1 and MARMA-2 models with respect to model order Q when m = 3 (GONEN) and Δ = 2 h.

3.2.2. Number of Samples

In order to solve the multichannel linear prediction filter coefficients, the selection of the number of previous samples, N, in Equation (10) is another critical parameter. It is observed that increasing the number of N after certain value do not improve the forecasting performances of the AR and VAR models as shown in Figure 10. On the other hand, MARMA-2's forecasting performance is better than the other models when relatively large number of previous samples are used. MARMA-2 uses different random noise processes for each channels and if the large number of previous samples are used, this model gives statistically efficient results.

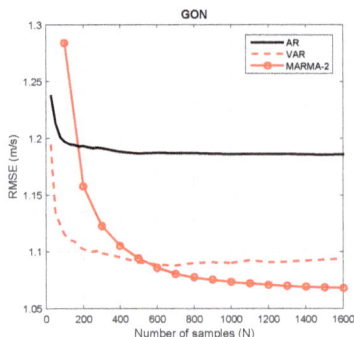

Figure 10. RMSE performances of the AR, VAR and MARMA models with respect to number of previous samples N when m = 3 (GONEN) and Δ = 2 h.

In the following section to make a fair comparison, N is selected as 1000 h for all models.

3.2.3. Number of Channels

In this part the effect of number of channels, M, in Equations (2), (3) and (4) is investigated. Table 2 shows RMSE performances of important channel selections. For $M = 4$, if we exclude BOZ or IPS from data set, RMSE increases which indicates the significance of these measurements for GON. However, if we exclude SIL, which has minimum correlation value with GON, RMSE value almost unchanged which verifies the correlation values in Figure 4.

Table 2. RMSE and MAE performances of MARMA-2 for different M when $m = 3$ (GON), Δ = 2 and $P = 4$.

Number of channel, M	MAE	RMSE
M = 5	0.7956	1.0721
M = 4 BOZ(1) is excluded	0.8071	1.0909
M = 4 IPS(2) is excluded	0.8035	1.0898
M = 4 BAN(4) is excluded	0.7994	1.0765
M = 4 SIL(5) is excluded	0.7968	1.0730
M = 3 BOZ(1), IPS(2) are excluded	0.8195	1.1146
M = 3 IPS(2), SIL(5) are excluded	0.8040	1.0891
M = 3 BAN(4), SIL(5) are excluded	0.8003	1.0763
M = 2 BOZ(1), IPS(2), BAN(4) are excluded	0.8355	1.1403
M = 2 IPS(2), BAN(4), SIL(5) are excluded	0.8142	1.1017
M = 1 All other channels are excluded	0.8480	1.1699

Figure 11 shows the RMSE performances with respect to the channel number, M. As it is seen, if the used channel is decreased the forecasting error demonstrate a steady increase.

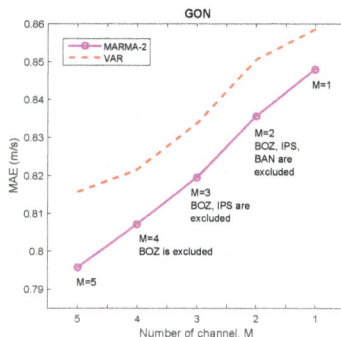

Figure 11. MAE performances of the VAR and MARMA-2 models with respect to number of channel M when m = 3 (GONEN) and Δ = 2 h.

3.2.4. Forecasting Results

Table 3 shows the RMSE and MAE of the target station (*m* = 3) according the forecast lead time (Δ). The multi-channel MARMA-2 has better RMSE and MAE performance than the persistence, AR and VAR models. Table 3 also show that when the lead time period is increased the MARMA models have much better performance than the others.

Table 3. RMSE and MAE performances of the persistent, AR, VAR and MARMA models according the forecast lead time (Δ).

Model	Δ = 1 h		Δ = 2 h		Δ = 3 h		Δ = 4 h	
	MAE	RMSE	MAE	RMSE	MAE	RMSE	MAE	RMSE
Persistent	0.8998	1.2443	1.0304	1.4148	1.1686	1.5842	1.2851	1.725
AR	0.7151	0.9947	0.8582	1.1909	0.9893	1.3578	1.1073	1.5081
VAR	0.6962	0.9438	0.8156	1.0984	0.9215	1.2252	1.0127	1.3348
MARMA-2	0.6854	0.9301	0.7956	1.0721	0.8879	1.1834	0.9588	1.2700

Table 4 shows the percentage improvements of the AR, VAR and MARMA-2 methods over persistence method. It is observed that the proposed MARMA-2 method has the best performance and approximately 2.6% more improvements on the average than the multichannel VAR method. It is also seen in Table 4, the multichannel (spatio-temporal) models (VAR and MARMA) which using the neighbouring measurements have significant improvements according the only temporal AR model.

Table 4. The percentage improvements in MAE of the models with respect to persistence model.

Model	Δ = 1	Δ = 2	Δ = 3	Δ = 4	Average
AR	20.52%	16.71%	15.34%	13.83%	16.60%
VAR	22.62%	20.84%	21.14%	21.19%	21.44%
MARMA-2	23.83%	22.79%	24.02%	25.40%	24.01%

The average execution times of the used and the proposed methods are given in Table 5 for a single Δ hour ahead forecasting. The used desktop computer has Intel Core(TM) i7-3770K CPU @ 3.50 GHz Processor and 16 GB RAM. Since all the single and multi-channel models are linear and uses efficient linear least square techniques, the observed execution times in table are less than one second with an ordinary desktop computer. It is possible to forecast very short term (in seconds) wind speed values with the proposed spatio-temporal linear MARMA model.

Table 5. The execution times of the used and the proposed models in milliseconds (ms).

Model	Prediction Time
Persistence	0.021 ms
AR	0.712 ms
VAR	1.500 ms
MARMA-2	31.20 ms

4. Conclusions

In this study, the spatio-temporal (multi-channel) linear models, which use the neighbouring measurements around the target location, are investigated for short-term wind speed forecasting problem. The problem formulation of the multi-channel ARMA models (called as MARMA) are presented and efficient multi-channel prediction coefficient estimation techniques are revised. The proposed MARMA models and solution techniques are tested using hourly averaged real wind values from the five station around Canakkale region of Turkey. The forecasting RMSE and MAE's of the MARMA-2 model is compared with the persistence, AR and multi-channel AR (VAR) methods. As a result, considerable improvements are observed compared to well known temporal persistence (24.01% improvement) and AR (7.41% improvement) methods. The proposed MARMA-2 model gives 2.6% better results than the spatio-temporal VAR method. It is shown that MARMA-2's performance is continuously improved when number of previous samples (N) and filter order are increased unlike the other models. It should be also noted that since the proposed MARMA model moves on the data set using the N previous available samples, it can also adapt the seasonal variations. It is also shown that the proposed multi-channel linear model can predict Δ hour wind speed value using an ordinary desktop computer in milliseconds which is suitable for very short term (in seconds) wind speed forecasting.

Acknowledgments: This work was supported by Anadolu University Scientific Research Projects Fund with projects 1505F512 and 1505F299. The received fund covers the costs to publish in open access. The authors would also like to acknowledge Turkish State Meteorological Service for providing the necessary wind data.

Conflicts of Interest: The author declares no conflict of interest.

References

1. Filik, Ü.B.; Filik, T.; Gerek, Ö.N. New electric transmission systems—Experiences from Turkey. In *The Handbook of Clean Energy Systems*; Yan, J.Y., Ed.; John Wiley and Sons Ltd: New York, NY, USA, 2015; Volume 4, pp. 1981–1994.
2. Filik, Ü.B.; Gerek, Ö.N.; Kurban, M. A novel modeling approach for hourly forecasting of long-term electric energy demand. *Energy Convers. Manag.* **2011**, *52*, 199–211.
3. Moldan, B.; Janouskova, S.; Hak, T. How to understand and measure environmental sustainability: Indicators and targets. *Ecol. Indic.* **2012**, *17*, 4–13.
4. Farhangi, H. The path of a smart grid. *IEEE Power Energy Mag.* **2010**, *8*, 18–28.
5. Tastu, J.; Pinson, P.; Trombe, P.J.; Madsen, H. Probabilistic forecasts of wind power generation accounting for geographically dispersed information. *IEEE Trans. Smart Grid* **2014**, *5*, 480–489.
6. Bracale, A.; De Falco, P. An Advanced Bayesian Method for Short-Term Probabilistic Forecasting of the Generation of Wind Power. *Energies* **2015**, *8*, 10293–10314.
7. Lei, M.; Shiyan, L.; Jiang, C.W.; Liu, H.L.; Zhang, Y. A review on the forecasting of wind speed and generated power. *Renew. Sustain. Energy Rev.* **2009**, *13*, 915–920.
8. Zhu, X.X.; Genton, M.G. Short Term Wind Speed Forecasting for Power System Operations. *Int. Stat. Rev.* **2012**, *80*, 2–23.
9. De Giorgi, M.G.; Campilongo, S.; Ficarella, A.; Congedo, P.M. Comparison Between Wind Power Prediction Models Based on Wavelet Decomposition with Least-Squares Support Vector Machine (LS-SVM) and Artificial Neural Network (ANN). *Energies* **2014**, *7*, 5251–5272.

10. Jung, J.; Broadwater, R.P. Current status and future advances for wind speed and power forecasting. *Renew. Sustain. Energy Rev.* **2014**, *31*, 762–777.

11. Zhang, Y.; Wang, J.X.; Wang, X.F. Review on probabilistic forecasting of wind power generation. *Renew. Sustain. Energy Rev.* **2014**, *32*, 255–270.

12. Tascikaraoglu, A.; Uzunoglu, M. A review of combined approaches for prediction of short-term wind speed and power. *Renew. Sustain. Energy Rev.* **2014**, *34*, 243–254.

13. Sun, W.; Liu, M.; Liang, Y. Wind Speed Forecasting Based on FEEMD and LSSVM Optimized by the Bat Algorithm. *Energies* **2015**, *8*, 6585–6607.

14. Potter, C.W.; Negnevitsky, M. Very short-term wind forecasting for Tasmanian power generation. *IEEE Trans. Power Syst.* **2006**, *21*, 965–972.

15. Gneiting, T. Larson, K.; Westrick, K.; Genton, M.G.; Aldrich, E. Calibrated probabilistic forecasting at the stateline wind energy center: The regime-switching space time method. *J. Am. Stat. Assoc.* **2006**, *101*, 968–979.

16. Hering, A.S.; Genton, M.G. Powering up with space-time wind forecasting. *J. Am. Stat. Assoc.* **2010**, *105*, 92–104.

17. He, M.; Yang, L.; Zhang, J.S.; Vittal, V. A spatio-temporal analysis approach for short-term forecast of wind farm generation. *IEEE Trans. Power Syst.* **2014**, *29*, 1–12.

18. Xie, L.; Gu, Y.Z.; Zhu, X.X.; Genton, M.G. Short-term spatio-temporal wind power forecast in robust look-ahead power system dispatch. *IEEE Trans. Smart Grid* **2014**, *5*, 511–520.

19. Dowell, J.; Weiss, S.; Hill, D.; Infield, D. Short term spatio temporal prediction of wind speed and direction. *Wind Energy* **2014**, *17*, 1945–1955.

20. Mohandes, M.A.; Rehman, S.; Rahman, S.M. Spatial estimation of wind speed. *Int. J. Energy Res.* **2012**, *36*, 545–552.

21. Pourhabib, A.; Huang, J.Z.; Ding, Y. Short-term Wind Speed Forecast Using Measurements from Multiple Turbines in a Wind Farm. *Technometrics* **2015**, *58*, 138–147.

22. Sfetsos, A. A comparison of various forecasting techniques applied to mean hourly wind speed time series. *Renew. Energy* **2000**, *21*, 23–35.

23. Riahy, G.H.; Abedi, M. Short term wind speed forecasting for wind turbine applications using linear prediction method. *Renew. energy* **2008**, *33*, 35–41.

24. Brown, B.G.; Katz, R.W.; Murphy, A.H. Time series models to simulate and forecast wind speed and wind power. *J. Clim. Appl. Meteorol.* **1984**, *23*, 1184–1195.

25. Hill, D.C.; McMillan, D.; Bell K.R.W.; Infield, D. Application of auto-regressive models to UK wind speed data for power system impact studies. *IEEE Trans. Sustain. Energy* **2012**, *3*, 134–141.

26. Shukur, O.B.; Lee, M.H. Daily wind speed forecasting through hybrid KF-ANN model based on ARIMA. *Renew. Energy* **2015**, *76*, 637–647.

27. Cadenas, E.; Rivera, W. Wind speed forecasting in three different regions of Mexico, using a hybrid ARIMA ANN model. *Renew. Energy* **2010**, *35*, 2732–2738.

28. Hernandez, L.; Baladron, C.; Aguiar, J.M.; Calavia, L.; Carro, B.; Sanchez-Esguevillas, A.; Perez, F.; Fernandez, A.; Lloret, J. Artificial Neural Network for Short-Term Load Forecasting in Distribution Systems. *Energies* **2014**, *7*, 1576–1598.

29. Swami, A.; Giannakis, G.; Shamsunder, S. Multichannel ARMA processes. *IEEE Trans. Signal Process.* **1994**, *42*, 898–913.

30. Moulines, E.; Duhamel, P.; Cardoso, J.F.; Mayrargue, S. Subspace methods for the blind identification of multichannel FIR filters. *IEEE Trans. Signal Process.* **1995**, *43*, 516–525.

31. Yu, C.P.; Zhang, C.S.; Xie, L.H. Blind identification of multi-channel ARMA models based on second-order statistics. *IEEE Trans. Signal Process.* **2012**, *60*, 4415–4420.

32. Filik, T. A Multidimensional Short-Term Wind Speed/Directional Analysis Based on Spatial Covariance Matrix Model. In Proceedings of the Turkish-German Conference on Energy Technologies, METU, Ankara, Turkey, 13–15 October 2014.

33. Filik, T. An Improved Multichannel Spatial Correlation Matrix Based ARMA Model for Short-Term Wind Forecasting. In Proceedings of the 5th International 100% Renewable Energy Conference (IRENEC-2015), Istanbul, Turkey, 28–30 May 2015.

34. Van Trees, H.L. *Detection, Estimation, and Modulation Theory, Optimum Array Processing*; John Wiley and Sons Ltd: New York, NY, USA, 2004.

35. Kay, S.M. *Fundamentals of Statistical Signal Processing, Volume I: Estimation Theory*; Prentice Hall: Upper Saddle River, NJ, USA, 1993.

36. Bjorck, A.; Elfving, T. Accelerated projection methods for computing pseudoinverse solutions of systems of linear equations. *BIT Numer. Math.* **1979**, *19*, 145–163.

37. Golub, G.H.; Van Loan, C.F. *Matrix Computations*; The Johns Hopkins University Press (JHU Press): Baltimore, MD, USA, 2012; Volume 3.

38. Filik, T. Wind Forecasting, Multichannel Wind Speed and Direction Data. Available online: http://www.eem.anadolu.edu.tr/tansufilik/EEM%20547/icerik/wind_forecasting.htm (accessed on 29 February 2016).

39. Parkes, J.; Tindal, A.; Works, S.V.S.; Lane, S. Forecasting short term wind farm production in complex terrain. In Proceedings of the 2004 European Wind Energy Conference, London, UK, 22–25 November 2004.

40. Akaike, H. A new look at the statistical model identification. *IEEE Trans. Autom. Control* **1974**, *19*, 716–723.

41. Rissanen, J. Modeling by shortest data description. *Automatica* **1978**, *14*, 465–471.

Article

Wind Power Generation Forecasting Using Least Squares Support Vector Machine Combined with Ensemble Empirical Mode Decomposition, Principal Component Analysis and a Bat Algorithm

Qunli Wu and Chenyang Peng *

Department of Economics and Management, North China Electric Power University, Baoding 071003, China; wuqunli2002@aliyun.com
* Correspondence: ncepupengchenyang@163.com; Tel.: +86-150-3128-0893

Academic Editor: José C. Riquelme
Received: 10 January 2016; Accepted: 29 March 2016; Published: 1 April 2016

Abstract: Regarding the non-stationary and stochastic nature of wind power, wind power generation forecasting plays an essential role in improving the stability and security of the power system when large-scale wind farms are integrated into the whole power grid. Accurate wind power forecasting can make an enormous contribution to the alleviation of the negative impacts on the power system. This study proposes a hybrid wind power generation forecasting model to enhance prediction performance. Ensemble empirical mode decomposition (EEMD) was applied to decompose the original wind power generation series into different sub-series with various frequencies. Principal component analysis (PCA) was employed to reduce the number of inputs without lowering the forecasting accuracy through identifying the variables deemed as significant that maintain most of the comprehensive variability present in the data set. A least squares support vector machine (LSSVM) model with the pertinent parameters being optimized by bat algorithm (BA) was established to forecast those sub-series extracted from EEMD. The forecasting performances of diverse models were compared, and the findings indicated that there was no accuracy loss when only PCA-selected inputs were utilized. Moreover, the simulation results and grey relational analysis reveal, overall, that the proposed model outperforms the other single or hybrid models.

Keywords: ensemble empirical mode decomposition (EEMD); least squares support vector machine (LSSVM); principal component analysis (PCA); bat algorithm (BA); grey relational analysis

---

1. Introduction

Wind power has been identified as one of the most important and efficient renewable energy and has been extensively utilized throughout the world [1–3]. With the rapid development of wind power, the proportion of wind power in the whole power system is becoming larger. However, wind power is a rolling source of electrical energy due to the variability of wind speed, temperature and other factors. The uncertainty of wind power undoubtedly affects the power system stability and increases the operation cost of power systems [2]. Therefore, accurate forecasting approaches with respect to wind power generation have positive implications on power system planning for unit commitment and dispatch, and electricity trading in certain electricity markets.

There is abundant literature on wind power forecasting, most of which has been published in recent years. In contrast to the wealth of studies on wind speed prediction, there has been less research looking at wind power generation forecasting. The approaches of these studies can be classified into three categories: time series models [4–8], artificial intelligent algorithm models [9–17] and time-series artificial intelligent algorithm models [18]. Most of these approaches utilize time series analysis models, including vector autoregressive (VAR) models [4,5], autoregressive moving average (AMRA) models, and autoregressive integrated moving average (ARIMA) models. Erdem [6] decomposed wind speed into lateral and longitudinal components with each component being represented by an ARMA model, then the predictive value results were obtained by accumulation. Liu [7] proposed an autoregressive moving average-generalized autoregressive conditional heteroscedasticity algorithm for modeling the mean and volatility of wind speed, with the model effectiveness being evaluated by multiple methods. The results suggested the proposed method effectively captured the characteristics of wind speed. Kavasseri [8] examined the use of ARIMA model to forecast wind speed. The simulation results indicated the forecasting accuracy of the proposed method outperformed the persistence models. Nevertheless, the wide implementation of time series models on wind power prediction can be problematic due to the poor nonlinear fitting capacity.

On the contrary, the adaptive and self-organized learning features of intelligent algorithms apparently facilitate the estimation of nonlinear time series. For instance, artificial neural network (ANN) [9,10] and least squares support vector machine (LSSVM) [11–14] are perceived to be highly effective methods in the field of wind power forecasting. Guo [11] successfully developed a hybrid seasonal auto-regression integrated moving average and LSSVM model to forecast the mean monthly wind speed. De Giorgi [12] developed a comparative study for the prediction of the power production of a wind farm using historical data and numerical weather predictions. The findings demonstrated that the hybrid approach based on wavelet decomposition with LSSVM significantly outperformed the hybrid artificial neural network (ANN)-based methods. Yuan [13] established a LSSVM model in the light of gravitational search algorithm (GSA) for short-term output power prediction of a wind farm. Compared with the back propagation (BP) neural network and support vector machine (SVM) model, the simulation results indicated that the GSA-LSSVM model had higher accuracy for short-term output power prediction. Wang [14] decomposed the non-stationary time series into several intrinsic mode functions (IMFs) and the corresponding, residue, then each sub-series was forecasted using diverse LSSVM models.

With the burgeoning use of artificial intelligence technology, many researchers have devoted increasing effort and time to delving into least squares support machine approaches. Since the performance of the prediction model depends on the regularization parameter and the kernel parameter of the LSSVM models, considerable research has established LSSVM models based on different intelligent algorithms for wind power prediction to attain satisfactory results [15–17]. Hu [15] introduced a modified quantum particle swarm optimization (QPSO) algorithm to select the optimal parameters of LSSVM, and the results suggested that the generalization capability and learning performance of LSSVM model were apparently enhanced. Sun [16] established a LSSVM model optimized by particle swarm optimization (PSO). The simulation results recognized that the proposed method can distinctly increase the predicting accuracy. Wang [17] constructed a LSSVM model where the parameters were tuned by a PSO method based on simulated annealing (PSOSA). A case study from four wind farms in Gansu Province, Northwest China was applied to corroborate the effectiveness of the hybrid model. Cai [18] utilized a time series model to select the input variables and multi-layer back propagation neural network and generalized regression neural network are applied it to conduct forecasting.

However, it can be concluded from the previous research that the PSO algorithm seems to suffer from the local optimum problem during the regularization parameter selection process. In order to overcome the weakness of existing algorithms, a novel global algorithm, namely, the bat algorithm (BA) originally was proposed by Yang in 2010, based on the echolocation behavior of bats [19]. With a good combination of the paramount advantages of PSO and genetic algorithm (GA), the superiority of the BA results from its simplification, powerful searching ability and fast convergence. Recently, a burgeoning number of studies focusing on the BA for parameter optimization have appeared [20–22]. Hafezi [20] explored a hybrid solution based on a BA to predict stock prices over a long term period. The model was examined through forecasting eight years of deutscher aktienindex (DAX) stock prices and conceived as an appropriate tool for predicting stock prices. Senthilkumar [21] selected the best set of features from the initial sets using a BA, perceived as a the recent optimization algorithm for reducing the time consumption in detecting record duplication. Yang [22] exploited an efficient multi-objective optimization method in accordance with the BA to suppress critical harmonics and determine power factors for passive power filters (PPFs). Considering the excellent capacity of the BA during the process of parameter optimization, it is the purpose of the current study to select the two pertinent parameters of the LSSVM model and obtain the global optimal strategy using the BA method.

From the previous literature, it can be seen that the original series tend to be regarded as the independent variables pertaining to wind power forecasting. However, it might be difficult to attain satisfactory results due to the stochastic nature and complexity of wind power generation. In order to explore a successful forecasting model, the necessity of analyzing the features of the raw time series should be increasingly highlighted. Therefore, the decomposition of wind power generation series appears to be an indispensable part in improving the forecasting accuracy. Empirical mode decomposition (EMD), perceived as an efficient decomposition method, is employed to decompose the wind power series into diverse IMFs for prediction [23,24]. Bao [23] presented a short term wind power output prediction model and the prediction of short-term wind power was implemented by differential EMD and relevance vector machine (RVM). In [24] a hybrid prediction model of wind farm power using EMD, chaotic theory and grey theory was constructed. The ultimate results indicated that the proposed method had good prediction accuracy. From the presented literature, it is possible to see that sometimes EMD cannot correctly decompose the raw data sequences. The IMFs extracted by EMD have lost their physical meaning and weaken the regularity. To address the mode mixing issue of the EMD technology, an improved method called ensemble empirical mode decomposition (EEMD) was introduced by Wu and Huang in 2009 [25]. Wang [26] selected EEMD as a data-cleaning method aiming to remove the high frequency noise embedded in the wind speed series. In this study, the EEMD is applied to decompose the original wind power generation series into several empirical modes, and the simulation results are encompassed in comparison with EMD.

Furthermore, a wealth of variables have great influence on the forecasting accuracy and efficiency, and the literature on the input selection gives this scant regard. These studies tend to select inputs using personal experience alone. However, in this research reported here, principal component analysis (PCA) was conducted to select inputs. PCA, a multivariate data analysis technology, can transform a set of correlated variables into new uncorrelated variables, namely principal components, containing most of the comprehensive variability of the original dataset. Lam [27] conducted PCA to extract a 2-component model from five raw variables for modelling the electricity use in office buildings. The literature on the importance of input dimensionality reduction and the appropriate selection of modelling variables has been widely reported. Ndiaye [28] applied PCA to select nine variables from all available variables to predict the electricity consumption in residential dwellings. Hong [29] proposed a hybrid PCA neural network model to forecast the day-ahead electricity price.

In this paper, the principal purpose of the experiment was to investigate a more accurate forecasting method for wind power generation. A hybrid model based on EEMD-principal component analysis (PCA)-least squares support vector machine (LSSVM)-bat algorithm (BA) was employed to forecast wind power generation. In addition, different models were developed using all available variables (least squares support machine-bat algorithm (LSSVM-BA), ensemble empirical mode decomposition-least squares support machine-bat algorithm (EEMD-LSSVM-BA)), and using only the variables deemed as significant by the PCA procedure (PCA-LSSCM-BA, ensemble empirical mode decomposition-principal component analysis-least squares support machine (EEMD-PCA-LSSVM), EEMD-PCA-LSSVM-BA). Therefore, a secondary aim of the present study was to determine whether an accuracy loss occurs when reducing the number of modelling variables using PCA. In comparison with the EMD method, the EEMD method can effectively mine the features of the original series through decomposing the series according to the difference of frequencies. First, the EEMD method was adopted to decompose the original wind power generation series to enhance the prediction performance. Then, PCA was utilized to reduce the number of modelling inputs by identifying the significant variables maintaining most of the information present in the data set. Finally, LSSVM models were developed to predict the sub-series. Noticeably, in this work the two parameters of LSSVM were fine-tuned by the BA to ensure the generalization and the learning ability of LSSVM. The wind power generation forecasting values can be obtained according to the accumulation of the prediction values of all sub-series. To demonstrate the effectiveness of the proposed method, a case study from China was examined and the grey relational analysis was applied to evaluate the rationality of the forecasting series stemmed from the hybrid model from the perspective of geometric shape.

The advantages of the proposed hybrid model, which result in the better forecasting performance, can be summed up in the following several aspects: in the beginning, many single methods are applied to implement wind prediction using the original series directly, but the forecasting accuracy is not very satisfactory due to the influence of random noise in the raw series. In this study, EEMD is employed to preprocess the original wind power generation series to reduce the effect of random noise. Then, the determination of inputs in the proposed model is more novel. From previous papers, the selection of inputs is usually based on personal experience. However, wind power generation may be affected by many factors such as temperature, wind speed and installed capacity. Thus, the innovation of this paper is the application of PCA to select the proper inputs. Moreover, since artificial neutral networks suffer from several disadvantages such as the occurrence of local minima, over fitting and slow convergence rate, LSSVM utilized in this study can improve the training speed for solving the problem. Unlike other LSSVM parameters optimization methods, which only utilize personal experience or traditional intelligent algorithms such as particle swarm optimization, the BA applied in this paper can avoid falling into local optimization and guarantee the generalization and the learning ability of LSSVM. Finally, grey relational analysis is utilized to demonstrate the superiority of the presented model considering the geometric shape of forecasting series and statistics. In brief, the novelty of the proposed model is described as follows: (a) a data preprocessing approach is explored to achieve the treatment of the original wind power generation series; (b) a PCA procedure is conducted to reduce the number of inputs without lowering the forecasting accuracy; (c) a LSSVM model with the relevant parameters optimized by BA is built to predict wind power generation; (d) grey relational analysis is adopted to cast light on the forecasting capacity of the proposed model.

The rest of this paper is organized as follows: Section 2 describes the modelling approaches of the proposed technique in detail. In Section 3 a hybrid model is constructed which is designed to predict wind power generation. Then, in Section 4 the proposed model is examined by a case study and an in depth comparison with other existing methods. Finally, Section 5 provides some conclusions of the whole research.

2. Methodology

2.1. Ensemble Empirical Mode Decomposition

EMD, originally proposed by Huang [30], is a powerful signal decomposition technology that aims to decompose complicated signals into several IMF components. However, sometimes EMD cannot correctly decompose the raw data sequences. These IMFs extracted by EMD have lost their physical meanings and weaken the regularity. Compared with EMD, EEMD has good performance in non-stationary signal decomposition. EEMD adds a white noise series to the raw signal $x(t)$ to eliminate the mode mixing, obtaining the IMFs through the EMD procedures. The computational steps of the EEMD algorithm are described as follows:

Step 1: Calculate $x^i(t) = x(t) + n^i(t)$, where $n^i(t)$ ($i = 1,2,3...,N$) represent the random white Gaussian noise series.

Step 2: Decompose the series $x^i(t)$ using the EMD technology to obtain IMF modes $imf^i_m(t)$ ($m = 1,2,3...,N$).

Step 3: Compute the mean of the corresponding series $imf^i_m(t)$ as follows:

$$\overline{imf_m(t)} = \frac{1}{N}\sum_{i=1}^{N} imf_m^i(t) \tag{1}$$

Step 4: Repeat the above mean procedure to complete the process of EEMD. The decomposed results of the original signal series $x(t)$ will be obtained as follows:

$$x(t) = \sum_{m=1}^{k} \overline{imf_m(t)} + r_k(t) \tag{2}$$

where $\overline{imf_m(t)}$, ($m = 1,2,3\ldots,k$) are the IMFs decomposed by EEMD, $r_k(t)$ denotes the corresponding residue.

2.2. Principal Component Analysis (PCA)

PCA based on population correlation coefficients is a statistical modelling technology which can identify the correlation among variables and generalize the data group in the light of particular linear combinations of variables, named principal components. In this study, PCA is employed to select the significant modelling inputs. Every principal component maintains interrelated variables resembling a data set. The first component indicates the paramount source of variance in the original data, and the other components account for the remaining variability. Details of the PCA method procedure are reported in [31].

2.3. Least Squares Support Vector Machine

The LSSVM, put forward by Suykens [32], is a variation of the standard support vector machine (SVM), adopting the loss function different from SVM and minimizing the square error. A quadratic programming problem can be transformed into linear equations through replacing inequality constraints with equality constraints, greatly reducing the computational complexity. In the LSSVM model, the training sample set $S = \{(x_i,y_i) \mid i = 1,2,3,\ldots,t\}$, $x_i = R^n$, $y_i = R$. Then, the optimal decision function is framed by using the high dimensional feature space. The decision function can be expressed as follows:

$$f(x) = w^T \varphi(x) + b \tag{3}$$

where $\varphi(x)$ represents the nonlinear mapping function from input space to high dimensional feature space, w is weight, b is bias.

The structural risk minimization can be described as follows:

$$R = \frac{1}{2}||w||^2 + cR_{emp} \tag{4}$$

where $||w||^2$ suggests the complex degree of the model, c is the regularization parameter, controlling the degree of punishment beyond the error samples, R_{emp} is the empirical risk function, the objective function of LSSVM is obtained as follows:

$$\min Z(w,\xi) = \frac{1}{2}||w||^2 + c\sum_{i=1}^{t} \xi_i^2 \tag{5}$$

$$s.t. \quad y_i = w\varphi(x_i) + \xi_i + b, \quad i = 1,2,3,\cdots,t$$

where ξ_i is the error, the Lagrange function can be defined as follows:

$$L(w,b,\xi,\lambda) = \frac{1}{2}||w||^2 + c\sum_{i=1}^{t} \xi_i^2 - \sum_{i=1}^{t} \lambda_i(w\varphi(x_i) + \xi_i + b - y_i) \tag{6}$$

where $\lambda_i(1,2,3,\ldots,t)$ are the Lagrange multipliers

According to the Karush-Kuhn-Tucker (KKT) conditions, Equation (7) is shown as follows:

$$\begin{cases} \omega - \sum_{i=1}^{t} \lambda_i \xi_i^2 = 0, \\ \sum_{i=1}^{t} \lambda_i = 0, \\ \lambda_i - c\xi_i = 0, \\ \omega\varphi(x_i) + \xi_i + b - y_i = 0. \end{cases} \quad (7)$$

In the light of Equation (7), the optimization problem can be converted into the process of solving linear equations, which is presented as follows:

$$\begin{bmatrix} 0 & I^T \\ I & J + \dfrac{1}{c} \end{bmatrix} \begin{bmatrix} b \\ \lambda \end{bmatrix} = \begin{bmatrix} 0 \\ y \end{bmatrix} \quad (8)$$

where $I = [11...1]^T$ is a $t \times 1$ dimensional column vector, $\lambda = [\lambda_1 \ \lambda_2 \ ...\lambda_t]^T$, $y = [y_1 \ y_2 \ ...y_t]^T$, $J_{ij} = \varphi(x_i)^T \varphi(x_j) = K(x_i, x_j)$, K is the kernel function which satisfies the condition of Mercer, the final form of LSSVM model emerges as follows:

$$f(x) = \sum_{i=1}^{t} \lambda_i K(x_i, x_j) + b \quad (9)$$

In this research, the radial basis function (RBF) is selected as the kernel function, as shown in Equation (10):

$$K(x_i, x_j) = \exp\left[\dfrac{-||x_i - x_j||^2}{2\sigma^2}\right] \quad (10)$$

where σ^2 is the parameter of the kernel function

Then, there are two parameters, the regularization parameter and the kernel parameter, determining the LSSVM model. In previous studies, experimental comparison, grid searching methods and cross validation methods were applied to optimize the two parameters, but they are time-consuming and inefficient. Therefore, this paper adopts a BA to optimize the two parameters, which can enhance and further the adaptability of the model and effectively improve the forecasting accuracy.

2.4. Bat Algorithm (BA)

The BA is a novel meta-heuristic algorithm inspired by the echolocation behavior of bats. The BA offers an excellent way for optimization and classification in a powerful selection of complicated problems [19]. The basic flow of the BA can be generalized by the pseudo code listed in Algorithm 1.

Algorithm 1. Pseudo code of the Bat Algorithm.

---

(1)	Initialize the position of bat population x_i ($i = 1, 2, ..., n$) and v_i
(2)	Initialize pulse frequency f_i at x_i, pulse rates r_i and the loudness A_i
(3)	**While** ($t <$ maximum number of iterations)
(4)	Generate new solutions by adjusting frequency
(5)	Update the velocities and solutions
(6)	**If** (rand $> r_i$)
(7)	Select a solution among the best solutions
(8)	Generate a local solution around the selected best solution
(9)	**End if**
(10)	Generate a new solution by flying randomly

---

(11)	**If** (rand $< A_i$ & $f(x_i) < f(x^*)$)
(12)	Accept the new solutions
(13)	Increase r_i and reduce A_i
(14)	**End if**
(15)	Rank the bats and find the current best x^*
(16)	**End while**

2.5. Grey Relational Analysis

Based on the proximity measure similarity, the grey relational analysis theory was first proposed by Deng [33]. The purpose of the grey relational analysis is to examine whether the various series have a close relationship on the basis of the similarity degree of the geometric shape of the series. The higher the similarity degree is, the greater the correlation is. The basic steps of grey relational analysis are as follows:

Step 1: Define reference and comparison series

A reference time series can be defined as follows:

$$Y_0 = (Y_0(1), Y_0(2), \cdots, Y_0(n)) \tag{11}$$

Then, t time series can be explained as follows:

$$Y_i = (Y_i(1), Y_i(2), \cdots, Y_i(n)), i = 1, 2, \cdots t \tag{12}$$

Step 2: Dimensionless processing of time series:

$$\overline{Y_i} = \frac{1}{n} \sum_{m=1}^{n} Y_i(m) \tag{13}$$

$$S_i = \sqrt{\frac{1}{n-1} \sum_{m=1}^{n} (Y_i(m) - \overline{Y_i})} \tag{14}$$

$$y_i(m) = \frac{Y_i(m) - \overline{Y_i}}{S_i} \tag{15}$$

Step 3: Compute the correlation coefficient:

$$r(y_0(k), y_i(k)) = \frac{\min_i \min_k |y_0(k) - y_i(k)| + \xi \max_i \max_k |y_0(k) - y_i(k)|}{|y_0(k) - y_i(k)| + \xi \max_i \max_k |y_0(k) - y_i(k)|} \tag{16}$$

where ξ is the distinguishing coefficient. In this work here, let $\xi = 0.5$.

Step 4: Calculate the grey relational degree:

$$r_i = \frac{1}{n} \sum_{k=1}^{n} r(y_0(k), y_i(k)) \tag{17}$$

where r_i is the grey relational degree of (y_0, y_i), representing the similarity degree of the geometric shape of the series.

Step 5: Sort the grey relational degree

The grey relational degrees of series are ranked according to the size of the grey relational degrees. If $r_i > r_k$, then the similarity of the curve of i series to the curve of the reference series is higher than that of the k series.

3. Wind Power Generation Forecasting Model

In this section, the proposed model (EEMD-PCA-LSSVM-BA) is constructed in detail. The flowchart of the presented model is given in Figure 1. In addition, the diverse LSSVM models are developed by using all variables from the data set, and using only the variables previously deemed significant by PCA procedure. The forecasting accuracy of both methods is compared to determine whether the PCA procedure is successful in selecting significant inputs. The following four parts constitute the hybrid model.

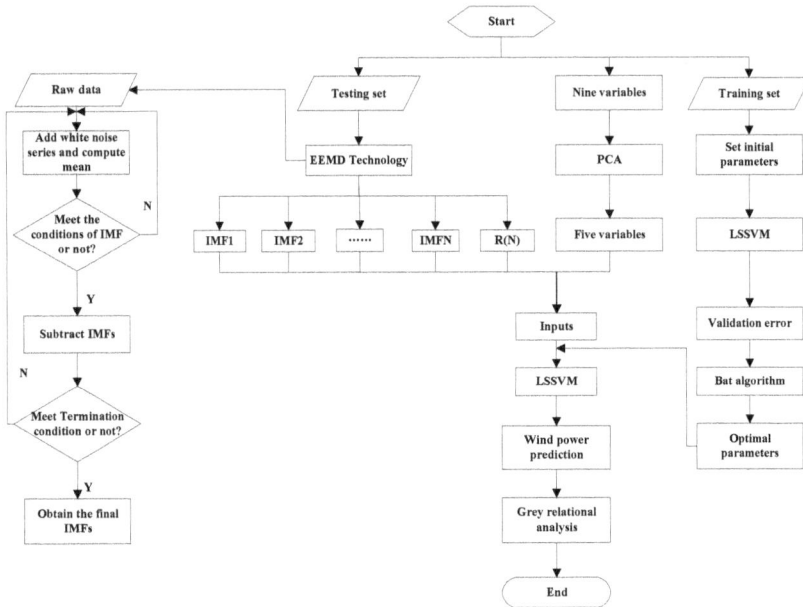

Figure 1. The flowchart of the proposed model.

Part one: Data preprocessing. The EEMD approach is adopted to decompose the original wind power generation series into different IMFs. The aim of this technology is to diminish the non-stationary character of the series for the high-precision prediction.

Part two: Input selection. Using the PCA to reduce the number of modelling inputs without lowering the prediction accuracy, the procedure can efficiently mine the significant variables containing most of the overall variability present in the data sets.

Part three: Training and validation of model. In this study, wind power generation forecasting approach is in the light of LSSVM-BA model, the basic steps can be described as follows:

Step 1: Parameter setting

The main parameters of BA are initial population size n, maximum iteration number N, original loudness A, pulse rate r, location vector x, speed vector v, respectively.

Step 2: Initialize population

Initialize the bat populations position, each bat location strategy is a component of (γ, σ^2), which can be defined as follows:

$$x = x_{\min} + \text{rand}(1, d) \times (x_{\max} - x_{\min}) \tag{18}$$

where the dimension of the bat population: $d = 2$.

Step 3: Update parameters

Calculate the fitness value of population, find the current optimal solution and update the pulse frequency, velocity and position of bats as follows:

$$f_i = f_{min} + (f_{max} - f_{min}) \times \beta \tag{19}$$

$$v_i^t = v_i^{t-1} + (x_i^t - x^*) \times f_i \tag{20}$$

$$x_i^t = x_i^{t-1} + v_i^t \tag{21}$$

where β denotes uniformly random numbers, $\beta \in [0,1]$; f_i is the search pulse frequency of the bat i, $f_i \in [f_{min}, f_{max}]$; v_i^t and v_i^{t-1} are the velocities of the bat i at time t and $t - 1$, respectively; further, x_i^t and x_i^{t-1} represent the location of the bat i at time t and $t - 1$, respectively; x^* is the present optimal solution for all bats.

Step 4: Update loudness and pulse frequency

Produce a uniformly random number rand, if rand $> r_i$, disturb the optimal strategy randomly and acquire a new strategy; if rand $< A_i$ and $f(x) > f(x^*)$, then the new strategy can be accepted, the r_i and A_i of the bat are updated as follows:

$$A_i^{t+1} = \alpha A_i^t \tag{22}$$

$$r_i^{t+1} = r_i^0 [1 - \exp(-\gamma t)] \tag{23}$$

where α and γ are constants.

Step 5: Output the global optimal solution

The current optimal solution can be obtained depending on the rank of all fitness values of the bat population. Repeat the steps of Equation (19) to Equation (21) till the maximum iterations are completed and output the global optimal solution. Therefore, a wind power generation prediction model can be generated.

In addition, the LSSVM approach is employed to model the training set, and the mean square errors of the true values and forecasting values are adopted as the fitness functions of the BA. Then, the group of parameters of LSSVM is optimized by BA for the minimum fitness value. Finally, the LSSVM model with optimal parameters can be applied to predict the wind power generation.

Part four: Wind power generation forecasting. In this part, the LSSVM approach with the parameters optimized by the BA is employed to predict each series decomposed by EEMD. Then, the forecasting series of wind power generation can be obtained by accumulating the prediction values of each subsequence. After obtaining the prediction values through the presented hybrid model, grey relational analysis was developed to determine the forecasting performance of the hybrid model.

4. Case Study

4.1. Study Area and Data Set

In this paper, the selected study area is a wind farm: Zhangjiakou, which is located in northwest China-Hebei Province, featuring an abundant wind energy source. In this work the daily wind power generation data from 1 January 2015 to 28 October 2015 are chosen as the samples to illustrate the effective performance of the proposed model. The daily measurements of the nine variables of this period are average wind speed, daily mean temperature, highest temperature, equivalent utilization hours, lowest temperature, availability of fan, maximum wind speed, minimum wind speed and installed capacity, respectively. The total number of daily wind power generation data is 301. The series are divided into two parts: training set and testing set. Data from 1 January 2015 to 8 August 2015 accounting for approximately 73% of the data are selected as training set. The rest of the data are regarded as the testing set.

4.2. Performance Criteria of Prediction Accuracy

In this paper, root mean square error (RMSE), mean absolute error (MAE) and mean absolute percentage error (MAPE) conceived as evaluation criteria are employed to assess the forecasting performance of the proposed model quantitatively:

$$\text{RMSE} = \sqrt{\frac{1}{n} \sum_{i=1}^{n} \left(x_i - \hat{x}_i \right)^2} \tag{24}$$

$$\text{MAE} = \frac{1}{n} \sum_{i=1}^{n} \left| x_i - \hat{x}_i \right| \tag{25}$$

$$\text{MAPE} = \frac{1}{n} \sum_{i=1}^{n} \left| \frac{x_i - \hat{x}_i}{x_i} \right| \tag{26}$$

where x_i is the actual value at i, and $\hat{x}_i$ is the corresponding predictive value.

4.3. Selection of Modelling Inputs

The selection of the variables utilized as the modelling inputs plays a pivotal role in exploring a powerful forecasting model. The convergence problem and poor forecasting performance may appear when redundant variables or ones that offer little contribution to the model are utilized. Moreover, the variables can increase the effort to develop models [34]. In addition, too few variables may result in lower prediction accuracy resulting from the inability of the available inputs to explain the model output behavior [35]. A successful model should employ relatively few inputs containing enough relevant information to attain satisfactory forecasting precision.

In this study, PCA was applied to measure all the variables except the wind power generation-the variable to be forecasted. According to the theory of PCA, each component is expressed by a linear equation involving all variables, and in this equation every variable can obtain a coefficient. Variables that comprise most information present in the data set and have large coefficients in the first components can be perceived as significant variables due to the fact they have the most contribution to the overall data variability. In the method proposed here, variables with coefficients having an absolute value larger than 0.1 in the components which cumulatively explain at least 95% of the overall variability were conceived to be significant. The first five components extracted from the data set approximately explain 97% of the result, but the other four components explain less than 3%. Thus, variables having coefficients with absolute values greater than 0.1 from the first five components were considered significant. The data variability explained by the top five components and the absolute values of these variable coefficients are described in Tables 1 and 2 respectively.

Table 1. Data variability explained by the top five principal components (%).

Components	Per Component	Cumulative
Comp.1	41.4457	41.4457
Comp.2	25.9709	67.4166
Comp.3	17.8437	85.2602
Comp.4	7.2593	92.5195
Comp.5	4.6729	97.1924

From Table 2, it can be seen that average wind speed, daily mean temperature, equivalent utilization hours, availability of fan and maximum wind speed have the largest coefficients contributing the most to the overall data variability. However, the phenomenon that some variables may have coefficients with absolute values slightly larger than 0.1 in the first principal components while they have much greater coefficients in the bottom components might occur. To address this issue, there needs to be a further analysis pertaining to the coefficients of the variables chosen previously. Specifically, a variable would not be perceived significant when the first five coefficients of the variable do not belong to the first five components. The ranking of the absolute values of variable correlation coefficients of the first five principal components can be noticeably indicated in Table 3. For instance, from Table 2 it can be seen that the variable including measurements of minimum wind speed has a coefficient superior to 0.1 in the component 4 which is one of the top five components. However, it can be seen from Table 3 that the all coefficients of this variable do not rank among the largest five in the first components. Therefore, this variable would not be regarded as significant.

Table 2. Variable correlation coefficients of the first five principal components.

Variables	Coefficients in Components				
	Comp.1	Comp.2	Comp.3	Comp.4	Comp.5
Average wind speed	0.3315	0.3095	0.4210	0.4513	0.0668
Daily mean temperature	0.4918	0.3049	0.0932	0.1211	0.0076
Highest temperature	0.0862	0.0893	0.0671	0.0276	0.0600
Equivalent utilization hours	0.2703	0.5167	0.1130	0.1994	0.5545
Lowest temperature	0.0800	0.0092	0.0405	0.0141	0.0137
Availability of fan	0.1835	0.3512	0.5498	0.4711	0.3210
Maximum wind speed	0.2655	0.4906	0.1342	0.1503	0.7618
Minimum wind speed	0.0652	0.0313	0.0544	0.1145	0.0300
Installed capacity	0.0000	0.0000	0.0000	0.0000	0.0000

Table 3. The ranking of the absolute values of variable correlation coefficients of the first five principal components.

Variables	Ranking of the Absolute Values of Coefficients in Components				
	Comp.1	Comp.2	Comp.3	Comp.4	Comp.5
Average wind speed	2	4	2	2	4
Daily mean temperature	1	5	5	5	8
Highest temperature	6	6	6	7	5
Equivalent utilization hours	3	1	4	3	2
Lowest temperature	7	8	8	8	7
Availability of fan	5	3	1	1	3
Maximum wind speed	4	2	3	4	1
Minimum wind speed	8	7	7	6	6
Installed capacity	9	9	9	9	9

Based on the above procedure, in this study five significant variables are identified and utilized as the inputs of the models to forecast the wind power generation. The result of selected variables through PCA can be shown in Table 4.

Table 4. The selected variables through principal component analysis (PCA).

Variables	
Average wind speed	√
Daily mean temperature	√
Highest temperature	
Equivalent utilization hours	√
Lowest temperature	
Availability of fan	√
Maximum wind speed	√
Minimum wind speed	
Installed capacity	

4.4. Ensemble Empirical Mode Decomposition Results

To improve the forecasting performance of wind power generation series, in this study, EEMD is devised for decomposing the raw series. To corroborate the performance of EEMD, EMD is employed to decompose the original series. EMD is similar to EEMD, and the two technologies both decompose the raw wind power generation series into seven IMFs and one residue. However, for the same IMF7 from Figures 2 and 3 it can be seen that EEMD can retain the true information of the original data sequence to the utmost, and effectively suppress the occurrence of mode mixing and eliminate the noise, in line with the actual situation.

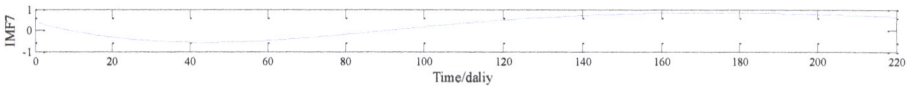

Figure 2. The ensemble empirical mode decomposition (EEMD) decomposed results of wind power generation of the training set.

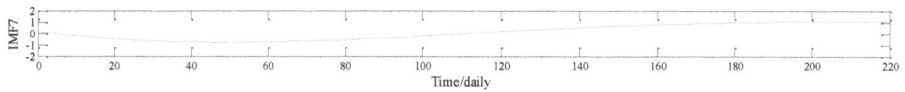

Figure 3. The empirical mode decomposition (EMD) decomposed results of wind power generation of the training set.

4.5. Selection of LSSVM Model

Previous studies on the LSSVM model for prediction demonstrate that the performance of the LSSVM approach relies on its parameters and the kernel function. The optimization of parameters is an indispensable part of any LSSVM model. The BA regarded as a population intelligent optimization algorithm offers a novel idea for searching the optimal parameters of LSSVM. In this paper, RBF is chosen as the kernel function of LSSVM algorithm, decreasing the complexity of the model and improving the training speed. Thus, the regularization parameter γ and kernel parameter σ^2 can obtain the optimal values using the powerful automatic searching ability of BA. The main parameters of the BA are listed in Table 5. Table 6 shows the optimal parameters (γ, σ^2) of the LSSVM models obtained using the BA approach.

Table 5. Main parameters of bat algorithm (BA).

Parameters	Values	Parameters	Values
Initial population size	10	Minimum frequency	0
Initial loudness	0.25	Maximum frequency	5
Pulse rate	0.5	Max-iteration number	50

Table 6. The optimal parameters in the LSSVM model. Intrinsic mode function: IMF.

Components	γ	σ^2
IMF1	0.8200	3.5274
IMF2	71.8112	5.1095
IMF3	81.2527	6.7157
IMF4	18.2928	78.1796
IMF5	13.5707	34.3623
IMF6	4.8853	54.2762
IMF7	2.2090	6.4192
RES	1.2663	11.1730

Then, the forecasting performance of the LSSVM model with the parameters tuned by the BA is examined by using the testing set. The prediction errors of LSSVM are presented in Figure 4. From Figure 4, it can be seen that the error change of LSSVM is relatively steady, and only four errors exceed 5. Furthermore, the maximum error is −5.5741 amongst all errors. Thus, it implies that the forecasting results can be considered acceptable.

In addition, in terms of EEMD-PCA-LSSVM-PSO, the max-iteration number of PSO is 200, the size of population is 20, the inertial factor (w) is 0.6, the learning factors ($c_1 = c_2$) are both 1.7. With respect to BPNN, the number of neuron in hidden layer is 13, the training number of BP is 200, and the learning rate is 0.04. Considering the EEMD-PCA-LSSVM approach, grid searching method and cross validation approach are applied to select the optimal regularization parameter (γ) and the kernel parameter (σ^2). For the grid searching method, $\gamma = 2^{-10}$–2^{15}, the step of γ is 1; $\sigma^2 = 2^{15}$–2^{-10}, the step of σ^2 is 1.

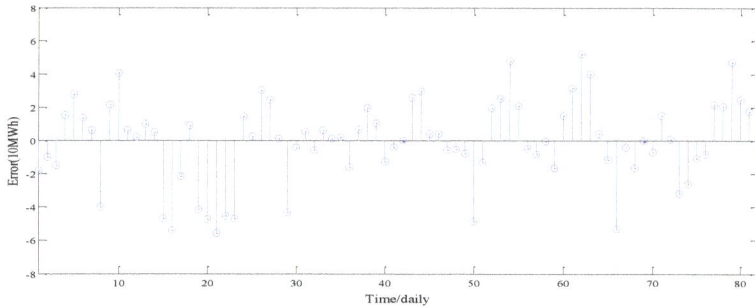

Figure 4. The prediction error of the proposed model.

4.6. Comparative Analysis of Different Methods

To illustrate the excellent performance of the proposed model, this paper employs the LSSVM-BA, principal component analysis-least squares support machine-bat algorithm (PCA-LSSVM-BA), EEMD-PCA-LSSVM, EEMD-LSSVM-BA, empirical mode decomposition-principal component analysis-least squares support machine-bat algorithm (EMD-PCA-LSSVM-BA) and ensemble empirical mode decomposition-principal component analysis-least squares support machine-particle swarm optimization (EEMD-PCA-LSSVM-PSO) for comparison. Meanwhile, the single LSSVM and ARIMA models are developed to predict wind power generation. In addition, the back propagation neutral network (BPNN) is utilized to forecast wind power generation. The comparison of prediction results with various models is shown in Table 7. Compared with other forecasting models, the proposed model displays better capacity on the prediction of wind power generation, capturing the characteristics of the wind power generation series, and achieving good forecasting performance.

Moreover, from Table 7, it can be seen that the hybrid model (EEMD-PCA-LSSVM-BA) has the highest accuracy compared with the ARIMA, BPNN, LSSVM, LSSVM-BA, PCA-LSSVM-BA, EEMD-PCA-LSSVM, EEMD-LSSVM-BA, EMD-PCA-LSSVM-BA and EEMD-PCA-LSSVM-PSO models. For instance, the proposed model achieves reductions of 44.44%, 42.99%, 41.64%, 38.14%, 28.42%, 16.58%, 20.98%, 10.93% and 9.58% in total MAPE compared with the ARIMA, BPNN, LSSVM, LSSVM-BA, PCA-LSSVM-BA, EEMD-PCA- LSSVM, EEMD-LSSVM-BA, EMD-PCA-LSSVM-BA and EEMD-PCA-LSSVM-PSO models. The abatements of MAE, RMSE and MAPE are evidently listed in Table 8. The computational formulas of the abatements of MAE, RMSE and MAPE can be defined as follows:

$$\frac{MAE_{comparative\,model} - MAE_{EEMD\text{-}PCA\text{-}LSSVM\text{-}BA}}{MAE_{comparative\,model}} \times 100\% \tag{27}$$

$$\frac{RMSE_{comparative\,model} - RMSE_{EEMD\text{-}PCA\text{-}LSSVM\text{-}BA}}{RMSE_{comparative\,model}} \times 100\% \tag{28}$$

$$\frac{\text{MAPE}_{\text{comparative model}} - \text{MAPE}_{\text{EEMD-PCA-LSSVM-BA}}}{\text{MAPE}_{\text{comparative model}}} \times 100\% \qquad (29)$$

where the comparative models represent ARIMA, BPNN, LSSVM, LSSVM-BA, PCA-LSSVM-BA, EEMD-PCA-LSSVM, EEMD-LSSVM-BA, EMD-PCA-LSSVM-BA and EEMD-PCA-LSSV-M-PSO models; the initial values of MAE, RMSE and MAPE can be obtained from Table 7.

In addition, the absolute errors between the real values and the estimated values may be captured on the basis of Figure 5. From subfigure (a) of Figure 5, it can be seen that the single ARIMA, BPNN, and LSSVM models have poor performance in forecasting wind power generation, revealing the inability of the single models to address the comprehensive features of the original wind power generation series. In contrast, the prediction values obtained by the hybrid LSSVM methods can be acceptable. Furthermore, subfigure (b) of Figure 5 shows that the hybrid models of LSSVM with the parameters optimized by intelligent algorithms have great advantages in wind power generation forecasting. Most importantly, the LSSVM-BA method outperforms the LSSVM-PSO model.

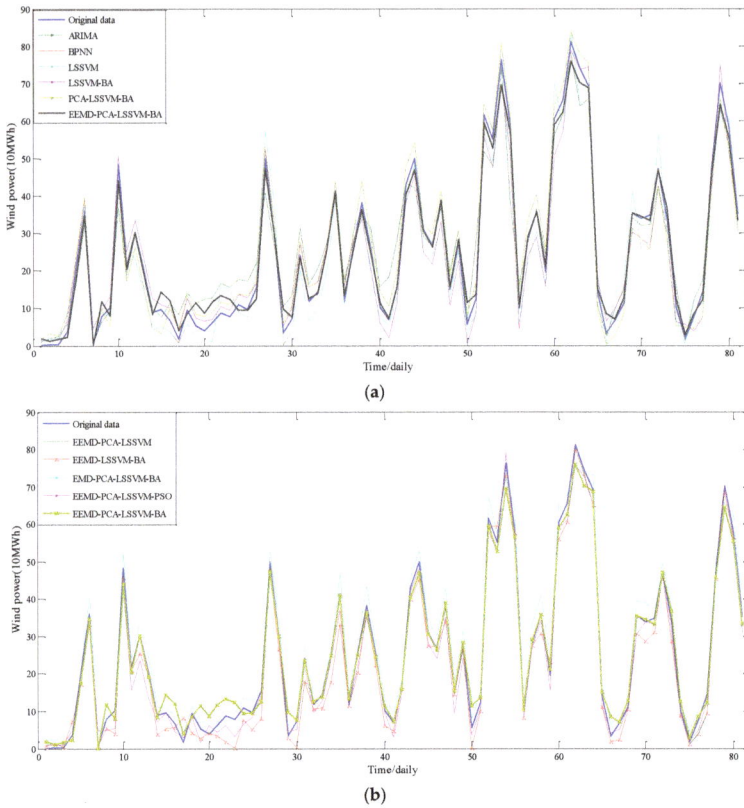

(a)

(b)

Figure 5. The forecasting results with different models. (**a**) autoregressive integrated moving average (ARIMA), BPNN, LSSVM, least squares support machine-bat algorithm (LSSVM-BA), PCA-LSSVM-BA and EEMD-PCA-LSSVM-BA models; (**b**) EEMD-PCA-LSSSVM, ensemble empirical mode decomposition-least squares support machine-bat algorithm (EEMD-LSSVM-BA), EMD-PCA-LSSVM-BA, EEMD-PCA-LSSVM-PSO and EEMD-PCA-LSSVM-BA models.

Form Table 7, it can be concluded that: (a) among all the forecasting models, the proposed EEMD-PCA-LSSVM-BA model achieves the best performance. Especially, compared with

PCA-LSSVM-BA and EMD-PCA-LSSVM-BA models, it can be found that EEMD technology can apparently enhance the forecasting ability of wind power generation series with regard to the evaluation indexes of MAE, RMSE and MAPE. For instance, the MAPE of EEMD-PCA-LSSVM-BA is 21.43%, but PCA-LSSVM-BA and EMD-PCA-LSSVM-BA models are 29.94% and 24.06%, respectively; (b) in comparison with EEMD-LSSVM-BA, the EEMD-PCA-LSSVM-BA model has better forecasting performance, demonstrating that the PCA procedure is successful in selecting significant inputs. Also, the PCA-LSSVM-BA model using the PCA-selected inputs slightly outperforms the LSSVM-BA model using all the inputs, corroborating the excellent ability of PCA procedure; (c) this study establishes two improved LSSVM models, and the performance of LSSVM based on BA is superior to the EEMD-PCA-LSSVM-PSO model concerning the three criteria of MAE, RMSE and MAPE. For instance, the MAE of EEMD-PCA-LSSVM-BA is 2.0298, while the MAE of EEMD-PCA-LSSVM-PSO is 2.4356. There seems to be a paramount reason for this phenomenon, namely that the BA adopts the major advantages of the existing intelligent algorithms in some way, combining the amazing echolocation behavior of bats, while particle swarm optimization is a special case of the BA in simplified form; (d) the improved LSSVM models have better performance than single LSSVM approach. The primary reason may be that the process of automatic searching is added to the improved LSSVM model, which equips the LSSVM model with better learning and generalization ability to acquire the global optimal solution easily; (e) in comparison with the ARIMA model merely using the raw wind power generation series, the improved LSSVM methods and the Neural network model(BPNN) are more powerful than the ARIMA model, which proves that the intelligent approaches have more research value and development space than the statistical models in the realm of wind power generation forecasting.

Furthermore, the time durations of the computing about different approaches are described in Table 9. In this study, a computer equipped with an Intel® Core™ i3-3110M processor CPU @ 2.40 GHz, 4 GB RAM and the 64 bit Windows 7 operating system (OS) was used. Also, MATLAB R2014a was applied to write all programs of this paper.

From Table 9, it can be seen that the single models such as ARIMA and LSSVM take less time compared with the hybrid models. However, the prediction accuracy of single models is lower than the hybrid models. Thus, it is reasonable to adopt more accurate wind power generation approaches taking a little more time for the security of the power system. In addition, the forecasting time of EEMD-PCA-LSSVM-BA is smaller than EEMD-PCA- LSSVM-PSO. It suggests that BA can reduce the time of parameter optimization of LSSVM effectively.

4.7. The Results of Grey Relational Analysis

In the current work, grey relational analysis is applied to verify whether the curve of the prediction result from the presented model has the highest similarity to the curve of the actual wind power generation series. Table 10 describes the grey relational degrees among the forecasting results from different models. From Table 10, it can be seen apparently that the proposed EEMD-PCA-LSSVM-BA model has the greatest similarity to the true wind power generation curve from the perspective of the geometric shape of the series.

Table 7. The comparison of prediction results with different models.

Indexes	Forecasting Methods									
	ARIMA	BPNN	LSSVM	LSSVM-BA	PCA-LSSVM-BA	EEMD-PCA-LSSVM	EEMD-LSSVM-BA	EMD-PCA-LSSVM-BA	EEMD-PCA-LSSVM-PSO	EEMD-PCA-LSSVM-BA
MAE (10 MWh)	4.5965	4.5038	4.4376	3.8998	3.8402	3.1324	3.3559	2.9335	2.4356	2.0298
RMSE (10 MWh)	5.8381	5.1245	4.9528	4.3537	4.0914	3.3859	3.9463	3.1906	2.8756	2.7117
MAPE (%)	38.57%	37.59%	36.72%	34.64%	29.94%	25.69%	27.12%	24.06%	23.70%	21.43%

Table 8. The abatements of MAE, RMSE and MAPE reductions in comparison with model except EEMD-PCA-LSSVM-BA.

Indexes	The Proportion of Reductions								
	ARIMA	BPNN	LSSVM	LSSVM-BA	PCA-LSSVM-BA	EEMD-PCA-LSSVM	EEMD-LSSVM-BA	EMD-PCA-LSSVM-BA	EEMD-PCA-LSSVM-PSO
MAE (10 MWh)	55.84%	54.93%	54.26%	47.95%	47.14%	35.20%	39.52%	30.81%	16.66%
RMSE (10 MWh)	53.55%	47.08%	45.25%	37.72%	33.72%	19.91%	31.29%	15.01%	5.70%
MAPE (%)	44.44%	42.99%	41.64%	38.14%	28.42%	16.58%	20.98%	10.93%	9.58%

Table 9. The time durations of the computing about different approaches.

CPU time (s)	Forecasting Methods									
	ARIMA	BPNN	LSSVM	LSSVM-BA	PCA-LSSVM-BA	EEMD-PCA-LSSVM	EEMD-LSSVM-BA	EMD-PCA-LSSVM-BA	EEMD-PCA-LSSVM-PSO	EEMD-PCA-LSSVM-BA
Time (s)	1.045 s	61.284 s	2.660 s	55.977 s	50.500 s	44.063 s	53.165 s	48.172 s	72.074 s	48.277 s

Table 10. The grey relational degree of various models.

Grey Relational Degree									
ARIMA	BPNN	LSSVM	LSSVM-BA	PCA-LSSVM-BA	EEMD-PCA-LSSVM-BA	EEMD-LSSVM-BA	EMD-PCA-LSSVM	EEMD-PCA-LSSVM-PSO	EEMD-PCA-LSSVM-BA
0.7954	0.7902	0.7898	0.7924	0.8064	0.8007	0.8045	0.8157	0.8209	0.8331

5. Conclusions

In order to enhance the forecasting accuracy wind power generation efficiently, a hybrid model is framed in this study. First, an EEMD technology was employed to decompose the original wind power generation series. Then, principal component (PCA) was applied to select the significant modelling inputs: five significant variables were selected from nine available inputs. Next, the relevant parameters of the proposed model were optimized by a BA. Finally, the presented method with favorable learning ability and generalization was developed to predict wind power generation. The simulation results and grey relational analysis indicate that the proposed hybrid model performs better than ARIMA, BPNN, LSSVM, LSSVM-BA, PCA-LSSVM-BA, EEMD-PCA-LSSVM, EEMD-LSSVM-BA, EMD-PCA-LSSVM-BA and EEMD-PCA-LSSVM-PSO models.

The superiority of the proposed hybrid model over other models may be accounted for by following aspects: (a) the forecasting performance of wind power generation series can be greatly augmented by using an EEMD method; (b) the simplified model using a reduced number of inputs selected by PCA procedure is more accurate than the models using all the inputs. This could suggest that variables not considered significant not only do not bring valuable information to the input set, but also add noise and unnecessary variability affecting the forecasting accuracy of models; (c) the parameters of the LSSVM models play an essential role in wind power generation prediction. Therefore, in this paper the BA is employed to optimize the parameters of the LSSVM model; (d) the hybrid model can comprehensively capture the characteristics of the raw wind power generation series, whilst the single models can only tap into the limited features of the original series. In this sense, it might be rational to see that the proposed hybrid model performs better than the other single or hybrid models regarding the criteria of MAE, RMSE and MAPE. In addition, the larger grey relational values also confirm that the proposed model outperforms the other models from the perspective of the geometric shape of forecasting series and statistics. Thus, the current method is a credible and promising algorithm for wind power generation prediction.

Regarding some limitations of this study, further research is necessary. Due to the unavailability of a reliable numerical weather prediction system, the meteorological data such as pressure, relative humidity and air density cannot be obtained. Therefore, this study only selected nine variables as alternative variables of the modelling inputs. The other relevant variables that affect wind power generation need to be investigated in further research.

Acknowledgments: This study is supported by the Fundamental Research Funds for the Central Universities (2015ZD33).

Author Contributions: Qunli Wu designed this paper and made overall guidance; Chenyang Peng wrote the whole manuscript.

Conflicts of Interest: The authors declare no conflict of interest.

References

1. Wan, C.; Xu, Z.; Pinson, P.; Dong, Z.Y.; Wong, K.P. Optimal prediction intervals of wind power generation. *IEEE Trans. Power Syst.* **2014**, *29*, 1166–1174. [CrossRef]
2. Saleh, A.E.; Moustafa, M.S.; Abo-Al-Ez, K.M.; Abdullah, A.A. A hybrid neuro-fuzzy power prediction system for wind energy generation. *Int. J. Electr. Power Energy Syst.* **2016**, *74*, 384–395. [CrossRef]
3. Foley, A.M.; Leahy, P.G.; Marvuglia, A.; McKeogh, E.J. Current methods and advances in forecasting of wind power generation. *Renew. Energy* **2012**, *37*, 1–8. [CrossRef]
4. Ewing, B.T.; Kruse, J.B.; Schroeder, J.L.; Smith, D.A. Time series analysis of wind speed using var and the generalized impulse response technique. *J. Wind Eng. Ind. Aerodyn.* **2007**, *95*, 209–219. [CrossRef]
5. Hill, D.; Bell, K.R.W.; McMillan, D.; Infield, D. A vector auto-regressive model for onshore and offshore wind synthesis incorporating meteorological model information. *Adv. Sci. Res.* **2014**, *11*, 35–39. [CrossRef]
6. Erdem, E.; Shi, J. ARMA based approaches for forecasting the tuple of wind speed and direction. *Appl. Energy* **2011**, *88*, 1405–1414. [CrossRef]

7.   Liu, H.; Erdem, E.; Shi, J. Comprehensive evaluation of ARMA–GARCH(-M) approaches for modeling the mean and volatility of wind speed. *Appl. Energy* **2011**, *88*, 724–732. [CrossRef]
8.   Kavasseri, R.G.; Seetharaman, K. Day-ahead wind speed forecasting using *f*-ARIMA models. *Renew. Energy* **2008**, *34*, 1388–1393. [CrossRef]
9.   Huang, D.Z.; Gong, R.X.; Gong, S. Prediction of wind power by chaos and BP artificial neural networks approach based on genetic algorithm. *J. Electr. Eng. Technol.* **2015**, *10*, 41–46. [CrossRef]
10. Li, G.; Shi, J. On comparing three artificial neural networks for wind speed forecasting. *Appl. Energy* **2010**, *87*, 2313–2320. [CrossRef]
11. Guo, Z.H.; Zhao, J.; Zhang, W.Y.; Wang, J.Z. A corrected hybrid approach for wind speed prediction in Hexi Corridor of China. *Energy* **2011**, *36*, 1668–1679. [CrossRef]
12. De Giorgi, M.; Campilongo, S.; Ficarella, A.; Congedo, P. Comparison between wind power prediction models based on wavelet decomposition with least-squares support vector machine (LS-SVM) and artificial neural network (ANN). *Energies* **2014**, *7*, 5251–5272. [CrossRef]
13. Yuan, X.; Chen, C.; Yuan, Y.; Huang, Y.; Tan, Q. Short-term wind power prediction based on LSSVM–GSA model. *Energy Convers. Manag.* **2015**, *101*, 393–401. [CrossRef]
14. Wang, X.; Li, H. One-month ahead prediction of wind speed and output power based on EMD and LSSVM. In Proceedings of the International Conference on Energy and Environment Technology, Guilin, China, 16–18 October 2009; pp. 439–442.
15. Hu, Z.Y.; Liu, Q.Y.; Tian, Y.X.; Liao, Y.F. A short-term wind speed forecasting model based on improved QPSO optimizing LSSVM. In Proceedings of the International Conference on Power System Technology (POWERCON), Chengdu, China, 20–22 October 2014.
16. Sun, B.; Yao, H.T. The short-term wind speed forecast analysis based on the PSO-LSSVM predict model. *Power Syst. Prot. Control* **2012**, *40*, 85–89.
17. Wang, J.Z.; Wang, Y.; Jiang, P. The study and application of a novel hybrid forecasting model—A case study of wind speed forecasting in china. *Appl. Energy* **2015**, *143*, 472–488. [CrossRef]
18. Cai, K.; Tan, L.N.; Li, C.L.; Tao, X.F. Short term wind speed forecasting combing time series and neural network method. *Power Syst. Technol.* **2008**, *32*, 82–85.
19. Yang, X.-S. A new metaheuristic bat-inspired algorithm. *Nat. Inspir. Coop. Strateg. Optim.* **2010**, *284*, 65–74.
20. Hafezi, R.; Shahrabi, J.; Hadavandi, E. A bat-neural network multi-agent system (BNNMAS) for stock price prediction: Case study of dax stock price. *Appl. Soft Comput.* **2015**, *29*, 196–210. [CrossRef]
21. Senthilkumar, P.; Vanitha, N.S. A unified approach to detect the record duplication using bat algorithm and fuzzy classifier for health informatics. *J. Med. Imaging Health Inform.* **2015**, *5*, 1121–1132. [CrossRef]
22. Yang, N.C.; Le, M.D. Optimal design of passive power filters based on multi-objective bat algorithm and pareto front. *Appl. Soft Comput.* **2015**, *35*, 257–266. [CrossRef]
23. Bao, Y.; Wang, H.; Wang, B.N. Short-term wind power prediction using differential EMD and relevance vector machine. *Neural Comput. Appl.* **2014**, *25*, 283–289. [CrossRef]
24. An, X.; Jiang, D.; Zhao, M.; Liu, C. Short-term prediction of wind power using EMD and chaotic theory. *Commun. Nonlinear Sci.* **2012**, *17*, 1036–1042. [CrossRef]
25. Wu, Z.H.; Huang, N.E. Ensemble empirical mode decomposition: A noise-assisted data analysis method. *Adv. Adapt. Data Anal.* **2009**, *1*, 1–41. [CrossRef]
26. Wang, J.; Jiang, H.; Han, B.; Zhou, Q. An experimental investigation of FNN model for wind speed forecasting using EEMD and CS. *Math. Probl. Eng.* **2015**, *2015*. [CrossRef]
27. Lam, J.C.; Wan, K.K.W.; Cheung, K.L.; Yang, L. Principal component analysis of electricity use in office buildings. *Energy Build.* **2008**, *40*, 828–836. [CrossRef]
28. Ndiaye, D.; Gabriel, K. Principal component analysis of the electricity consumption in residential dwellings. *Energy Build.* **2011**, *43*, 446–453. [CrossRef]
29. Hong, Y.-Y.; Wu, C.-P. Day-ahead electricity price forecasting using a hybrid principal component analysis network. *Energies* **2012**, *5*, 4711–4725. [CrossRef]
30. Huang, N.E.; Shen, Z.; Long, S.R.; Wu, M.C.; Shih, H.H.; Zheng, Q.; Yen, N.C.; Tung, C.C.; Liu, H.H. The empirical mode decomposition and the hilbert spectrum for nonlinear andnon-stationary time series analysis. *Proc. R. Soc. Math. Phys. Eng. Sci.* **1998**, *454*, 903–995. [CrossRef]
31. Jolliffe, I. *Principal Component Analysis*, 2nd ed.; Springer: New York, NY, USA, 2002; pp. 299–335.

32. Suykens, J.A.K.; Vandewalle, J. Recurrent least squares support vector machines. *IEEE Trans. Circuits Syst. I Fundam. Theory Appl.* **2000**, *47*, 1109–1114. [CrossRef]

33. Deng, J.L. *Introduction to Grey System Theory*; International Academic Publishers: Beijing, China, 1989; pp. 1–24.

34. Back, A.D.; Trappenberg, T.P. Selecting inputs for modeling using normalized higher order statistics and independent component analysis. *IEEE Trans. Neural Netw.* **2001**, *12*, 612–617. [CrossRef] [PubMed]

35. May, R.; Dandy, G.; Maier, H. Review of input variable selection methods for artificial neural networks. In *Artifical Netural Networks-Methodological Adances and Biomedical Applications*; Suzuki, K., Ed.; INTECH Open Access Publisher: Rijeka, Croatia, 2011; pp. 19–44.

energies

MDPI

Article

A Large-Eddy Simulation Study of Vertical Axis Wind Turbine Wakes in the Atmospheric Boundary Layer

Sina Shamsoddin and Fernando Porté-Agel *

Wind Engineering and Renewable Energy Laboratory (WIRE),
École Polytechnique Fédérale de Lausanne (EPFL), EPFL-ENAC-IIE-WIRE, Lausanne 1015, Switzerland;
sina.shamsoddin@epfl.ch
* Correspondence: fernando.porte-agel@epfl.ch; Tel.: +41-21-693-6138; Fax: +41-21-693-6135

Academic Editor: Frede Blaabjerg
Received: 29 March 2016; Accepted: 29 April 2016; Published: 13 May 2016

Abstract: In a future sustainable energy vision, in which diversified conversion of renewable energies is essential, vertical axis wind turbines (VAWTs) exhibit some potential as a reliable means of wind energy extraction alongside conventional horizontal axis wind turbines (HAWTs). Nevertheless, there is currently a relative shortage of scientific, academic and technical investigations of VAWTs as compared to HAWTs. Having this in mind, in this work, we aim to, for the first time, study the wake of a single VAWT placed in the atmospheric boundary layer using large-eddy simulation (LES). To do this, we use a previously-validated LES framework in which an actuator line model (ALM) is incorporated. First, for a typical three- and straight-bladed 1-MW VAWT design, the variation of the power coefficient with both the chord length of the blades and the tip-speed ratio is analyzed by performing 117 simulations using LES-ALM. The optimum combination of solidity (defined as Nc/R, where N is the number of blades, c is the chord length and R is the rotor radius) and tip-speed ratio is found to be 0.18 and 4.5, respectively. Subsequently, the wake of a VAWT with these optimum specifications is thoroughly examined by showing different relevant mean and turbulence wake flow statistics. It is found that for this case, the maximum velocity deficit at the equator height of the turbine occurs 2.7 rotor diameters downstream of the center of the turbine, and only after that point, the wake starts to recover. Moreover, it is observed that the maximum turbulence intensity (TI) at the equator height of the turbine occurs at a distance of about 3.8 rotor diameters downstream of the turbine. As we move towards the upper and lower edges of the turbine, the maximum TI (at a certain height) increases, and its location moves relatively closer to the turbine. Furthermore, whereas both TI and turbulent momentum flux fields show clear vertical asymmetries (with larger magnitudes at the upper wake edge compared to the ones at the lower edge), only slight lateral asymmetries were observed at the optimum tip-speed ratio for which the simulations were performed.

Keywords: vertical-axis wind turbines (VAWTs); VAWT wake; atmospheric boundary layer (ABL); large-eddy simulation (LES); actuator line model (ALM); turbulence

1. Introduction

Vertical axis wind turbines (VAWTs) offer some advantages over their horizontal axis counterparts and are being considered as a viable alternative to horizontal axis wind turbines (HAWTs). The research on VAWT technology started in the 1970s, and while the main focus of the performed studies has been on the overall turbine performance (quantities such as power and torque) and on the mechanical loading on the blades, relatively few studies have attempted to analyze the wake of a VAWT (for a comprehensive and chronological review of the studies on VAWTs before 2000, see Paraschivoiu [1] (Chapters 4–7)). Having a thorough understanding of VAWT wakes is especially crucial in designing VAWT wind farms, where downstream turbines can potentially be located in the wake of upstream ones, and consequently, the performance of the whole wind farm could be significantly affected by the wake flow characteristics. Among the experimental works investigating VAWT wakes, one can find a relatively larger number of studies that have focused only on the near wake region (e.g., [2–5]), compared to those that have considered also the far wake region (e.g., [6–8]). Nevertheless, from a wind farm design point of view, it is the far wake behavior of the flow that has more relevance and importance.

In the numerical flow simulation domain, the studies performed on the flow through VAWTs can be divided into two main categories: (1) the simulations in which the blades of the turbine (and consequently, the boundary layer around them) are resolved; and (2) the simulations in which the blades are modeled by an actuator-type technique, which uses immersed-body forces to take into account the effects of the blades on the flow. While the first approach (for instance, the work of Castelli *et al.* [9]) can be highly valuable to calculate the loading on the blades and the flow characteristics inside the rotor and in the near wake, to simulate the far wake of VAWTs and especially VAWT wind farms, the second approach is deemed to be more feasible and attainable [10,11]. The use of actuator-type techniques for VAWTs dates back to the 1980s, when Rajagopalan and Fanucci [12] for the first time modeled the VAWT rotor by a porous surface, swept by the blades, on which time-averaged blade forces are distributed and continuously act on the flow (which has also been called the actuator swept-surface model [11]). An extension of this work to three dimensions was made by Rajagopalan *et al.* [13]. Later on, Shen *et al.* [14] introduced the actuator surface model and employed it to obtain the flow field past a VAWT in two dimensions. More recently, Shamsoddin and Porté-Agel [11] used large-eddy simulation (LES) coupled with both the actuator-swept surface model (ASSM) and the actuator line model (ALM) to simulate the flow through a VAWT placed in a water channel and compared the resulting wake profiles with experimental data.

Acknowledging the fact that any given real VAWT is likely to be working in the atmospheric boundary layer (ABL) and benefiting from the helpful experience gained from the extensive research on HAWT wakes, it is imperative to study in detail the characteristics of the wake of VAWTs placed in boundary layer flows, especially if VAWT farms are to be envisaged as a viable source of power in future energy outlooks. Having this in mind, the present study is a step in this direction and attempts to use a previously-validated LES framework, in which an actuator line model is incorporated, to analyze the wake of a typical straight-bladed VAWT in a relatively long downstream range. Moreover, before the wake study, using the same framework, the power production performance of the VAWT for different combinations of blade chord lengths and tip-speed ratios is studied to find the optimum combination for the aforementioned wake analysis. To the best knowledge of the authors, this study is the first attempt to characterize the wake of a VAWT in ABL using LES.

The LES framework is presented in Section 2, and the numerical setups and techniques are described in Section 3. Next, the results for both the power production parametric study and the wake analysis are presented and discussed in Section 4. Finally, a summary of the study is given in Section 5.

2. Large-Eddy Simulation Framework

In the LES framework used for the simulations of this paper, the filtered incompressible Navier–Stokes equations (for a neutrally-stratified ABL) are solved. These equations can be written in rotational form as:

$$\frac{\partial \tilde{u}_i}{\partial x_i} = 0 \tag{1}$$

$$\frac{\partial \tilde{u}_i}{\partial t} + \tilde{u}_j \left(\frac{\partial \tilde{u}_i}{\partial x_j} - \frac{\partial \tilde{u}_j}{\partial x_i} \right) = -\frac{\partial \tilde{p}^*}{\partial x_i} - \frac{\partial \tau_{ij}}{\partial x_j} - \frac{f_i}{\rho} + F_p \delta_{i1} \tag{2}$$

where the tilde represents a three-dimensional spatial filtering operation at scale $\tilde{\Delta}$, $\tilde{u}_i$ is the filtered velocity in the i-th direction (with $i = 1, 2, 3$ corresponding to the streamwise (x), spanwise (y) and vertical (z) directions, respectively), $\tilde{p}^* = \frac{\tilde{p}}{\rho} + \frac{1}{2}\tilde{u}_i\tilde{u}_i$ is the modified kinematic pressure where $\tilde{p}$ is the filtered pressure, $\tau_{ij} = \widetilde{u_i u_j} - \tilde{u}_i \tilde{u}_j$ is the kinematic subgrid-scale (SGS) stress, f_i is a body force (per unit volume) representing the force exerted by the flow on the turbine blades (observe the minus sign), F_p is an imposed pressure gradient and ρ is the constant fluid density. In this paper, u, v and w notations are also used for the u_1, u_2 and u_3 velocity components, respectively. Regarding the parametrization of the SGS stresses, in these simulations, the Lagrangian scale-dependent dynamic model [15] is used.

To parameterize the VAWT-induced forces on the flow (*i.e.*, to model the term f_i/ρ in Equation (2)), an actuator line model is used. According to the ALM, each blade of the turbine is represented by an actuator line on which the turbine forces, calculated based on the blade-element theory, are distributed. This method has the advantage of being capable of tracking the rotation of the blades at each time step. For a detailed explanation of the application of the ALM for VAWTs, the reader can refer to Shamsoddin and Porté-Agel [11] (Section 2.2).

3. Numerical Setup

In this section, the techniques used to numerically solve Equations (1) and (2), as well as the configuration of the performed numerical experiments are presented.

The LES code, which is used to realize the simulations in this study, is a modified version of the code described by Albertson and Parlange [16], Porté-Agel *et al.* [17] and Porté-Agel *et al.* [18]. The computational mesh is a 3D structured one, which has N_x, N_y and N_z nodes in the x, y and z directions, respectively. The mesh is staggered in the z direction in a way that the layers in which the vertical component of velocity (w) is stored are located halfway between the layers in which all of the other main flow variables (u, v, p) are stored. The first w-nodes are located on the $z = 0$ plane, while the first uvp-nodes are located on the $z = \Delta z/2$ plane.

To compute the spatial derivatives, a Fourier-based pseudospectral scheme is used in the horizontal directions, and a second-order finite difference method is used in the vertical direction. The governing equations for conservation of momentum are integrated in time with the second-order Adams–Bashforth scheme.

The pressure term in Equation (2) is not a thermodynamic quantity, and it only serves to have a divergence-free (*i.e.*, incompressible) velocity field. Therefore, by taking the divergence of the momentum Equation (2) and using the continuity Equation (1), we can solve the arising Poisson equation for the modified pressure, $\tilde{p}^*$, using the spectral method in the horizontal directions and finite differences in the vertical direction.

The boundary conditions (BCs) in the horizontal directions are mathematically (and implicitly through using the spectral method) periodic. For the bottom BC, the instantaneous surface shear stress is calculated using the Monin–Obukhov similarity theory [19] as a function of the local horizontal velocities at the nearest (to the surface) vertical grid points ($z = \Delta z/2$) (see, for instance, Moeng [20], Stoll and Porté-Agel [21]). For the upper boundary, an impermeable stress-free BC is applied, *i.e.*, $\partial \tilde{u}_1/\partial z = \partial \tilde{u}_2/\partial z = \tilde{u}_3 = 0$.

Since the study of the flow through a single turbine is desired, we need to numerically enforce an inflow BC to practically override the implicitly-imposed periodic BC in the x direction. For this purpose, a buffer zone upstream of the VAWT is employed to adjust the flow to an undisturbed ABL inflow condition. The inflow field is obtained by saving the instantaneous velocity components in a specific y-z plane in a similar precursory simulation of ABL over a flat terrain (with the same surface roughness) with no turbine on it. The use of this technique, *i.e.*, using an inflow boundary condition in a direction in which the flow variables are discretized using Fourier series, has been shown to be successful in the works of Tseng *et al.* [22], Wu and Porté-Agel [23] and Porté-Agel *et al.* [24].

To implement the ALM, values of the airfoil's lift and drag coefficients (C_L and C_D, respectively) as a function of Reynolds number (Re) and angle of attack (α) (*i.e.*, $C_{L(D)} = f(Re, \alpha)$) are needed. This information was obtained from the tabulated data provided by Sheldahl and Klimas [25]. Moreover, the dynamic stall phenomenon, which is known to have a considerable effect on the performance of VAWTs [14,26], is accounted for using the modified MIT model [27]. A detailed explanation of the implementation of the dynamic stall model is provided in Appendix A.

Figures 1 and 2 show the geometrical specifications of the VAWT and the computational domain in which it is placed. The turbine rotor is made of three straight blades and has a diameter (*D*) of 50 m and a height of 100 m. The blades' airfoil is selected to be the symmetrical NACA 0018 airfoil, which is widely used for VAWTs. It is attempted that these chosen turbine specifications are representative of those of real VAWTs with a nominal capacity of 1 MW (this fact will be reaffirmed by the results of the simulations). For example, a curve-bladed (or Φ-rotor) VAWT of similar size and capacity (96 m high and an equatorial diameter of 64 m) with two NACA 0018 blades of a 2.4-m chord length was operational as part of Project Éole in Cap Chat, Quebec, Canada, between 1987 and 1993 [28]. This turbine was designed to deliver a maximum power of about 4 MW (at high winds and high rotational speeds), and its maximum measured power of about 1.3 MW is hitherto one of the greatest measured power outputs for a VAWT ([1] Section 7.3.4).

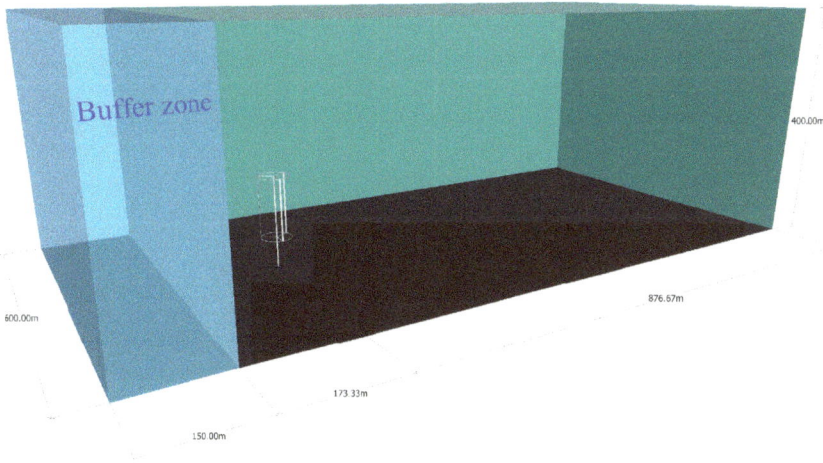

Figure 1. Schematic of the computational domain, including the simulated VAWT.

The buffer zone occupies about 12% of the domain length. The domain dimensions are $L_x = 1200$ m (=24*D*), $L_y = 600$ m (=12*D*) and $L_z = 400$ m (=8*D*) in the streamwise, spanwise and vertical directions, respectively. The blockage ratio of the turbine in the computational domain is 2.08%, which is well below the value of 10%, which is reported by Chen and Liou [29] as the threshold below which it is acceptable to neglect the blockage effect. Regarding the computational mesh, the number of grid points in each of the three directions is $N_x = 360$, $N_y = 180$ and $N_z = 240$. The code has been shown to yield grid-independent results provided that a minimum number of grid points is used to resolve the rotor [11]. In this study, we chose a resolution (15 points in each horizontal direction covering the rotor area) that falls within the grid-independent range. The time resolution for all of the simulations is 0.0155 s. For the wake study simulation, the total physical time of the simulation is 90.4 min, and for mean velocity and turbulence statistics results, we have time-averaged the quantities in question over the final 77.5-min time span.

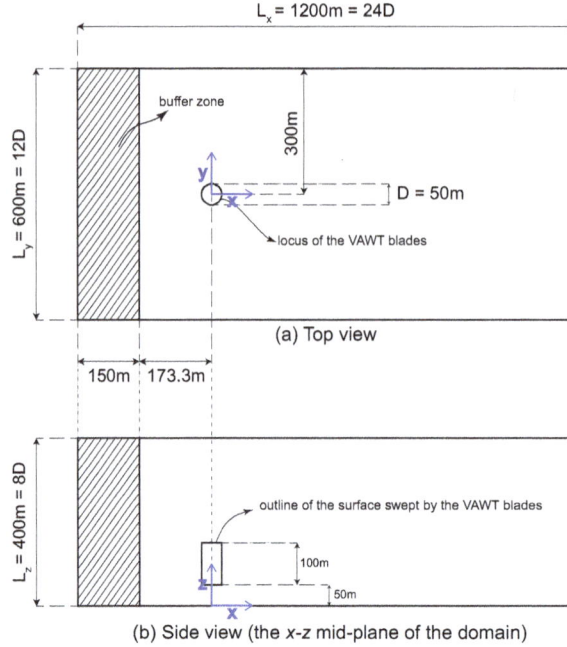

Figure 2. Plane views of the geometrical configuration of the simulations: (**a**) top view of the domain; (**b**) side view of the domain, seen in the *x-z* mid-plane of the domain.

Figure 3 shows mean and standard deviation profiles of the inflow streamwise velocity. As mentioned earlier, the inflow field is generated by using the flow field of a precursory simulation of the neutrally-stratified ABL on a flat terrain. The surface roughness, z_0, and the friction velocity, u_*, used in this precursory simulation are 0.1 m and 0.52 m/s, respectively. In Figure 3a, it can be seen that the mean streamwise velocity profile approximately follows the log law in the surface layer. The mean inflow streamwise velocity at the equator height of the turbine (*i.e.*, $z = 100$ m in this case), U_{eq}, and the turbulence intensity of the inflow at the same height (σ_u / U_{eq}) are 9.6 m/s and 8.3%, respectively. It should be noted that the above-mentioned inflow field is used for all of the simulations of this paper (*i.e.*, both Subsections 4.1 and 4.2).

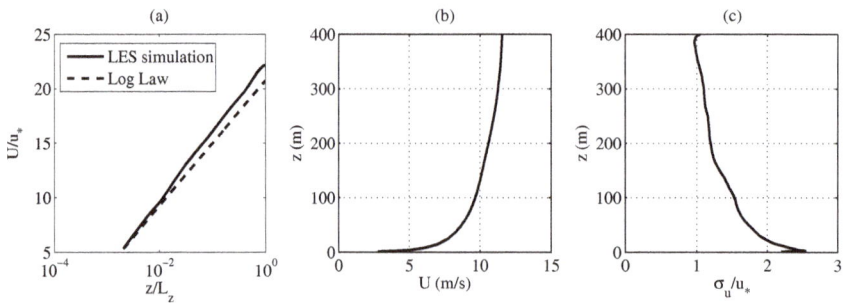

Figure 3. Inflow characteristics: (**a**) vertical profile of the mean streamwise velocity compared to a log-law profile (horizontal axis in logarithmic scale); (**b**) vertical profile of the mean streamwise velocity (linear scale); (**c**) vertical profile of the standard deviation of the streamwise velocity.

4. Results and Discussion

In this section, the results of the simulations are presented and discussed. First, we examine the turbine's energy-extraction performance, and next, we study the wake flow of a VAWT placed in the ABL.

4.1. Turbine Performance and Power Extraction

In this subsection, we are interested in how the power production of the turbine is affected by different combinations of tip-speed ratio, TSR, and chord length, c. For this purpose, 117 simulations have been performed to obtain the power coefficient, C_P, of the turbine as a function of both TSR and c, i.e., $C_P(TSR, c)$. Figure 4 shows how C_P varies with different values of TSR and c. Figure 4a is generated with a resolution of 0.5 m for chord length and 0.5 for TSR. It can be seen that, as we increase the chord length, the useful TSR range (a range in which C_P is higher than a certain value) decreases. Moreover, the figure shows that the maximum power coefficient of the turbine occurs for a TSR of 4.5 and a chord length of 1.5 m (which corresponds to a solidity of $Nc/R = 0.18$, where N is the number of blades and R is the rotor radius). This combination results in a power extraction, P, of 1.3 MW and a C_P of 0.47 (C_P is defined as $C_P = P/(0.5\rho DHU_{eq}^3)$), where ρ is the fluid density and considered equal to 1.225 kg/m^3, D is the rotor diameter and H is the rotor height).

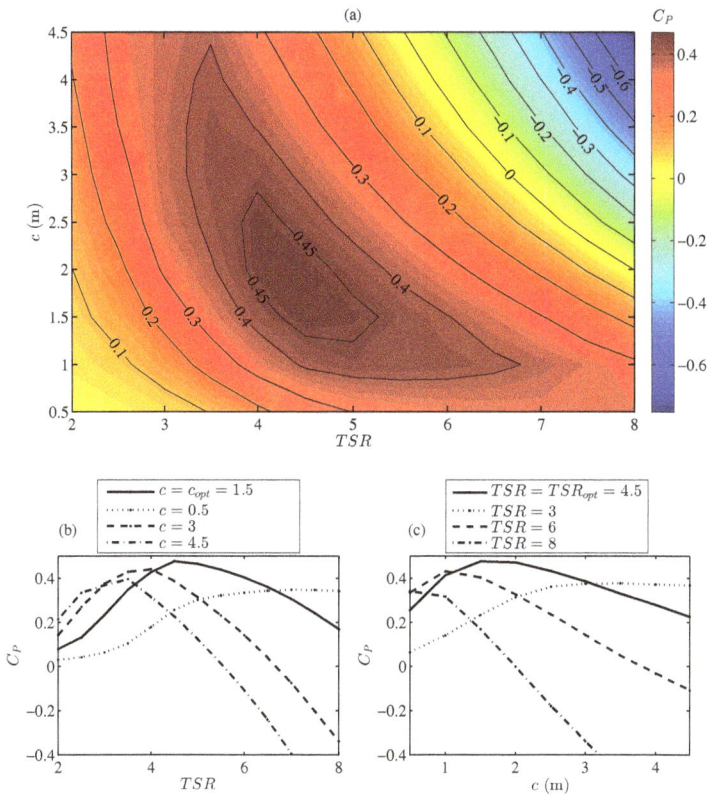

Figure 4. Variation of the power coefficient of a three-bladed VAWT with tip-speed ratio (TSR) and chord length: (**a**) C_P as a function of both TSR and chord length; (**b**) C_P as a function of TSR for four different chord lengths; (**c**) C_P as a function of chord length for four different TSR values.

4.2. VAWT Wake

In this subsection, we have picked the optimum combination of TSR and chord length ($TSR = 4.5$ and $c = 1.5$ m) for the VAWT rotor and studied the wake flow behind it. Figure 5 shows the instantaneous streamwise velocity field of the flow in three different orthogonal planes. In all of the following figures in this section containing contour plots, the black circles and rectangles represent the outline of the locus of the blades. The sense of the rotation of the turbine blades is counterclockwise when seen from above. The wake of the VAWT and the highly turbulent nature of the flow are obvious in this figure and in the Videos S1 and S2 included in the Supplementary Material. It should be noted that the average thrust coefficient of the turbine (defined as $C_T = T/(0.5\rho DHU_{eq}^2)$, where T is the total thrust force of the turbine in the x direction) in this case is found to be 0.8.

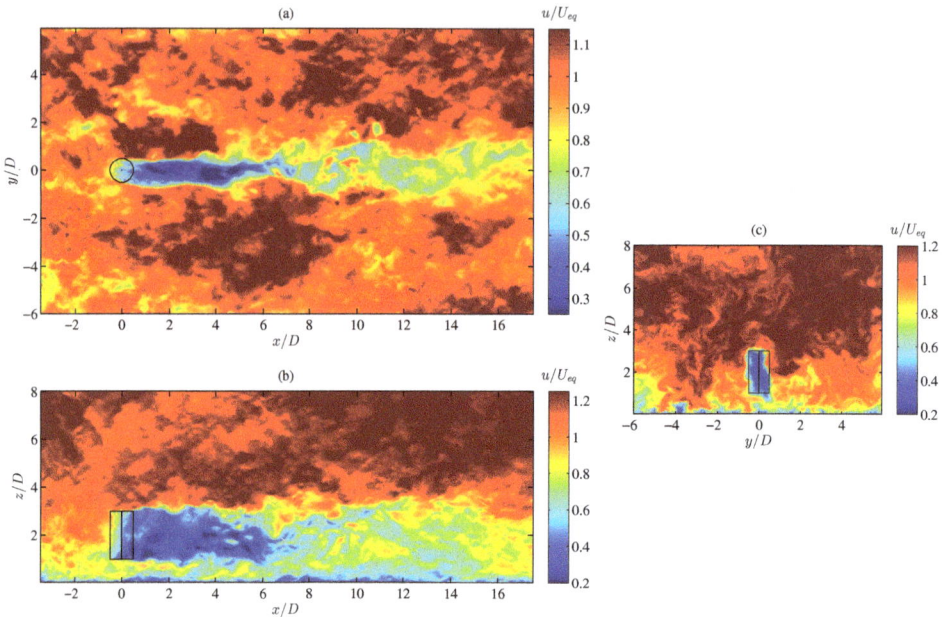

Figure 5. Contour plots of the instantaneous normalized streamwise velocity (u/U_{eq}) in three different planes: (**a**) the *x-y* plane at the equator height of the turbine; (**b**) the *x-z* plane going through the center of the turbine; (**c**) the *y-z* plane which is 2D downstream of the center of the turbine.

Figures 6 and 7 show contour plots of the mean streamwise velocity in the *x-y* plane at the equator height of the turbine and in the *x-z* mid-plane of the turbine. It can be seen in these figures that it takes a long distance for the wake to recover; at a downwind distance as large as 14 rotor diameters, the wake center velocity reaches only 85% of the incoming velocity. Moreover, Figure 8 shows the mean velocity contours in six *y-z* planes downstream of the turbine. In all of these figures (Figures 6–8), one can observe how the wake recovers in the streamwise direction (after a certain distance) and how it expands in the spanwise direction as it advances farther downstream.

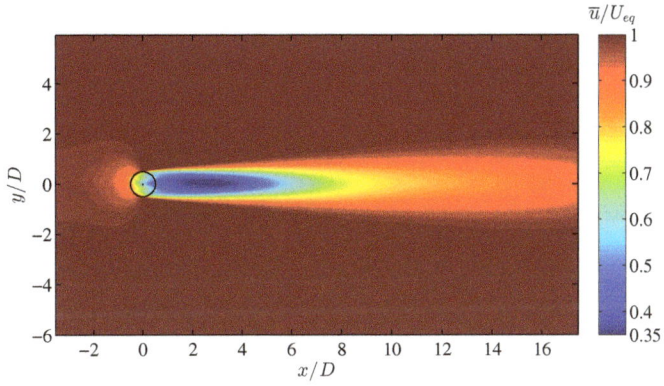

Figure 6. Contours of the normalized mean streamwise velocity ($\bar{u}/U_{eq}$) in the x-y plane at the equator height of the turbine.

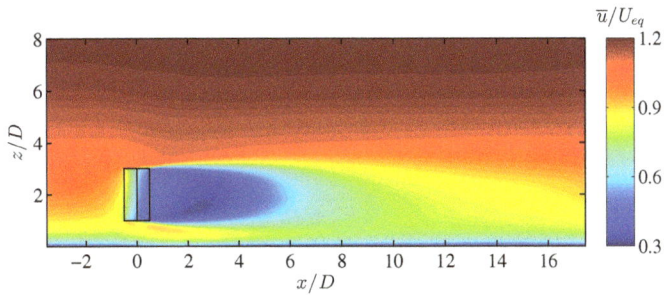

Figure 7. Contours of the normalized mean streamwise velocity ($\bar{u}/U_{eq}$) in the x-z plane going through the center of the turbine.

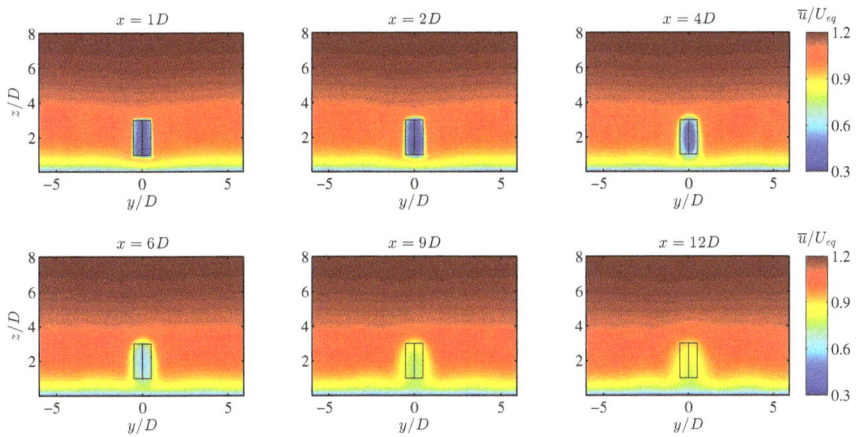

Figure 8. Contour plots of the normalized mean streamwise velocity ($\bar{u}/U_{eq}$) in six different y-z planes at different distances downstream of the center of the turbine.

To have a more quantitative and precise insight about the VAWT wake, Figures 9 and 10 can be consulted. Figure 9 shows spanwise profiles of the mean streamwise velocity in a horizontal plane at the equator height of the turbine in eight downstream positions. Besides, Figure 10 presents vertical profiles of the mean streamwise velocity in the x-z mid-plane of the turbine at different downstream positions. An interesting observation that can be made from Figures 6, 9 and 10 is that the maximum velocity deficit occurs at a downstream distance of about 2.7 rotor diameters; this distance is significantly larger than the equivalent one for the case of HAWT wakes [23]. After the point where the maximum velocity deficit (more than 65% of U_{eq} in this case) occurs has been reached, the wake starts to recover with a relatively high recovery rate (defined here as the magnitude of the rate of change of the maximum velocity deficit with streamwise distance). As we go farther downstream, the rate of the wake recovery decreases considerably; so that in the distances as large as 17 rotor diameters, where the velocity deficit reaches values of about 90%, the recovery rate is comparably very small.

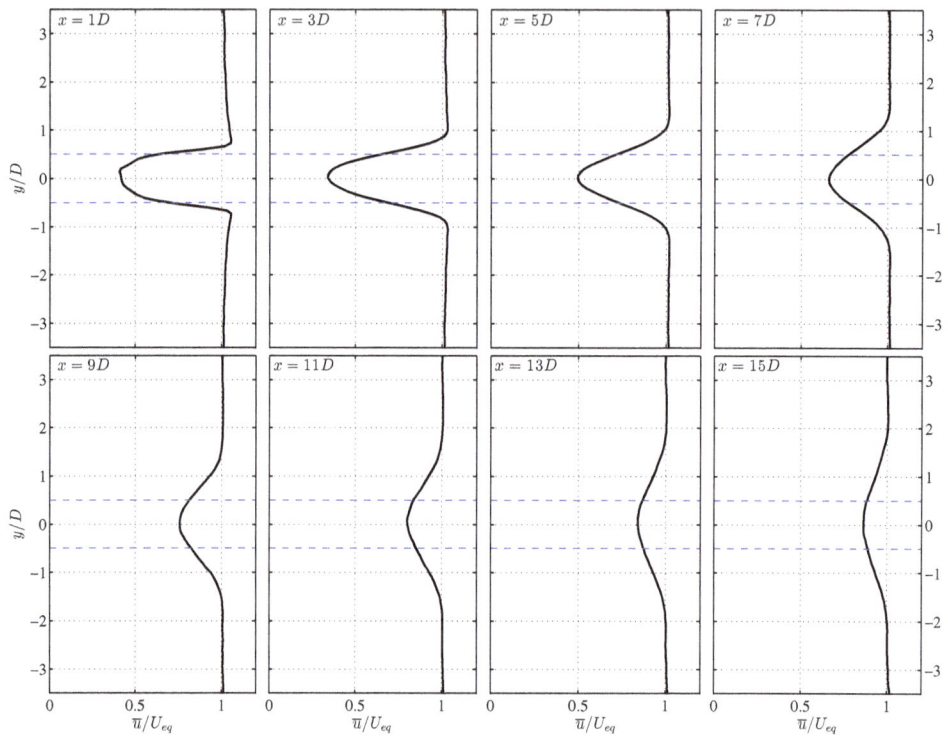

Figure 9. Horizontal-spanwise profiles of the normalized mean streamwise velocity ($\bar{u}/U_{eq}$) in the x-y plane at the equator height of the turbine at different downstream positions. The blue horizontal dashed lines show the extent of the turbine.

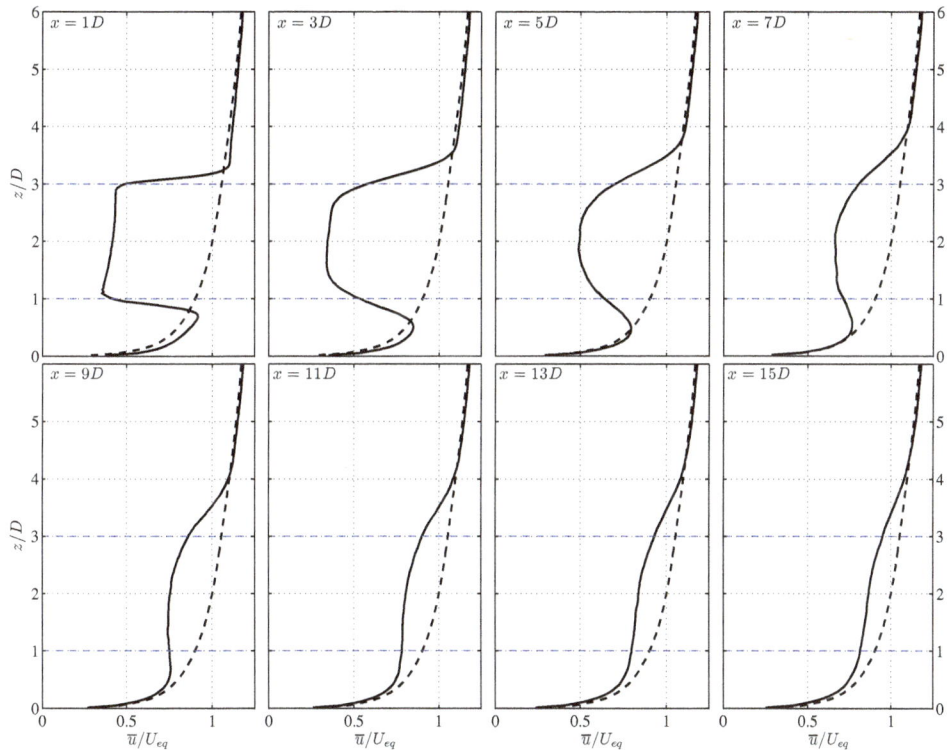

Figure 10. Vertical profiles of the normalized mean streamwise velocity ($\bar{u}/U_{eq}$) in the x-z plane going through the center of the turbine at different downstream positions. The black dashed line represents the inflow profile, and the blue horizontal dashed lines show the extent of the turbine.

Another group of crucial quantities that has a significant importance in characterizing turbine wakes is the turbulence-related statistics, such as turbulence intensity and turbulent fluxes. These quantities are especially important for the design of wind farms, due to their role in both wake recovery and mechanical loads on turbine blades. Figure 11 shows contours of turbulence intensity (TI) in two different orthogonal planes (x-y and x-z) in the wake of the turbine. Here, the turbulence intensity is defined as $TI = \sigma_u/U_{eq}$. In addition, Figure 12 shows the distribution of TI in y-z planes at different downstream locations. In Figure 11a, it can be seen that two branches of high TI regions start to develop from the two spanwise extremities of the rotor swept surface (the black circle in the figure). These two branches grow in spanwise width as we go further downstream, until the point where they meet each other (for this case, in about 3.5 rotor diameters downstream of the turbine in the horizontal mid-plane of the turbine). Starting from the turbine area, the TI in each of these branches increases, until a point where the maximum TI occurs (about 3.8 rotor diameters downstream in this case); after this maximum point, the TI starts to decrease as the flow advances downstream, while the width of the branches continues to expand. Figure 13 examines the previous figure quantitatively, by showing the spanwise profiles of the TI in the equator height of the turbine. One can readily see that at each downstream position, the horizontal TI profiles have two maxima at two spanwise positions, which correspond to the two aforementioned TI branches. Although slight asymmetries can still be seen in the TI values of the two branches, the lateral asymmetry is not significantly pronounced. It should be noted that the degree to which the VAWT wake is laterally asymmetric is influenced by parameters, such as TSR, airfoil type and the Reynolds number in which the turbine is working.

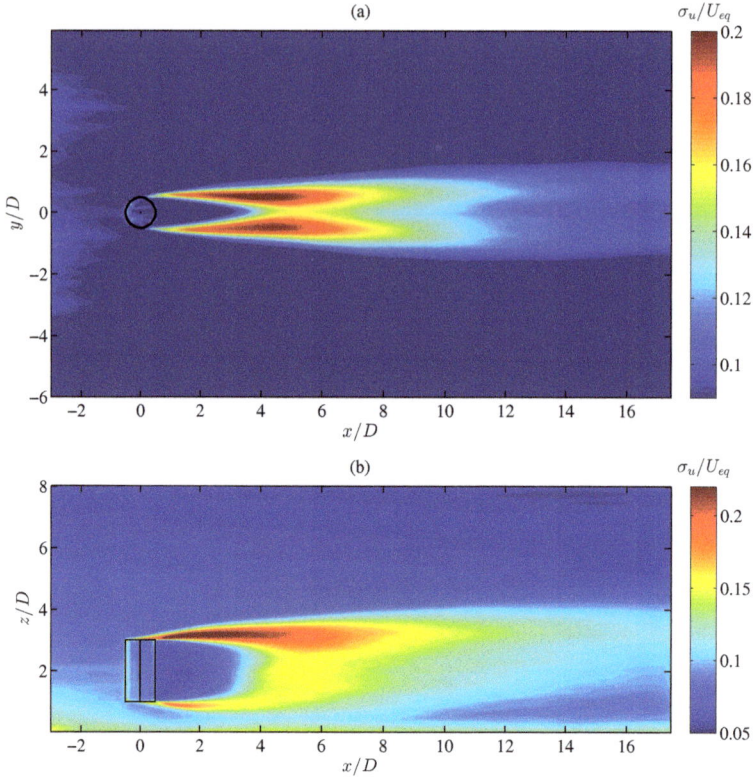

Figure 11. Contour plots of the streamwise turbulence intensity, σ_u / U_{eq}: (**a**) in the x-y plane at the equator height of the turbine; (**b**) in the x-z plane going through the center of the turbine.

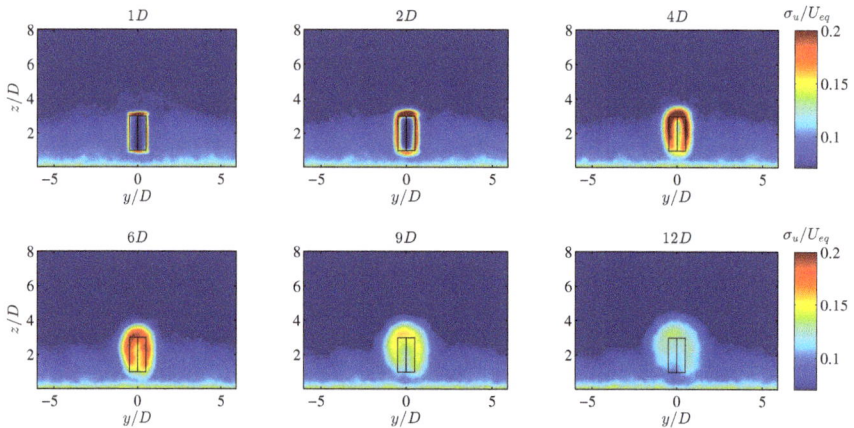

Figure 12. Contour plots of the streamwise turbulence intensity, σ_u / U_{eq}, in six different y-z planes at different distances downstream of the center of the turbine.

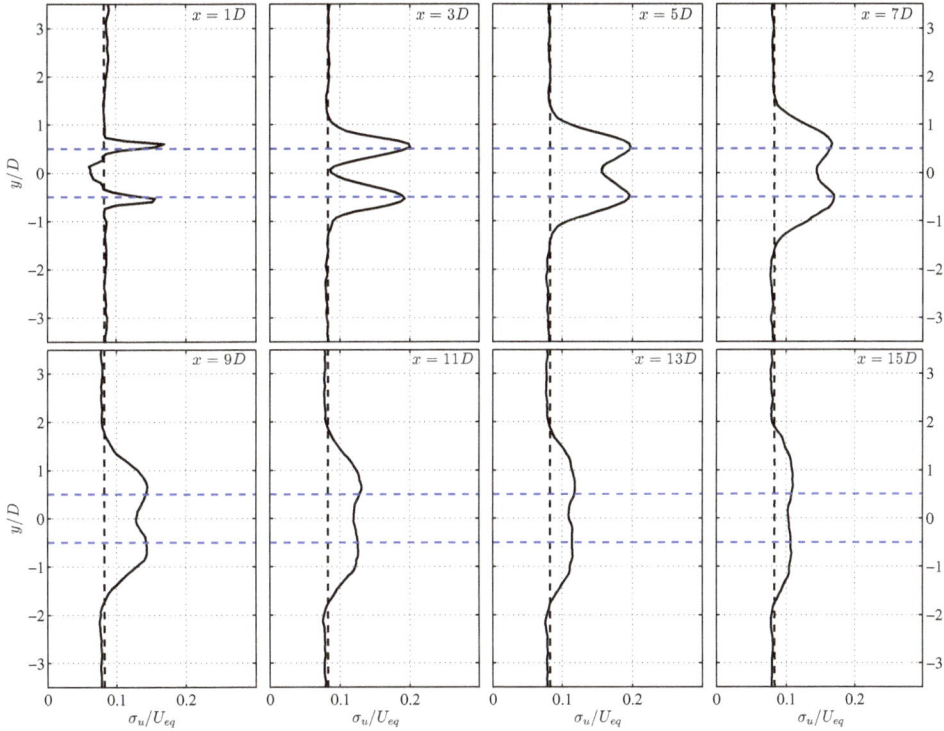

Figure 13. Horizontal profiles of the streamwise turbulence intensity in the *x*-*y* plane at the equator height of the turbine at different downstream positions. The black dashed line represents the inflow profile, and the blue horizontal dashed lines show the extent of the turbine.

In Figure 11b, one can observe a similar behavior by noticing the two high TI regions originating from the upper and lower extremities of the turbine; however, in this case, the TI originating from the upper edge of the blades is clearly larger than the one originating from the lower edge. To further quantify this, and to have a better understanding of the vertical variation of the TI in a VAWT wake, one can study Figure 14, in which vertical profiles of TI are shown at different downstream positions. In this figure, it can also be seen that in the region below the turbine blades' lower edge, the turbulence intensity has even decreased to values lower than the inflow TI; this behavior has also been observed in HAWT wakes, as well (e.g., [23]).

Furthermore, turbulent momentum fluxes in the VAWT wake are believed to be worthy of inspection, as they quantify the rate of flow entrainment into the wake, which is responsible for the recovery and lateral expansion of the wake. Figure 15 shows the normalized lateral turbulent flux ($\overline{u'v'}$) at the horizontal mid-plane of the turbine. The positive and negative regions of $\overline{u'v'}$, which are located on the two lateral edges of the wake, show an inward entrainment of momentum into the wake region. This lateral entrainment can also be seen in the *y*-*z* planes in Figure 16. Figure 17 shows the spanwise profiles of $\overline{u'v'}$ at the equator height of the turbine at different downstream distances from the turbine. We notice that for this case, the maximum absolute value of $\overline{u'v'}$ at the equator height of the turbine occurs between 4.5 and 5*D* (4.9*D* for positive values and 4.7*D* for negative values) downstream of the turbine, which is about 1*D* farther downwind compared to the maximum TI point. It should be noted that, again in this figure, only a slight lateral asymmetry (in terms of $|\overline{u'v'}|$) can be observed.

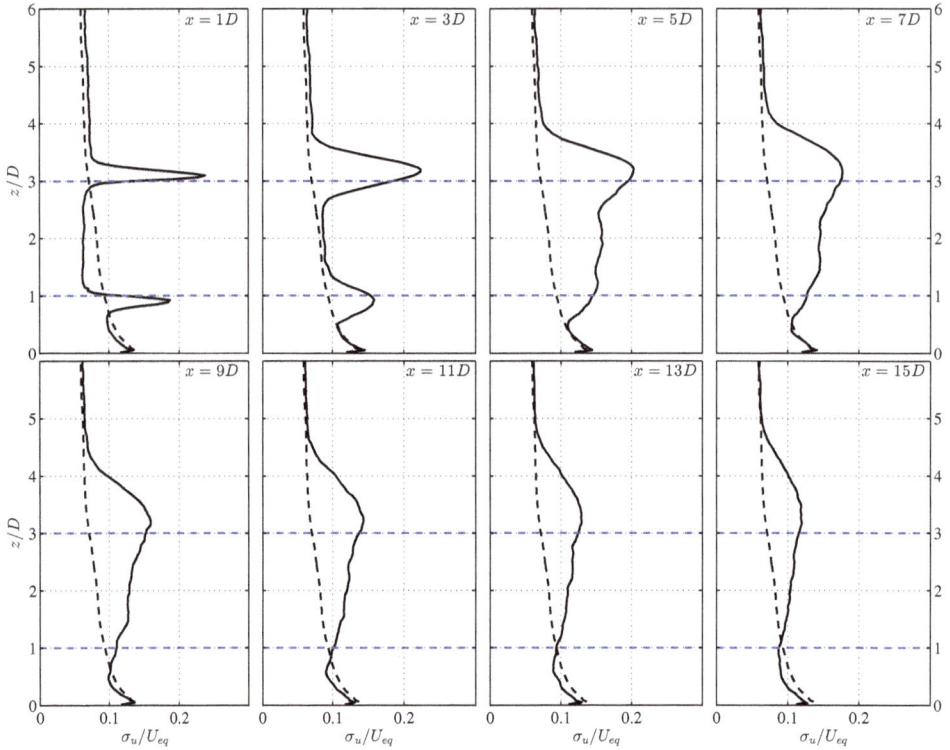

Figure 14. Vertical profiles of the streamwise turbulence intensity in the x-z plane going through the center of the turbine at different downstream positions. The black dashed line represents the inflow profile, and the blue horizontal dashed lines show the extent of the turbine.

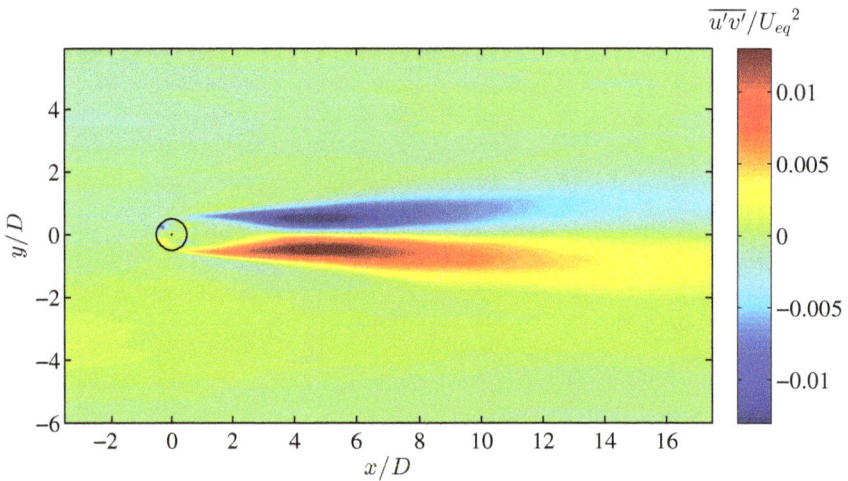

Figure 15. Contour plot of the normalized lateral turbulent flux, $\overline{u'v'}/U_{eq}{}^2$, in the x-y plane at the equator height of the turbine.

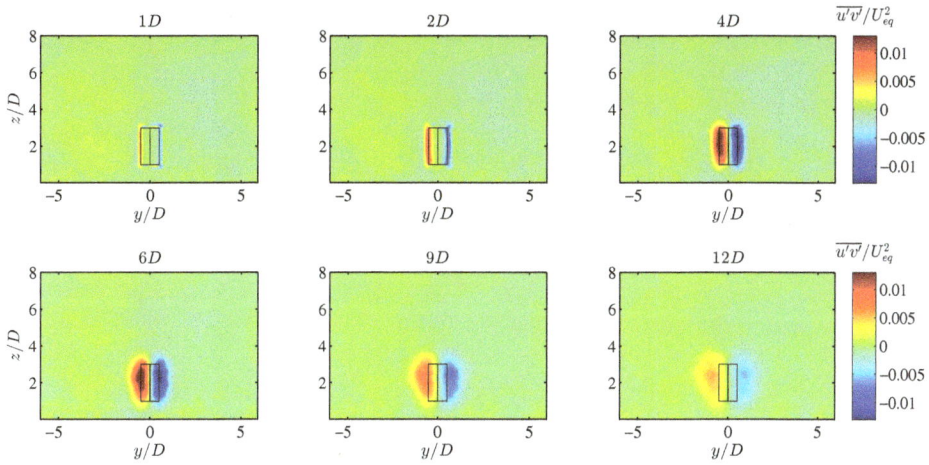

Figure 16. Contour plots of the normalized lateral turbulent flux, $\overline{u'v'}/U_{eq}^2$, in six different y-z planes at different distances downstream of the center of the turbine.

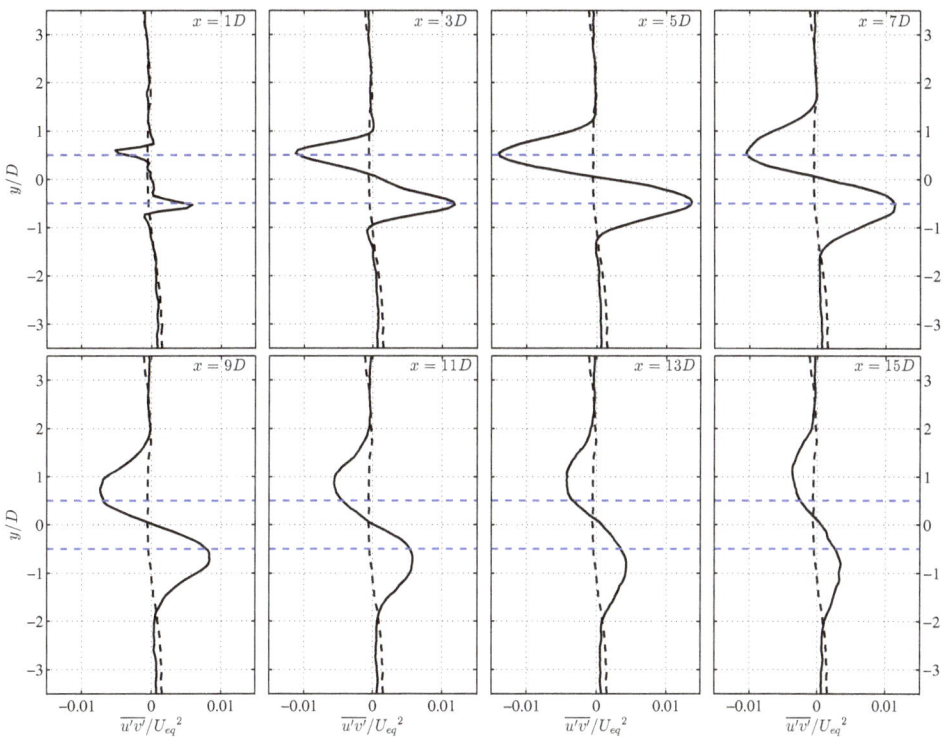

Figure 17. Horizontal profiles of the normalized lateral turbulent flux $(\overline{u'v'}/U_{eq}^2)$ in the x-y plane at the equator height of the turbine at different downstream positions. The black dashed line represents the inflow profile, and the blue horizontal dashed lines show the extent of the turbine.

Figures 18 and 19 show the normalized vertical turbulent flux ($\overline{u'w'}$) in the wake flow. The vertical inward entrainment from both above and below the wake region is clear in these figures. Figure 20 displays the vertical profiles of $\overline{u'w'}$ in the x-z plane going through the center of the turbine. It can be seen in this figure that the values of the vertical turbulent flux are higher in upper edge of the wake with respect to the lower edge. Here, we can observe that the magnitude of $\overline{u'w'}$ (in the aforesaid vertical plane) peaks relatively close to the turbine (1.9D for positive values and 0.5D for negative values) at heights near to the ones of the upper and lower edges of the blades.

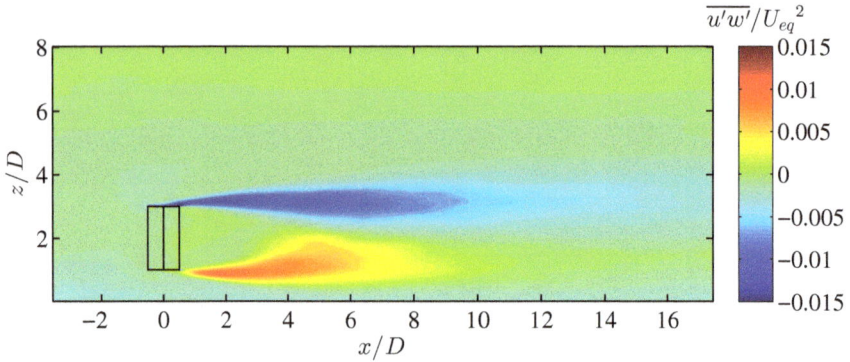

Figure 18. Contour plots of the normalized vertical turbulent flux, $\overline{u'w'}/U_{eq}^2$, in the $x-z$ plane going through the center of the turbine.

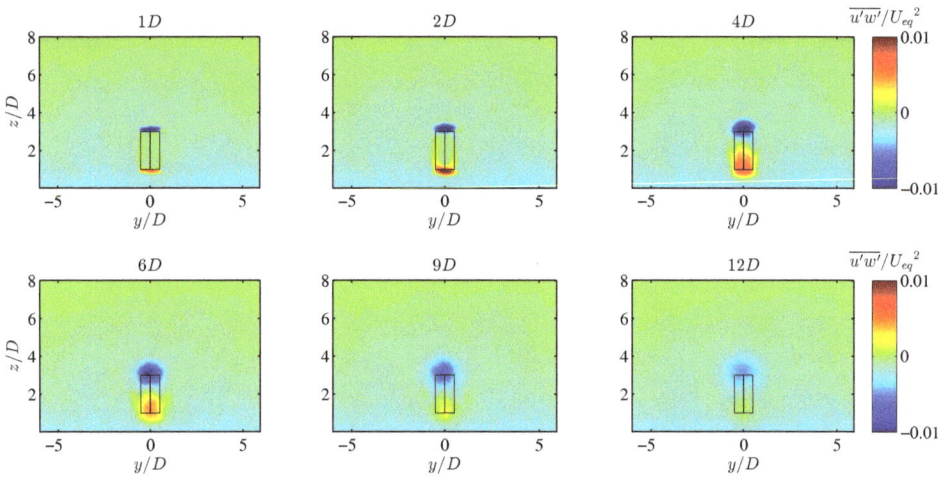

Figure 19. Contour plots of the normalized vertical turbulent flux, $\overline{u'w'}/U_{eq}^2$, in six different y-z planes at different distances downstream of the center of the turbine.

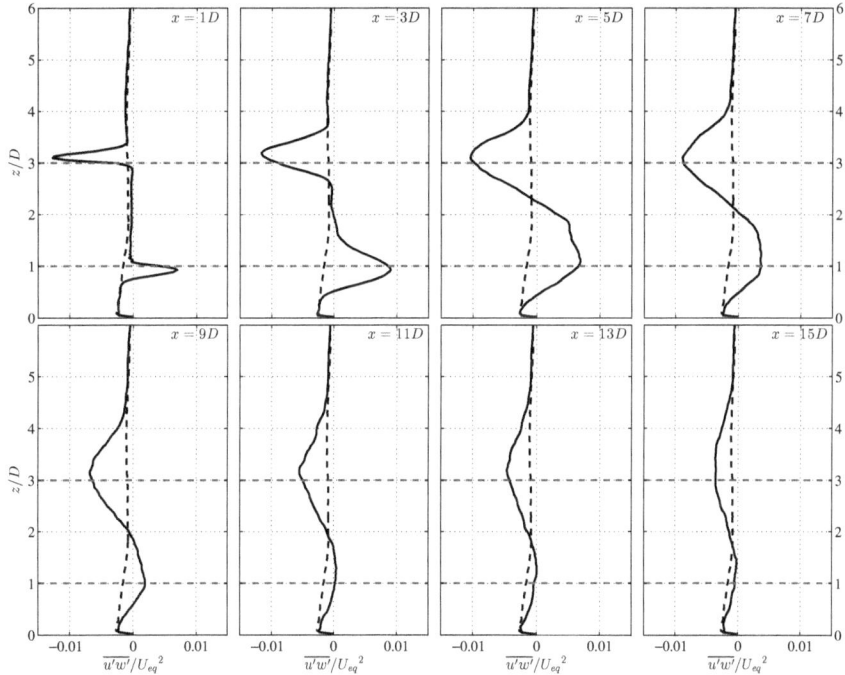

Figure 20. Vertical profiles of the normalized vertical turbulent flux $(\overline{u'w'}/U_{eq}^2)$ in the *x*-*z* plane going through the center of the turbine at different downstream positions. The black dashed line represents the inflow profile, and the blue horizontal dashed lines show the extent of the turbine.

5. Summary

Acknowledging the prospects of VAWTs as alternative wind energy extractors along with HAWTs in a future clean-energy outlook, which is likely to be marked by diversity, targeted research on VAWTs' performance is deemed to be highly useful and necessary. One of the research targets, which is especially crucial in designing potential VAWT farms, is to characterize VAWT wakes; a target which is still considerably underachieved for VAWTs, particularly with respect to HAWTs. In this view, one of the approaches that can greatly contribute to the cause is to use turbulence-resolving numerical simulation techniques, which can provide plenitude of high-resolution spatial and temporal information about the flow field and lead to valuable insight into the behavior of the turbine wake.

In this study, we used a previously-validated large-eddy simulation framework, in which an actuator line model is employed to parameterize the blade forces on the flow, to simulate the atmospheric boundary layer flow through stand-alone VAWTs placed on a flat terrain. For a typical straight-bladed 1-MW VAWT rotor design, first, the variation of the power coefficient with the tip-speed ratio and the chord length of the blades was studied. In doing so, the optimum combination of TSR and solidity (Nc/R), which yielded the maximum power coefficient of 0.47, was found to be 4.5 and 0.18, respectively. Second, for a VAWT with this optimum combination, a detailed study on the characteristics of its wake was performed, in which different mean and turbulence statistics were inspected. The mean velocity in the wake was found to need a long distance to recover; for example, the wake requires a distance of 14 rotor diameters to recover its center velocity to 85% of the incoming velocity. It was also seen that for this case, the point with the maximum velocity deficit is located 2.7 rotor diameters downstream of the center of the turbine (at the equator height of the turbine), and only after this point, the wake recovery starts with a rate (based on the change of the maximum velocity deficit) that is decreasing with streamwise distance. The turbulence intensity was observed to reach its maximum value (at the equator height of the turbine) 3.8 rotor diameters downstream of the VAWT. As we go towards the upper and lower extremities of the rotor, the height-specific maximum of the TI moves closer to the turbine and its value also increases. Turbulent momentum fluxes, which are a gauge for flow entrainment and, as a consequence, are responsible for the recovery of the wake, were also quantified, and it was shown that in the equator height of the turbine, the magnitude of the lateral flux peaks about $1D$ farther downwind of the maximum TI point. The above-mentioned mean and turbulence statistics corresponding to the optimum tip-speed ratio show only slight lateral asymmetries in the wake. However, significant vertical asymmetries were observed in terms of both the TI and magnitude of momentum fluxes, with higher values at the upper edge of the blades compared to the ones at the lower edge.

This study paves the way to further explore VAWT wakes and to discover the effects of different relevant parameters on the wake behavior. Moreover, it can serve as a solid foundation for future studies on performance, characteristics and optimization of VAWT farms.

Supplementary Materials: Zenodo DOI:10.5281/zenodo.51387 (https://zenodo.org/record/112316). Video S1: Normalized instantaneous streamwise velocity field both on a vertical plane (*x-z*) going through the center of the turbine and on a horizontal plane at the equator height of the turbine (Note: the physical time corresponding to this video is 1 minute and 17 seconds, and the size of the blades is magnified for illustration purposes). Video S2: Normalized instantaneous streamwise velocity field on a horizontal plane at the equator height of the turbine for two cases: when the turbine starts to operate (top) and when the flow has reached statistically steady condition (bottom) (Note: the physical time corresponding to both videos is 1 minute and 17 seconds, and the size of the blades is magnified for illustration purposes).

Acknowledgments: This research was supported by EOS (Energie Ouest Suisse) Holding, the Swiss Federal Office of Energy (Grant SI/501337-01) and the Swiss Innovation and Technology Committee (CTI) within the context of the Swiss Competence Center for Energy Research "FURIES: Future Swiss Electrical Infrastructure". Computing resources were provided by the Swiss National Supercomputing Centre (CSCS) under Project IDs s599 and s542.

Author Contributions: This study was done as part of Sina Shamsoddin's doctoral studies supervised by Fernando Porté-Agel.

Conflicts of Interest: The authors declare no conflict of interest.

Appendix A

In this Appendix, the procedure of the method with which the dynamic stall phenomenon is modeled is described in detail. The dynamic stall model is based on the modified MIT model developed by Noll and Ham [27], which is a practical modification of the original MIT model [30]. This model has the advantage of being simple and easy to use and also has been found to work better for VAWTs compared to other available models [1]. It is noteworthy that the following procedure can be implemented for both VAWTs and HAWTs.

Dynamic stall is a phenomenon that occurs for an airfoil when the angle of attack of the incident flow keeps changing with time and its rate of change (*i.e.*, $\dot{\alpha} = \frac{d\alpha}{dt}$) is sufficiently large. For a blade element of a turbine (either VAWT or HAWT) (placed in a turbulent flow), the change of α with time can be originated by three main sources: (1) the turbulent fluctuations of the incident flow; (2) the changes (spatial or temporal) in the mean incident flow; and (3) the rotation of the blades. Of these three reasons, the second one is normally specific to HAWTs, since an HAWT blade element experiences the variation of the boundary layer mean velocity profile at different heights; which is not the case for a VAWT blade element, as it moves at a constant height. However, the third reason is specific to VAWTs, because the geometry of a VAWT rotor is such that α (for a given blade element) oscillates between a maximum positive value and a minimum negative value in each revolution (even with a uniform inflow); however, for an HAWT blade element, assuming a uniform mean inflow, α remains constant during one revolution. Since the MIT model (and other similar practical models) is (are) only appropriate for the large-scale behavior of α in time, in our implementation of this model, the dynamic stall effects arising from the above-mentioned second and third sources, as well as the relatively large-scale turbulent fluctuations (from the first source) are modeled, while the changes of α arising from the relatively small-scale turbulent fluctuations of the incident flow are filtered out.

In order to implement the above-mentioned procedure, $\dot{\alpha}$ is calculated from a time-averaged and smoothed curve of $\alpha_f = \alpha_{fit}(\theta)$ during one revolution. For this purpose, the angle of attack at each azimuthal angle is time-averaged during each N_{rev} revolutions of the blades, and then, a polynomial curve, $\alpha_{fit}(\theta)$, is fitted on the time-averaged curve, $\alpha_{avg}(\theta)$. For the rest of the dynamic stall calculations, it is the $\alpha_{fit}(\theta)$ curve that is used. Figure A1 shows an example for this procedure for $TSR = 2$ and $c = 2$ m. The azimuthal angle, θ, is considered to increase counterclockwise (when seen from above) from $-90°$ to $270°$, in a way that $\theta = 0°$ and $\theta = 180°$ correspond to the most downstream and the most upstream points of the rotor, respectively. It is also noteworthy to mention again that the sense of the rotation of the turbine blades is counterclockwise when seen from above. Here, for the curve fitting, an eighth order polynomial is used to detect the two extrema accurately.

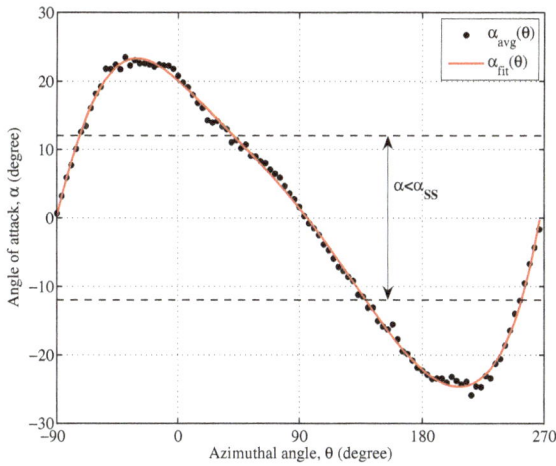

Figure A1. The time-averaged (black circles) and curve-fitted (red line) behavior of the variation of angle of attack as a function of azimuthal angle in one revolution of a blade element.

Subsequently, we implement the modified MIT model on the $\alpha_{fit}(\theta)$ curve and construct $C_{L,DS}(\alpha)$ and $C_{D,DS}(\alpha)$ curves, which are lift and drag coefficients as a function of the angle of attack considering dynamic stall. In the modified MIT model, we use the tabulated airfoil data for lift and drag coefficients, and based on that, $C_{L,DS}(\alpha)$ and $C_{D,DS}(\alpha)$ are constructed. Based on the tabulated airfoil data, we can determine the static stall angle, $\alpha_{SS} > 0$, and the lift coefficient at static stall, $C_{L,SS} > 0$. The static lift and drag coefficient functions derived from the tabulated airfoil data are designated as $C_{L,table}(\alpha)$ and $C_{D,table}(\alpha)$ hereafter. Moreover, the slope of the $C_{L,table}(\alpha)$ curve before the static stall can be calculated as $a_s = C_{L,SS}/\alpha_{SS}$, considering that in this region, normally, $C_{L,table}(\alpha)$ is linear.

As can be seen in Figure A1, the global (*i.e.*, the curve-fitted) behavior of $|\alpha|$ in one revolution of a blade element is such that $|\alpha|$ twice (once for positive α values and once for negative α values) increases from zero to a maximum value and then decreases to zero again. In each of these increase-decrease cycles of $|\alpha|$, the MIT dynamic stall model casts the flow in one of the four below dynamic stall states:

State 1 occurs when $|\alpha| \le \alpha_{SS}$. In this state, both lift and drag coefficients are extracted directly from the static tabulated airfoil data:

$$C_{L,DS}(\alpha) = C_{L,table}(|\alpha|) \tag{A1}$$

$$C_{D,DS}(\alpha) = C_{D,table}(|\alpha|) \tag{A2}$$

State 2 occurs when $\alpha_{SS} < |\alpha| < \alpha_{DS}$ and $\alpha_f \dot{\alpha} > 0$ (*i.e.*, $|\alpha|$ is increasing in time). α_{DS} is calculated with the following formula:

$$\alpha_{DS} = \alpha_{SS} + \gamma \sqrt{\left(\frac{|\dot{\alpha}|c}{2V_{rel}} \right)} \qquad (A3)$$

where c is the blade chord length, V_{rel} is the magnitude of the relative velocity (which is also a function of the azimuthal angle), $\dot{\alpha} = \Omega d\alpha_{fit}/d\theta$, Ω is the angular velocity of the blade and γ is a constant that has a dimension of an angle and is weakly a function of the airfoil type and is determined experimentally [27]. If an experimental value for γ is not available, a value of one radian is recommended [31]. We keep calculating α_{DS} in this state, until the point at which $|\alpha|$ is on the verge of becoming larger than α_{DS} (*i.e.*, the point at which the model goes to State 3). We designate this last value of α_{DS} as $\alpha_{DS,final}$, and with this value, we calculate the maximum value of $C_{L,DS}$ (*i.e.*, $C_{L,max}$):

$$C_{L,max} = C_{L,SS} + 40(\frac{|\dot{\alpha}|c}{V_{rel}}) \qquad (A4)$$

and we apply the following clipping conditions on $C_{L,max}$:

$$\begin{aligned} &\text{If} \quad C_{L,max} > 3.0 \quad \text{then} \quad C_{L,max} = 3.0 \\ &\text{If} \quad C_{L,max} < a_s \sin(\alpha_{DS,final}) \quad \text{then} \quad C_{L,max} = a_s \sin(\alpha_{DS,final}) \end{aligned} \qquad (A5)$$

Throughout this state, the lift coefficient is extrapolated from static values, and the drag coefficient is still directly extracted from the static tabulated data:

$$C_{L,DS}(\alpha) = a_s \sin(|\alpha|) \qquad (A6)$$

$$C_{D,DS}(\alpha) = C_{D,table}(|\alpha|) \qquad (A7)$$

where (as in Noll and Ham [27]) a sine function is used for extrapolation (noting that in the range of angles of attack, on which we normally apply the model, $|\alpha|$ is small, and we have $\sin(|\alpha|) \approx |\alpha|$).

State 3 occurs when $\alpha_{DS,final} < |\alpha|$ and $\alpha_f \dot{\alpha} > 0$ (*i.e.*, $|\alpha|$ is still increasing in time). As soon as the model enters State 3, we start to calculate the elapsed time from the moment in which State 3 is triggered; in other words, we start to calculate the time elapsed after the $\alpha_{DS,final}$ value has been reached; we call this time t_{DS}.

In this state, the lift and drag coefficients are calculated as:

$$C_{L,DS}(\alpha) = a_s \sin(|\alpha|) \qquad (A8)$$

$$C_{D,DS}(\alpha) = C_{L,DS} \tan(|\alpha|) \qquad (A9)$$

However, in this state, we only keep using Equations (A8) and (A9) as long as these conditions are both satisfied: $C_{L,DS} \leq C_{L,max}$ and $t_{DS}V_{rel}/c < 1$; otherwise, we set the lift coefficient to the $C_{L,max}$ value and calculate the drag coefficient accordingly (as shown below). We designate the value of $|\alpha|$ of the moment in which either of the aforesaid conditions is on the verge of being violated as $\alpha_{C_{L,max}}$.

$$\text{If} \quad C_{L,DS} > C_{L,max} \quad \text{Or} \quad t_{DS} \frac{V_{rel}}{c} \geq 1:$$

$$\begin{aligned} C_{L,DS} &= C_{L,max} \\ C_{D,DS} &= C_{L,max} \tan(\alpha_{C_{L,max}}) \end{aligned} \qquad (A10)$$

State 4 occurs when $|\alpha| > \alpha_{SS}$ and $\alpha_f \dot{\alpha} \leq 0$ (*i.e.*, when $|\alpha|$ starts to decrease with time). We designate the azimuthal angle of the moment in which $|\alpha|$ starts to decrease as $\theta_{\alpha_{max}}$. At this stage, $C_{L,DS}$ is lowered exponentially (in time) from $C_{L,max}$ to $C_{L,SS}$.

$$C_{L,DS} = (C_{L,max} - C_{L,SS}) \exp\left(-(\theta - \theta_{\alpha_{max}})\frac{2R}{c}\right) + C_{L,SS} \tag{A11}$$

$$C_{D,DS}(\alpha) = C_{L,DS} \tan(|\alpha|) \tag{A12}$$

where R is the radius of the blade element about the axis of rotation (in the case of a VAWT, R is simply the radius of the VAWT rotor).

As can be noticed in the above procedure, $\alpha(t)$ (*i.e.*, $\alpha(\theta)$) needs to be a smooth function for the above model to work. Because of this, we use $\alpha_f = \alpha_{fit}(\theta)$ (*i.e.*, the time-averaged and curve-fitted value of α) in the above procedure instead of α. Thus, at the end of each N_{rev} revolution and after getting the $\alpha_{fit}(\theta)$ function, we apply the MIT model on this curve, and we construct the $C_{L,DS}(\alpha)$ and $C_{D,DS}(\alpha)$ functions, which will be used in the next N_{rev} revolutions. For the first N_{rev} revolutions (for which we still do not have $\alpha_{fit}(\theta)$), one can preliminarily just use the static tabulated airfoil data.

Figure A2 shows an example of a constructed $C_{L,DS}(\alpha)$ curve under dynamic stall, which corresponds to the $\alpha_{fit}(\theta)$ shown in Figure A1. The aforesaid states of the model are shown in the figure. As can be seen in this figure, to construct this curve, we need both the tabulated airfoil data and some parameters, which we should obtain from the MIT model. As a summary, all of the necessary data and parameters required to construct the $C_{L,DS}(\alpha)$ and $C_{D,DS}(\alpha)$ curves are listed below:

(1) Tabulated airfoil data: $C_{L,table}(\alpha)$ and $C_{D,table}(\alpha)$; (2) α_{SS}; (3) $C_{L,SS}$; (4) a_s; (5) γ; (6) $\alpha_{DS,final}$; (7) $\alpha_{C_{L,max}}$; (8) $C_{L,max}$; (9) $\theta_{\alpha_{max}}$

It should be noted that for the last four items, two values are obtained for each revolution: one for the $\alpha \geq 0$ cycle and one for the $\alpha < 0$ cycle. In our simulations, we have used $\gamma = 1$ radian and $N_{rev} = 15$.

A comprehensive and step-by-step procedure to implement the MIT dynamic stall model is given in the flowchart of Figure A3; the flowchart is deliberately given in a way that can be followed for coding purposes in any common programming language.

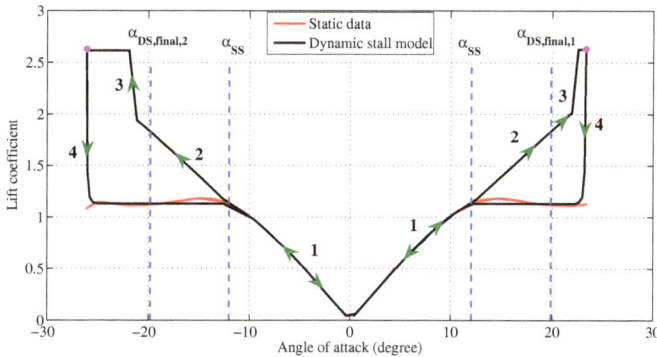

Figure A2. Lift coefficient curve under the dynamic stall model (black line) as compared to the static lift coefficient curve (red line). The four states in the dynamic stall model are indicated in the graph.

Start

Provide these input parameters:
$\Delta\theta$, α_{SS}, $C_{L,SS}$, a_s, γ

Initialize these variables: $\theta = -90°$, f_DS = 0,
f_$C_{L,max}$ = 0, f_α_{max} = 0, t_{DS} = 0

$\theta = \theta + \Delta\theta$
$\alpha_f = \alpha_{fit}(\theta)$

$\theta \geq 270°$

Check if $\theta < 270°$

$\theta < 270°$

Check if $|\alpha_f| > \alpha_{SS}$

$|\alpha_f| \leq \alpha_{SS}$

$C_{L,DS}(\alpha_f) = C_{L,table}(\alpha_f)$ and
$C_{D,DS}(\alpha_f) = C_{D,table}(\alpha_f)$

$|\alpha_f| > \alpha_{SS}$

$\dot{\alpha} = \Omega d\alpha_{fit}/d\theta$

Check if $\alpha_f\dot{\alpha} > 0$

$\alpha_f\dot{\alpha} \leq 0$

Check if f_α_{max} = 0

f_α_{max} $\neq$ 0

f_α_{max} = 1
$C_{L,DS} = (C_{L,max} - C_{L,SS})\exp\left(-(\theta - \theta_{\alpha_{max}})\frac{2R}{c}\right) + C_{L,SS}$
$C_{D,DS}(\alpha) = C_{L,DS}\tan(|\alpha|)$

$\alpha_f\dot{\alpha} > 0$

f_α_{max} = 0

$\theta_{\alpha_{max}} = \theta$

Check if f_DS = 0

f_DS = 0

$\alpha_{DS} = \alpha_{SS} + \gamma\sqrt{(|\dot{\alpha}|c/2V_{rel})}$

$C_{L,max} = C_{L,SS} + 40(|\dot{\alpha}|c/V_{rel})$
If $C_{L,max} > 3.0$ then $C_{L,max} = 3.0$
If $C_{L,max} < a_s \sin(\alpha_{DS})$
then $C_{L,max} = a_s \sin(\alpha_{DS})$

f_DS $\neq$ 0

Check if $|\alpha_f| < \alpha_{DS}$

$|\alpha_f| \geq \alpha_{DS}$

f_DS = 1
$t_{DS} = t_{DS} + \Delta\theta/\Omega$
$C_{L,DS} = a_s \sin(|\alpha_f|)$
$C_{D,DS} = C_{L,DS}\tan(|\alpha_f|)$

Check if
$C_{L,DS} \leq C_{L,max}$
And
$t_{DS}V_{rel}/c < 1$

$C_{L,DS} \leq C_{L,max}$ And $t_{DS}V_{rel}/c < 1$

$|\alpha_f| < \alpha_{DS}$

$C_{L,DS} = a_s \sin(|\alpha_f|)$
$C_{D,DS} = C_{D,table}(|\alpha_f|)$

$C_{L,DS} > C_{L,max}$ Or $t_{DS}V_{rel}/c \geq 1$

If f_$C_{L,max}$ = 0 Then $\alpha_{C_{L,max}} = \alpha_f$

End

f_$C_{L,max}$ = 1
$C_{L,DS} = C_{L,max}$
$C_{D,DS} = C_{L,max}\tan(\alpha_{C_{L,max}})$

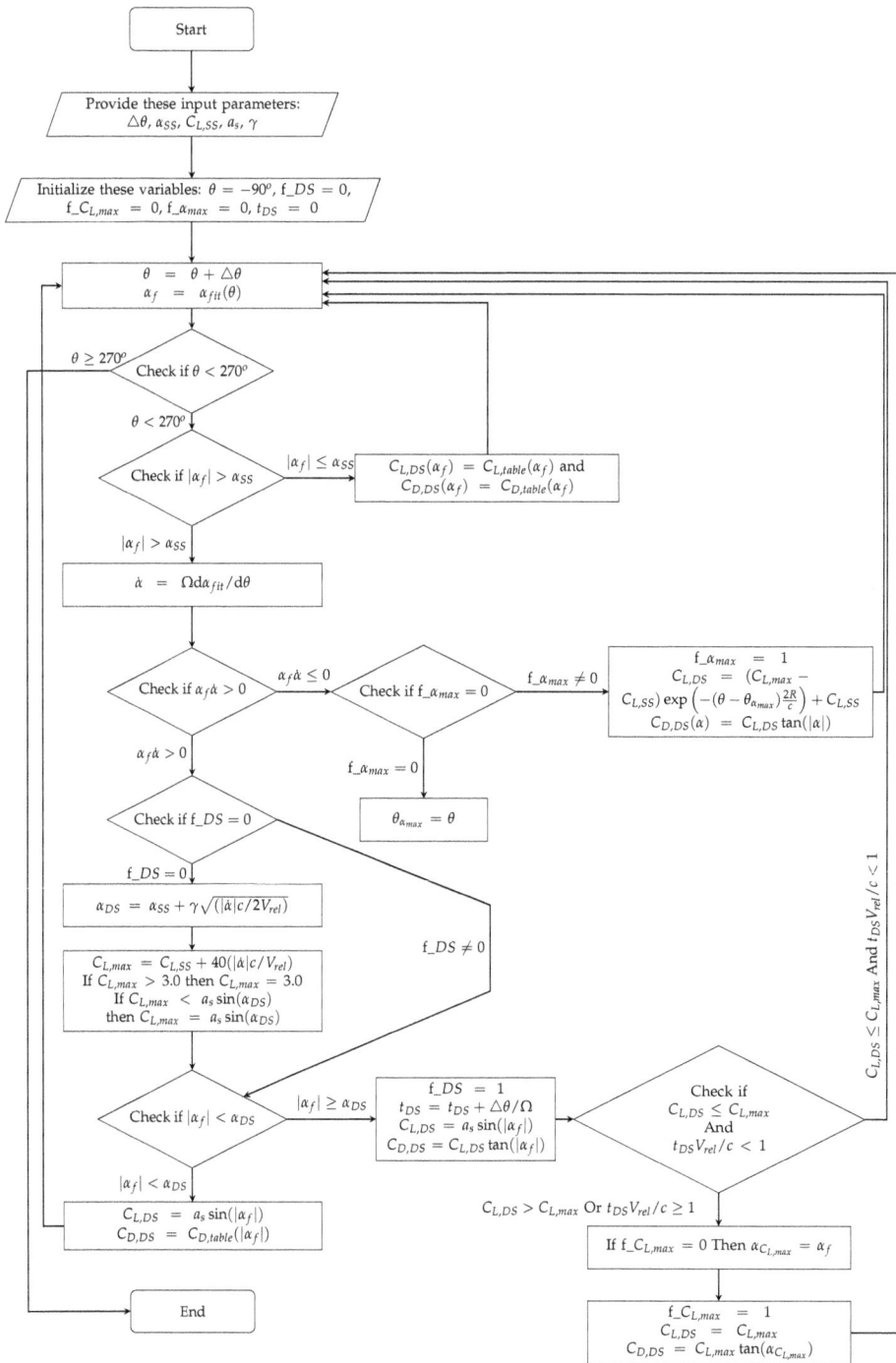

Figure A3. Flowchart to implement the modified MIT dynamic stall model for a VAWT in turbulent flow.

References

1. Paraschivoiu, I. *Wind Turbine Design—With Emphasis on Darrieus Concept*; Polytechnic International Press: Montreal, QC, Canada, 2002.

2. Tescione, G.; Ragni, D.; He, C.; Simao Ferreira, C.J.; van Bussel, G.J. Experimental and numerical aerodynamic analysis of vertical axis wind turbine wake. In Proceedings of the International Conference on Aerodynamics of Offshore Wind Energy Systems and Wakes, Lyngby, Denmark, 17–19 June 2013.

3. Battisti, L.; Zanne, L.; Dell'Anna, S.; Dossena, V.; Persico, G.; Paradiso, B. Aerodynamic measurements on a vertical axis wind turbine in a large scale wind tunnel. *J. Energy Resour. Technol.* **2011**, *133*, doi:10.1115/1.4004360.

4. Bachant, P.; Wosnik, M. Characterising the near-wake of a cross-flow turbine. *J. Turbul.* **2015**, *16*, 392–410.

5. Araya, D.B.; Dabiri, J.O. A comparison of wake measurements in motor-driven and flow-driven turbine experiments. *Exp. Fluids* **2015**, *56*, 1–15.

6. Brochier, G.; Fraunie, P.; Beguier, C.; Paraschivoiu, I. Water channel experiments of dynamic stall on darrieus wind turbine blades. *AIAA J. Propuls. Power* **1986**, *2*, 445–449.

7. Rolin, V.F.C.; Porté-Agel, F. Wind-tunnel study of the wake behind a vertical axis wind turbine in a boundary layer flow using stereoscopic particle image velocimetry. *J. Phys. Conf. Ser.* **2015**, *625*, 012012.

8. Ryan, K.J.; Coletti, F.; Elkins, C.J.; Dabiri, J.O.; Eaton, J.K. Three-dimensional flow field around and downstream of a subscale model rotating vertical axis wind turbine. *Exp. Fluids* **2016**, *57*, 1–15.

9. Castelli, M.R.; Englaro, A.; Benini, E. The darrieus wind turbine: Proposal for a new performance prediction model based on CFD. *Energy* **2011**, *36*, 4919–4934.

10. Pierce, B.; Moin, P.; Dabiri, J.O. *Evaluation of Point-Forcing Models with Application to Vertical Axis Wind Turbine Farms*; Annual Research Briefs; Center for Turbulence Research, Stanford University: Stanford, CA, USA, 2013.

11. Shamsoddin, S.; Porté-Agel, F. Large eddy simulation of vertical axis wind turbine wakes. *Energies* **2014**, *7*, 890–912.

12. Rajagopalan, R.G.; Fanucci, J.B. Finite difference model for vertical-axis wind turbines. *AIAA J. Propuls. Power* **1985**, *1*, 432–436.

13. Rajagopalan, R.G.; Berg, D.E.; Klimas, P.C. Development of a three-dimensional model for the darrieus rotor and its wake. *AIAA J. Propuls. Power* **1995**, *11*, 185–195.

14. Shen, W.; Zhang, J.; Sørensen, J. The actuator surface model: A new navier-stokes based model for rotor computations. *J. Sol. Energy Eng.* **2009**, *131*, doi:10.1115/1.3027502.

15. Stoll, R.; Porté-Agel, F. Dynamic subgrid-scale models for momentum and scalar fluxes in large-eddy simulation of neutrally stratified atmospheric boundary layers over heterogeneous terrain. *Water Resour. Res.* **2006**, *42*, doi:10.1029/2005WR003989.

16. Albertson, J.D.; Parlange, M.B. Surfaces length scales and shear stress: implications for land-atmosphere interactions over complex terrain. *Water Resour. Res.* **1999**, *35*, 2121–2132.

17. Porté-Agel, F.; Meneveau, C.; Parlange, M.B. A scale-dependent dynamic model for large-eddy simulation: Application to a neutral atmospheric boundary layer. *J. Fluid Mech.* **2000**, *415*, 261–284.

18. Porté-Agel, F.; Wu, Y.T.; Lu, H.; Conzemius, R.J. Large-eddy simulation of atmospheric boundary layer flow through wind turbines and wind farms. *J. Wind Eng. Ind. Aerodyn.* **2011**, *99*, 154–168.

19. Monin, A.; Obukhov, M. Basic laws of turbulent mixing in the ground layer of the atmosphere. *Tr. Akad. Nauk SSSR Geophiz. Inst.* **1954**, *24*, 163–187.

20. Moeng, C. A large-eddy simulation model for the study of planetary boundary-layer turbulence. *J. Atmos. Sci.* **1984**, *46*, 2311–2330.

21. Stoll, R.; Porté-Agel, F. Effect of roughness on surface boundary conditions for large-eddy simulation. *Bound. Layer Meteorol.* **2006**, *118*, 169–187.

22. Tseng, Y.H.; Meneveau, C.; Parlange, M.B. Modeling flow around bluff bodies and predicting urban dispersion using large eddy simulation. *Environ. Sci. Technol.* **2006**, *40*, 2653–2662.

23. Wu, Y.T.; Porté-Agel, F. Large-eddy simulation of wind-turbine wakes: Evaluation of turbine parametrisations. *Bound. Layer Meteorol.* **2011**, *138*, 345–366.

24. Porté-Agel, F.; Wu, Y.T.; Chen, C.H. A numerical study of the effects of wind direction on turbine wakes and power losses in a large wind farm. *Energies* **2013**, *6*, 5297–5313.

25. Sheldahl, R.E.; Klimas, P.C. *Aerodynamic Characteristics of Seven Airfoil Sections through 180 Degrees Angle of Attack for Use in Aerodynamic Analysis of Vertical Axis Wind Turbines*; Technical Report SAND80-2114; Sandia National Laboratories: Albuquerque, NM, USA, 1981.

26. Scheurich, F.; Brown, R. Effect of dynamic stall on the aerodynamics of vertical-axis wind turbines. *AIAA J.* **2011**, *49*, 2511–2521.

27. Noll, R.B.; Ham, N.D. *Dynamic Stall of Small Wind Systems*; Technical Report; Aerospace Systems Inc.: Burlington, MA, USA, 1983.

28. Templin, R.J.; Rangi, R.S. Vertical-axis wind turbine development in Canada. *IEE Proc. A Phys. Sci. Meas. Instrum. Manag. Educ. Rev.* **1983**, *130*, 555–561.

29. Chen, T.; Liou, L. Blockage corrections in wind tunnel tests of small horizontal-axis wind turbines. *Exp. Therm. Fluid Sci.* **2011**, *35*, 565–569.

30. Ham, N.D. Aerodynamic loading on a two-dimensional airfoil during dynamic stall. *AIAA J.* **1968**, *6*, 1927–1934.

31. Hibbs, B.D. *HAWTPerformance with Dynamic Stall*; Technical Report SERI/STR 217-2732; AeroVironment, Inc.: Monrovia, CA, USA, 1986.

energies

MDPI

Article

High Spatial Resolution Simulation of Annual Wind Energy Yield Using Near-Surface Wind Speed Time Series

Christopher Jung

Environmental Meteorology, Albert-Ludwigs-University of Freiburg, Werthmannstrasse 10, Freiburg D-79085, Germany; christopher.jung@mail.unr.uni-freiburg.de; Tel.: +49-761-203-6822

Academic Editor: Frede Blaabjerg
Received: 15 February 2016; Accepted: 22 April 2016; Published: 6 May 2016

Abstract: In this paper a methodology is presented that can be used to model the annual wind energy yield (AEY_{mod}) on a high spatial resolution (50 m $\times$ 50 m) grid based on long-term (1979–2010) near-surface wind speed (U_S) time series measured at 58 stations of the German Weather Service (DWD). The study area for which AEY_{mod} is quantified is the German federal state of Baden-Wuerttemberg. Comparability of the wind speed time series was ensured by gap filling, homogenization and detrending. The U_S values were extrapolated to the height 100 m ($U_{100m,emp}$) above ground level (AGL) by the Hellman power law. All $U_{100m,emp}$ time series were then converted to empirical cumulative distribution functions (CDF_{emp}). 67 theoretical cumulative distribution functions (CDF) were fitted to all CDF_{emp} and their goodness of fit (GoF) was evaluated. It turned out that the five-parameter Wakeby distribution (WK5) is universally applicable in the study area. Prior to the least squares boosting (LSBoost)-based modeling of WK5 parameters, 92 predictor variables were obtained from: (i) a digital terrain model (DTM), (ii) the European Centre for Medium-Range Weather Forecasts re-analysis (ERA)-Interim reanalysis wind speed data available at the 850 hPa pressure level (U_{850hPa}), and (iii) the Coordination of Information on the Environment (CORINE) Land Cover (CLC) data. On the basis of predictor importance (PI) and the evaluation of model accuracy, the combination of predictor variables that provides the best discrimination between $U_{100m,emp}$ and the modeled wind speed at 100 m AGL ($U_{100m,mod}$), was identified. Results from relative PI-evaluation demonstrate that the most important predictor variables are relative elevation (Φ) and topographic exposure (τ) in the main wind direction. Since all WK5 parameters are available, any manufacturer power curve can easily be applied to quantify AEY_{mod}.

Keywords: annual wind energy yield (AEY); Wakeby distribution (WK5); least squares boosting (LSBoost); predictor importance (PI); wind speed extrapolation

1. Introduction

The world's energy supply is facing multiple challenges. The depletion of conventional fuels is unavoidable [1,2], greenhouse gas emissions from the burning of fossil fuels most significantly contributes to global warming [3,4] and the emissions of air pollutants affect human health [3,5]. Although nuclear energy production enables the reduction of carbon dioxide (CO_2) emissions [6], nuclear power plants bear great short- and long-term risk of accidents [7]. In order to reduce and avoid negative impacts of the current use of energy resources on the environment and human health, alternative forms of energy utilization must be found.

Renewable energies provide a clean, environmentally friendly and health-compatible alternative to fossil energies and nuclear energy [2,5]. One major renewable energy resource is the kinetic energy contained in the atmosphere, commonly known as wind energy. The potential for wind energy utilization to play a key role in the future global energy mix is enormous. Wind energy could supply more than 40 times [8] the annual global electricity consumption. Consequently, the wind power generation capacity of the world is growing constantly with an average annual rate of about 30% over the last decade [2]. In the European Union, directive 2009/28/EC [9] aims to cover 20% of the primary energy demand by renewable energies in 2020, including wind energy. The leading wind power producer in Europe is Germany. Germany aims to supply at least 30% of the energy consumption in 2020 by renewable energies [4]. Yet, in some German federal states the utilization of wind energy is still far from being exhausted. For instance, the ministry of Environment, Climate Protection and the Energy Sector of the southwestern German federal state of Baden-Wuerttemberg plans to increase the share of wind energy in the energy mix from ~1% in 2015 to 10% in 2020 [10]. In order to achieve this political target, up to 1200 new wind turbines with an average output power between 2.5 MW and 3.0 MW must be installed in a period of only five years [10].

The first step in the onshore assessment of potential wind turbine sites is to quantify the site-specific atmospheric wind energy resource at the wind turbine hub height (~80–100 m) [11]. The wind resource is predetermined by the large-scale atmospheric circulation and modified by characteristics of surface roughness [12] and terrain [13]. As a result, the local wind resource can vary significantly over short distances [8]. In contrast to this, ground-based measurements of long-term wind speed at the landscape level are rare and only available for heights near the surface (10 m above ground level (AGL)). Because of the high spatiotemporal variability of the local wind resource [14,15], the low number of available near-surface wind speed measurement sites alone often limits the detailed assessment of the site-specific wind resource.

To overcome the problem of the low number of wind speed measurements and the strong influence of surface and terrain characteristics on the local wind resource, one option is highly resolved statistical modeling of wind speed at hub height. However, mapping of average wind speed alone is insufficient [16], since not only the central tendencies of wind speed distributions determine the wind resource. Therefore, fitting an appropriate theoretical wind speed distribution to empirical wind speed distributions is crucial [17]. Which theoretical distribution fits empirical wind speed distributions best is currently under discussion [18,19].

Due to the limited availability of wind speed measurement sites in Southwest Germany, a region with highly complex topography and mosaic-like land cover pattern, the goals of this study are (i) the quantification of the annual wind energy yield (*AEY*) on a high spatial resolution grid and (ii) the identification of the most important factors influencing the local wind resource.

2. Materials and Methods

2.1. Study Area and Wind Speed Measurements

The study area is the German federal state of Baden-Wuerttemberg (Figure 1). The low mountain ranges Black Forest (length ~150 km, width ~30–50 km, highest elevations >1400 m) and Swabian Alb (length ~180 km, width ~35 km, highest elevations >1000 m) are the most complex topographical features with the strongest impact on the wind resource over the study area [20]. The top of the Feldberg (1493 m) is the highest elevation in the study area. Approximately 38% (13,700 km^2) of the study area is covered with forests [21]. More details about land cover and topographical features in the study area are summarized in [20].

The wind speed database used in this study consists of time series of the daily mean wind speed measured from 1 January 1979 to 31 December 2010 at 58 meteorological stations by the German Weather Service (DWD). The height (h_S) of wind speed measurements varies between 3 m AGL (stations Bad Wildbad-Sommerberg, Isny) and 48 m AGL (station Karlsruhe). Data preparation included gap filling, testing for homogeneity and detrending according to [20].

The median wind speed near the surface values ($\tilde{U}_s$) vary in the range 0.5 m/s (station Triberg) to 7.5 m/s (station Feldberg) (Table 1). To extend the database, four measurement stations located in the bordering federal states Hesse and Bavaria were included in this study. Out of the 58 wind speed time series, 48 time series were put into a parameterization dataset (DS1). The remaining 10 time series, for which the original length was less than 10 years, belong to the validation dataset (DS2).

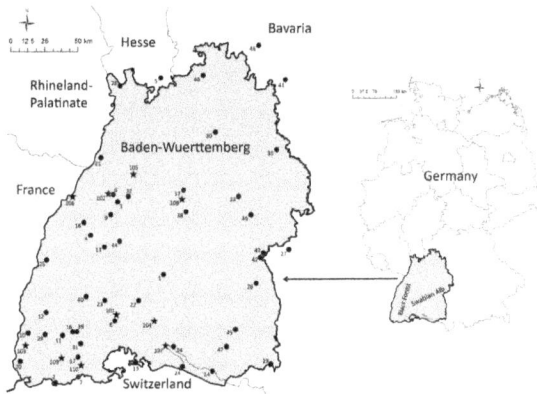

Figure 1. The study area Baden-Wuerttemberg in Southwest Germany and locations of German Weather Service (DWD) stations. Dots indicate parameterization dataset (DS1) stations; stars indicate validation dataset (DS2) stations.

Table 1. List of DWD stations and corresponding data features. DS1 stations are indicated by identification numbers (ID) 1–48; DS2 stations are indicated by ID values 101–110. $\tilde{U}_s$: median wind speed near the surface values; h_S: height.

ID	Station	$\tilde{U}_s$	h_s	ID	Station	$\tilde{U}_s$	h_s
1	Albstadt-Onstmettingen	1.7	17	30	Öhringen	2.3	16
2	Bad Säckingen	0.9	10	31	Schluchsee	1.6	10
3	Bad Wildbad-Sommerberg	0.7	3	32	Schömberg	0.6	10
4	Baiersbronn-Obertal	0.9	10	33	Schwäbisch-Gmünd	0.7	10
5	Beerfelden	1.7	10	34	Sipplingen	2.5	16
6	Dobel	2.3	10	35	Stimpfach-Weiptershofen	2.1	10
7	Dogern	1.7	10	36	Stötten	4.1	12
8	Donaueschingen	2.5	10	37	Stuttgart (Schnarrenberg)	2.5	12
9	Enzklösterle	1.0	10	38	Stuttgart-Echterdingen	2.5	10
10	Eschbach	2.4	10	39	Titisee	0.7	11
11	Feldberg	7.5	19	40	Triberg	0.5	15
12	Freiburg	2.4	12	41	Uffenheim	1.3	10
13	Freudenstadt	3.7	34	42	Ulm	2.2	10
14	Friedrichshafen	3.1	10	43	Ulm-Wilhelmsburg	2.7	15
15	Gailingen	1.6	12	44	Waldachtal-Lützenhardt	1.2	6
16	Hinterzarten	1.2	6	45	Waldsee, Bad-Reute	2.2	10
17	Höchenschwand	1.2	6	46	Walldürn	2.9	10
18	Hornisgrinde	5.9	10	47	Weingarten	1.9	12
19	Isny	2.1	3	48	Würzburg	2.8	12
20	Kandern-Gupf	2.0	10	101	Bad Dürrheim	1.3	10
21	Karlsruhe	3.3	48	102	Bad Herrenalb	0.6	10
22	Klippeneck	3.9	16	103	Müllheim	1.3	6
23	Königsfeld	1.0	6	104	Neuhausen ob Eck	2.5	10
24	Konstanz	1.7	17	105	Pforzheim-Ispringen	2.6	12
25	Lahr	2.3	10	106	Söllingen	2.6	10
26	Laupheim	2.5	10	107	Stockach-Espasingen	1.5	12
27	Leipheim	2.3	10	108	Stuttgart-Stadt	1.7	26
28	Mannheim	2.7	22	109	Todtmoos	1.1	10
29	Münstertal	1.3	10	110	Weilheim-Bierbronnen	2.6	10

2.2. Wind Speed Extrapolation

All wind speed near the surface (U_S) time series were extrapolated to 100 m AGL using the Hellman power law [22–24]. It was demonstrated by [11] that the power law performs well compared to similar wind speed extrapolation methods. According to [22], the accuracy of the power law increases when stratification effects and the influence of the wind speed are considered. Therefore, the Hellmann exponent (E) was computed on a daily basis.

As has previously been done by [20,25], daily mean wind speed at the 850 hPa pressure level (U_{850hPa}) and the height of the 850 hPa pressure level AGL (h_{850hPa}), both available from the European Centre for Medium-Range Weather Forecast [26], were used to calculate daily, station-specific E-values:

$$E = \frac{\ln\left(\frac{U_{850hPa}}{U_s}\right)}{\ln\left(\frac{h_{850hPa}}{h_s}\right)} \tag{1}$$

After the E-values were determined, daily, station-specific U_S-values were extrapolated to 100 m AGL yielding $U_{100m,emp}$:

$$U_{100m,emp} = U_s \times \left(\frac{100m}{h_s}\right)^E \tag{2}$$

2.3. Probability Distribution Fitting

Prior to the probability distribution fitting, $U_{100m,emp}$ time series were transformed to empirical cumulative distribution functions (CDF$_{emp}$). Afterwards, 67 CDF were fitted to each CDF$_{emp}$. The goodness of fit (GoF) of each CDF was quantified by calculating the coefficient of determination (R^2) from probability plots [19,27] and the Kolmogorov-Smirnov statistic (D) [28–30] to the fits. The D-values were obtained by measuring the largest vertical difference between CDF and CDF$_{emp}$. The transformation of time series, fitting and GoF evaluation were done by EasyFit software (Version 5.5, MathWave Technologies, Dnepropetrovsk, Ukraine) and Matlab® Software Optimization Toolbox (Release 2015a; The Math Works Inc., Natick, MA, USA).

According to D- and R^2-value evaluation, which will be presented in detail in the results section, the five-parameter Wakeby distribution (WK5) [31] is clearly the best-fitting distribution. It can be defined by its quantile function [20,25,31,32]:

$$U_{100m,distr}(F) = \varepsilon + \frac{\alpha}{\beta} \times \left[1 - (1-F)^\beta\right] - \frac{\gamma}{\delta} \times \left[1 - (1-F)^{-\delta}\right] \tag{3}$$

where F is the cumulative probability with $U_{100m,distr}(F)$ being the associated wind speed value. The four parameters α, β, γ, and δ are distribution parameters and the fifth parameter, ε, is the location parameter. WK5 can be interpreted as a mixed distribution [33] consisting of a left and right part [31,32]. This enables WK5 to reproduce shapes of wind speed distributions that other distributions cannot reproduce [25,31].

2.4. Predictor Variable Building

A total number of 92 predictor variables (50 m × 50 m) covering the study area were built by using the ArcGIS® 10.2 software (Esri, Redlands, CA, USA). All predictor variables originate from a digital terrain model (DTM), CORINE Land Cover (CLC) data [34] or ERA-Interim reanalysis U_{850hPa} [26].

The DTM was used to map Φ, τ [35,36], curvature, aspect and slope. The Φ-values were calculated by subtracting the mean elevation of an outer circle around each grid point from the grid point-specific elevation. Five different Φ variants with outer-circle radii of 250 m, 500 m, 1000 m (Φ_{1000m}), 2500 m (Φ_{2500m}) and 5000 m (Φ_{5000m}) were created.

The τ-maps were built for eight main compass directions (northeast (22.5°–67.4°), east (67.5°–112.4°), southeast (112.5°–157.4°), south (157.5°–202.4°), southwest (202.5°–247.4°), west

(247.5°–292.4°), northwest (292.5°–337.4°), north(337.5°–22.4°)) at 200 m radius intervals. This was done by summing angles up to a distance limited to 1000 m. Curvature, aspect and slope were calculated by using the Spatial Analyst Toolbox in ArcGIS.

Roughness length (z_0) was derived from CLC data with an original spatial resolution of 100 m × 100 m. Roughness length values were assigned to land cover types according to [20] yielding the local roughness length ($z_{0,l}$). Additionally, "effective" roughness length values ($z_{0,eff}$) for the eight main compass directions were calculated. This was done for four different radii around each grid point (100 m, 200 m, 300 m, 400 m). In the end, all z_0-values were interpolated to 50 m × 50 m resolution grids.

U_{850hPa} data (0.125° × 0.125° resolution) were included into model building because it represents large-scale airflow undisturbed by the surface [37]. The 0.01, 0.30, 0.50, 0.75 and 0.99 percentiles of U_{850hPa} time series covering the period from 01 January 1979 to 31 December 2010 were calculated ($U_{850hPa,0.01}$, $U_{850hPa,0.30}$, $U_{850hPa,0.50}$, $U_{850hPa,0.75}$ and $U_{850hPa,0.99}$) and mapped in ArcGIS®. A spline interpolation was applied to convert the U_{850hPa} layers to 50 m × 50 m resolution grids.

2.5. Wakeby Parameter Estimation and Modeling

The procedure applied to obtain the Wakeby parameters at every grid point in the study area comprised the following work steps: (1) estimating the Wakeby parameters of every CDF_{emp} based on L-moments [38,39]; (2) analyzing the obtained Wakeby parameters and identifying common characteristics of all distributions ; and (3) modeling target variables (Y) that enable the calculation of all WK5 parameters at every grid point in the study area. To make the WK5 parameter modeling more robust, the WK5 parameters estimated by L-moments were modeled indirectly according to [20,25].

Analyzing the estimated distributions led to the following parameter modeling and calculation approach: First, the estimated left-hand tail of WK5 (Y_L), which is represented by α, β and ε, was modeled:

$$Y_L = \varepsilon + \frac{10}{\beta} \times \left[1 - (1 - 0.25)^\beta\right] \tag{5}$$

The estimated location parameter ε, which represents the lower bound of the distribution, was directly modeled. Because the L-moment-based WK5 parameter estimation showed that $\alpha = 10$ at nearly all stations, it was set to this value. The use of a fixed α-value enabled the subsequent calculation of β.

Since Y_L affected WK5 parameter estimation up to $F = 0.25$, exactly as described by [31,32], the percentiles $F = 0.30$ (Y_{R1}), $F = 0.50$ (Y_{R2}), $F = 0.75$ (Y_{R3}) and $F = 0.99$ (Y_{R4}) were modeled to build the right-hand tail of WK5 (Y_R). A system of non-linear equations was solved at every grid point yielding γ and δ:

$$\begin{cases} \frac{\gamma}{\delta} \times \left[1 - (1 - 0.30)^{-\delta}\right] + [Y_{R1} - Y_L] = 0 \\ \frac{\gamma}{\delta} \times \left[1 - (1 - 0.50)^{-\delta}\right] + [Y_{R2} - Y_L] = 0 \\ \frac{\gamma}{\delta} \times \left[1 - (1 - 0.75)^{-\delta}\right] + [Y_{R3} - Y_L] = 0 \\ \frac{\gamma}{\delta} \times \left[1 - (1 - 0.99)^{-\delta}\right] + [Y_{R4} - Y_L] = 0 \end{cases} \tag{6}$$

In order to calculate $U_{100m,mod}$, Y_L and Y_R were recombined yielding WK5 with modeled parameters (WK5$_{mod}$).

All Y were computed for every grid point by least squares boosting (LSBoost) [40]. This was done by using the Ensemble Learning algorithm LSBoost implemented in the Matlab®Software Statistics Toolbox (Release 2015a; The Math Works Inc.). LSBoost is basically a sequence of simple regression trees, which are called weak learners (B). The objective of LSBoost is to minimize the mean squared error (*MSE*) between Y and the aggregated prediction of the weak learners (Y_{pred}). In the beginning, the median of the target variables ($\tilde{Y}$) is calculated. Afterwards, multiple regression trees $B_1, \ldots, B_m$

are combined in a weighted manner [41] to improve model accuracy. The individual regression trees are a function of selected predictor variables (X):

$$Y_{\text{pred}}(X) = \widetilde{Y}(X) + v \sum_{m=1}^{M} p_m \times B_m(X) \tag{4}$$

with p_m being the weight for model m, M is the total number of weak learners, and v with $0 < v \leqslant 1$ is the learning rate [20,41].

The predictor variable selection process comprised several steps. First, the most appropriate length of outer-circle radii for τ and $z_{0,\text{eff}}$ were determined by the correlation coefficient (r) between τ respectively $z_{0,\text{eff}}$ and Y. Secondly, the importance of the remaining predictor variables was evaluated by predictor importance (PI) which quantifies the relative contribution of individual predictor variables to the model output [21]. The PI-values were determined by summing up changes in MSE due to splits on every predictor and dividing the sum by the number of branch nodes. All predictor variables with $PI = 0.00$ were sorted out.

After PI-evaluation, combinations of predictor variables were tested for their predictive power. Starting with one predictor variable, further predictor variables were added to the model and kept when the model accuracy measures R^2, mean error (ME), mean absolute error (MAE), MSE and mean absolute percentage error ($MAPE$) improved [42–44]. For model parameterization, DS1 data were used. Model validation was done with both DS1 and DS2 data.

Multicollinearity among the predictor variables was investigated by assessing the variance inflation and the condition index in combination with variance decomposition proportions according to [45].

2.6. Annual Wind Energy Yield Estimation

The relationship between wind speed and the electrical power output (P) of wind turbines is typically established by a power curve [46]. Power curve values are developed from field measurements and can be used for studies involving energy calculations [47]. There are three important points characterizing a typical power curve (Figure 2): (1) at the cut-in speed the wind turbine starts to generate usable power; (2) after exceeding the rated output speed the maximum output power (rated power) is generated; and (3) after exceeding the cut-out speed turbines cease power generation and shut down [46]. A standard 2.5 MW power curve [48] for onshore wind power plants was applied to calculate the AEY.

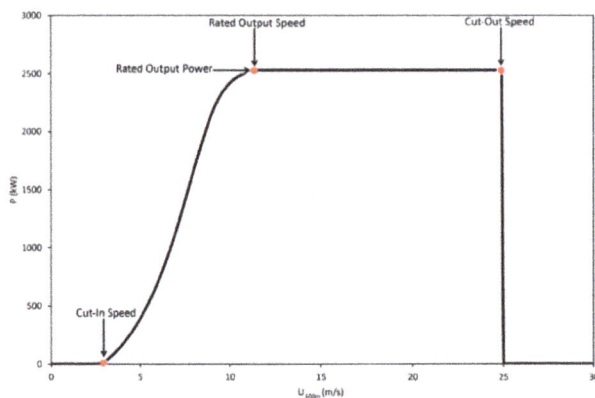

Figure 2. Power curve used to calculate empirical annual wind energy yield (AEY_{emp}) and modeled annual wind energy yield (AEY_{mod}) depending on wind speed in 100 m above ground level (AGL) ($U_{100\text{m}}$).

The discrete *P*-values from the manufacturer power curve were interpolated by a spline to obtain a continuous power curve. The basic attributes of the applied power curve are: cut-in speed U_{100m} = 3.0 m/s; cut-out speed U_{100m} = 25.0 m/s; rated output speed U_{100m} = 13.0 m/s and; rated output power *P* = 2580 kW. The empirical annual wind energy yield (AEY_{emp}) was calculated for each station in DS1 and DS2 following [49]:

$$AEY_{emp} = (\sum_{i=1}^{N} P(U_{100m,emp,i})/Z_1 \tag{7}$$

with *N* = 11,688 being the total number of days in the investigation period and the number of years in the investigation period (Z_1).

The average electrical power output ($\overline{P}$) was calculated according to [19,50]:

$$\overline{P} = \int_0^{\infty} P(U_{100m,mod}) \times f(U_{100m,mod}) \, dU_{100m,mod} \tag{8}$$

The above equation describes the electrical power produced at each wind speed class multiplied by the probability of the specified wind speed class and integrated over all possible wind speed classes [50] with $f(U_{100m,mod})$ being the probability density of $U_{100m,mod}$. After $\overline{P}$ is calculated modeled annual wind energy yield (AEY_{mod}) can be computed by multiplying $\overline{P}$ with the respective number of days per year (Z_2):

$$AEY_{mod} = \overline{P} \times Z_2 \tag{9}$$

2.7. Summary of the Methodology

The methodology for the quantification of *AEY* in the study area is summarized in Figure 3. The basic steps are:

(1) Extrapolation of near-surface wind speed time series to hub height;
(2) Identification of a theoretical distribution that is capable of reproducing various shapes of empirical wind speed distributions;
(3) Modeling the estimated parameters of the identified theoretical distribution, based on large-scale airflow, surface roughness and topographic features;
(4) Mapping of distribution parameters in the study area; and
(5) Calculation of the *AEY* using a wind turbine-specific power curve.

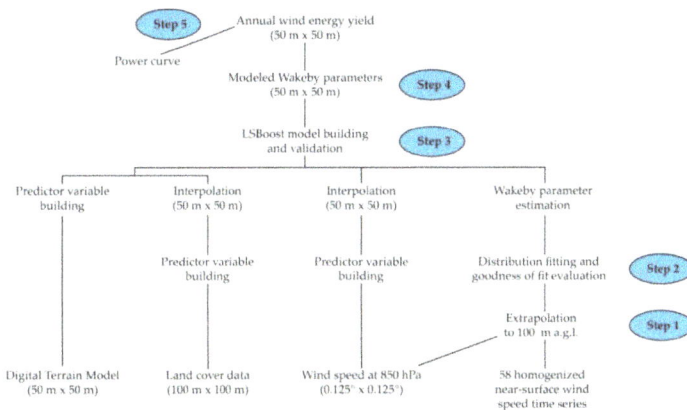

Figure 3. Schematic representation of the workflow applied to obtain annual wind energy yield (*AEY*).

3. Results and Discussion

3.1. Distribution Fitting

According to results from the D-evaluation, WK5 fits 23 CDF_{emp} best. As can be seen in Table 2, the D-value averaged over all stations for WK5 (0.02) is lower than the average D-value of all other theoretical distributions. Another well-fitting distribution is the four-parameter Johnson SB distribution ($D = 0.03$). The best fitting three-parameter distribution is the inverse Gaussian distribution ($D = 0.03$). In general, the performance of theoretical distributions defined by three or more parameters is better than the performance of two- and one-parameter distributions. In the case of eight theoretical distributions (Johnson SU, Log-Gamma, Log-Pearson 3, Nakagami, Pareto, Reciprocal, Phased Bi-Exponential, Phased Bi-Weibull) no fit to CDF_{emp} could be achieved and therefore the parameter estimation procedure failed.

Table 2. Distributions ranked (RK) by Kolmogorov-Smirnov statistic (D)-values with their number of parameters (NP). D- and coefficient of determination (R^2)-values are averages over all meteorological stations.

RK	Distribution	D	R^2	NP	RK	Distribution	D	R^2	NP
1	Wakeby	0.02	0.9992	5	35	Weibull	0.10	0.9768	2
2	Johnson SB	0.03	0.9991	4	36	Pert	0.11	0.9732	3
3	Inv. Gaussian	0.03	0.9981	3	37	Rayleigh	0.12	0.9721	2
4	Pearson 6	0.03	0.9984	4	38	Erlang	0.12	0.9866	3
5	Pearson 6	0.03	0.9983	3	39	Normal	0.13	0.9492	2
6	Lognormal	0.03	0.9982	3	40	Rice	0.13	0.9653	2
7	Dagum	0.03	0.9978	3	41	Logistic	0.13	0.9511	2
8	Fatigue Life	0.03	0.9978	3	42	Hypersecant	0.14	0.9497	2
9	Gen. Extreme	0.04	0.9975	3	43	Uniform	0.14	0.9351	2
10	Burr	0.04	0.9974	4	44	Cauchy	0.15	0.9845	2
11	Log-Logistic	0.04	0.9971	3	45	Erlang	0.15	0.9909	2
12	Burr	0.04	0.9971	3	46	Chi-Squared	0.16	0.9908	2
13	Lognormal	0.04	0.9973	2	47	Error	0.16	0.9437	3
14	Bimodal Weibull	0.04	0.9985	5	48	Laplace	0.17	0.9412	2
15	Fatigue Life	0.04	0.9970	2	49	Chi-Squared	0.19	0.9920	1
16	Inv. Gaussian	0.04	0.9974	2	50	Gumbel Min	0.20	0.8976	2
17	Pearson 5	0.04	0.9958	3	51	Exponential	0.23	0.9863	2
18	Bimodal Normal	0.04	0.9956	5	52	Exponential	0.27	0.9833	1
19	Gen. Pareto	0.04	0.9976	3	53	Pareto 2	0.28	0.9841	2
20	Gen. Gamma	0.04	0.9954	4	54	Triangular	0.31	0.9132	3
21	Dagum	0.04	0.9957	4	55	Power Func.	0.31	0.9138	3
22	Pearson 5	0.05	0.9954	2	56	Levy	0.36	0.9777	2
23	Gen. Logistic	0.05	0.9947	3	57	Levy	0.39	0.9769	1
24	Log-Logistic	0.05	0.9952	2	58	Error Func.	0.70	0.9103	1
25	Gamma	0.06	0.9931	3	59	Student's t	0.82	0.7991	1
26	Beta	0.06	0.9922	4	-	Johnson SU	No Fit		4
27	Gen. Gamma	0.06	0.9904	3	-	Log-Gamma	No Fit		2
28	Gamma	0.06	0.9910	2	-	Log-Pearson 3	No Fit		3
29	Frechet	0.06	0.9909	3	-	Nakagami	No Fit		2
30	Gumbel Max	0.07	0.9879	2	-	Pareto	No Fit		2
31	Weibull	0.07	0.9851	3	-	Reciprocal	No Fit		2
32	Frechet	0.07	0.9892	2	-	Phased Bi-Exp.	No Fit		4
33	Kumuaraswamy	0.09	0.9805	4	-	Phased Bi-Wei.	No Fit		6
34	Rayleigh	0.10	0.9776	1	-	-	-		-

A widely used theoretical distribution applied to empirical wind speed distributions is the two-parameter Weibull distribution [30,51–58]. However, in this study, the fit of the Weibull distribution is poor ($D = 0.10$) compared to many other theoretical distributions. These results are in accordance with similar studies where the GoF of various theoretical distributions to empirical distributions was compared [59–61]. The Weibull distribution is not even the best-fitting two-parameter distribution, which is the lognormal distribution. The best GoF of a one-parameter distribution was achieved by the also widely used Rayleigh distribution [62,63]. However, compared to many distributions defined by more parameters the GoF of the Rayleigh distribution was rather poor ($D = 0.10$). An explanation for the poor fit of distributions with less than three parameters might be that their capacity for reproducing irregular shapes of empirical distributions is limited. Irregularly shaped empirical wind speed distributions often result from complex topography [64].

The evaluation of averaged R^2-values confirms results of the D-values evaluation. The best-fitting distribution is WK5 ($R^2 = 0.9992$), followed by Johnson SB ($R^2 = 0.9991$).

The superior fit of WK5 is in accordance to GoF measures of empirical near-surface (10 m AGL) wind speed distributions in the study area [20]. Based on the results presented in this study it is concluded that WK5 is a universal wind speed distribution for the study area.

3.2. Predictor Variable Selection and Importance

The screening of r-values showed that the most appropriate length of outer-circle radius was 1000 m for τ and 200 m for z_{0eff}. Table 3 lists the predictor variables used for all six least squares boosting models (LSBM) and their relative impact to the model outputs. From the large set of predictor variables, predictor selection finally reduced their number to 14.

Table 3. Relative importance of predictor variables used for final least squares boosting models (LSBM) in percent. The top three important predictor variables are highlighted in red.

ID	Predictor variable	Symbol	Y_L	ε	Y_{R1}	Y_{R2}	Y_{R3}	Y_{R4}
1	Wind speed at 850 hPa level ($F = 0.75$)	$U_{850hPa,0.75}$	-	-	-	-	21.4	-
2	Wind speed at 850 hPa level ($F = 0.99$)	$U_{850hPa,0.99}$	-	-	-	-	-	14.8
3	Roughness length, local	$z_{0,l}$	1.9	0.1	0.4	-	-	1.5
4	Roughness length, effective, W	$z_{0eff,W}$	-	-	-	9.2	-	-
5	Roughness length, effective, SW	$z_{0eff,SW}$	-	-	0.6	-	-	-
6	Roughness length, effective, S	$z_{0eff,S}$	-	-	7.8	2.2	1.5	6.2
7	Roughness length, effective, N	$z_{0eff,N}$	-	-	1.1	-	6.9	-
8	Topographic exposure, NW	τ_{NW}	-	-	-	9.8	-	5.2
9	Topographic exposure, W	τ_W	-	21.4	-	-	-	20.8
10	Topographic exposure, SW	τ_{SW}	24.4	23.5	9.6	-	20.8	-
11	Topographic exposure, SE	τ_{SE}	-	-	-	-	-	2.4
12	Relative elevation, 1000 m	Φ_{1000m}	73.7	55.0	-	-	49.4	-
13	Relative elevation, 2500 m	Φ_{2500m}	-	-	80.5	-	-	-
14	Relative elevation, 5000 m	Φ_{5000m}	-	-	-	78.8	-	49.1

The main wind directions in the study area are west and southwest. It is therefore reasonable that southwesterly and westerly oriented τ- and z_{0eff}-predictor variables have a distinct impact to the model outputs. The highest PI-values for any roughness length predictor variable are found for the LSBM output Y_{R2} and the western sector ($PI = 9.2\%$). However, the PI-value for Y_{R1} and the southwestern sector is relatively low ($PI = 0.6\%$). The topographic exposure for the southwestern sector, respectively the western sector, is one of the most important predictor variables for modeling ε, Y_L, Y_{R3} and Y_{R4}.

It is important to note that U_{850hPa} was not used to model the left-hand tail of WK5, which represents $U_{100m,mod}$-values. Low wind speed values mostly occur when the atmosphere is stably stratified [22]. Thus, the influence of U_{850hPa} on $U_{100m,mod}$ is rather small.

When modeling Y_{R3} and Y_{R4}, the large-scale airflow becomes more important $PI = \{21.4\%, 14.8\%\}$ because high $U_{100m,mod}$-values usually occur when the atmosphere is neutrally stratified [22].

Results from PI-evaluation indicate the fundamental role of relative elevation in wind turbine site assessment. The high PI-values for Φ indicate the great importance of Φ for model outputs. The highest PI-value is 80.5% for Φ_{2500m} when modeling Y_{R1}. In contrast, the absolute elevation (ψ)

was never used as predictor variable. This is reasonable because sites with high ψ-values are not necessarily exposed to high wind speeds.

3.3. Wind Speed Mapping

Median $U_{100m,mod}$-values ($\tilde{U}_{100m,mod}$) are shown in Figure 4. In large parts (75%) of the study area, $\tilde{U}_{100m,mod}$-values are in the range between 3.0 m/s and 4.0 m/s. In only 0.2% of the study area, $\tilde{U}_{100m,mod}$-values are above 4.9 m/s. Due to the complex topography, high and low $\tilde{U}_{100m,mod}$-values can occur within small distances (<500 m). For example, in the Black Forest, which is characterized by narrow, forested valleys, $\tilde{U}_{100m,mod}$-values are very low. However, there are many exposed mountaintops in close proximity to these valleys where $\tilde{U}_{100m,mod}$-values are high. Beside narrow, forested valleys, lowest $\tilde{U}_{100m,mod}$-values (<3.1 m/s) occur in large cities. In the entire study area the effect of topographic exposure on the modeling results is evident by predominantly higher $\tilde{U}_{100m,mod}$-values at sites exposed to the West and Southwest.

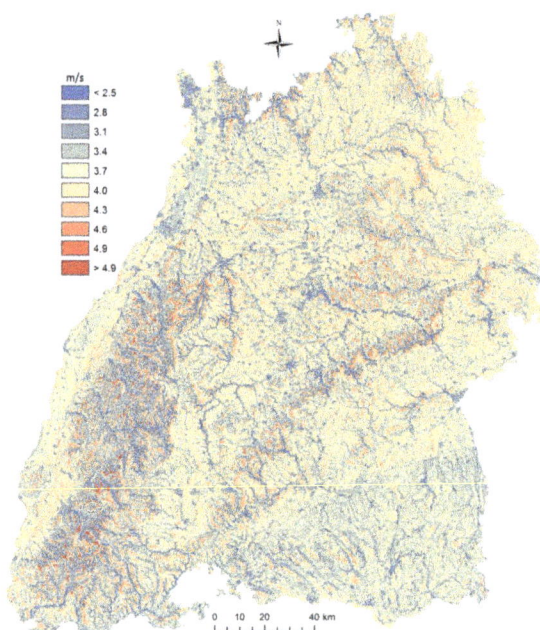

Figure 4. Median of modeled wind speed in 100 m AGL ($\tilde{U}_{100m,mod}$) in the study area. The legend values indicate highest class values.

3.4. Annual Wind Energy Yield

In Figure 5, the empirical *AEY* per wind speed class (ΔAEY_{emp}), the modeled *AEY* per wind speed class (ΔAEY_{mod}), the probability density distributions of WK5$_{mod}$ and the probability density distributions fitted to U_S-values ($U_{S,distr}$) are presented as a function of wind speed classes (intervals of 0.1 m/s) for the stations Hornisgrinde (Figure 5a) and Laupheim (Figure 5b).

It is clear that percentiles ($F = \{0.30\text{--}0.99\}$) from the right-hand tail of WK5$_{mod}$ contribute more to *AEY* and are thus more important for the total amount of AEY_{mod}. In Laupheim the mode of $U_{100,mod}$ is 2.3 m/s, whereas highest ΔAEY_{mod} is obtained at 8.0–8.1 m/s. Even at the top of the Hornisgrinde, which is one of the windiest places in the study area, the $U_{100m,mod}$ mode value at 4.2 m/s is clearly lower than the wind speed class assigned to the highest ΔAEY_{mod}-value (9.0–9.1 m/s). Overall, the ΔAEY_{mod}-curves fit ΔAEY_{emp}-values obtained for both stations well.

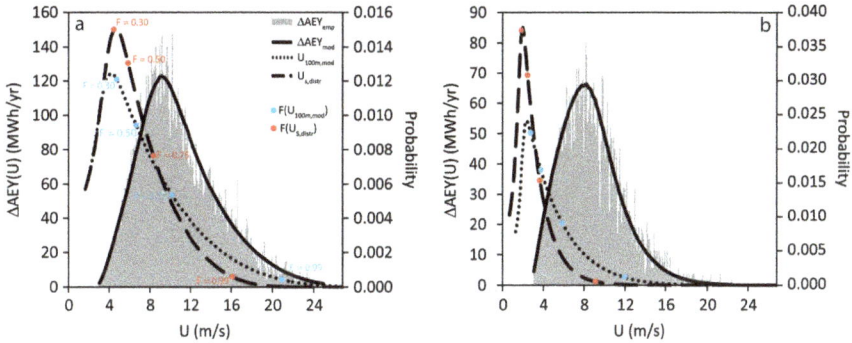

Figure 5. ΔAEY_{emp} and ΔAEY_{mod} as a function of wind speed classes (U) (intervals of 0.1 m/s) as well as the probability of wind speed near the surface fitted to a WK5 distribution $(U_{S,distr})$- and modeled wind speed in 100 m AGL $(U_{100m,mod})$-classes for stations: (**a**) Hornisgrinde; and (**b**) Laupheim.

The map of AEY_{mod} (Figure 6) shows similar patterns like the $\tilde{U}_{100m,mod}$-map. By applying the power curve to $U_{100m,mod}$, the mean AEY_{mod}-value in the study area is 3.4 GWh/yr. The highest AEY_{mod}-value (13.6 GWh/yr) occurs at the top of the Feldberg. Only in 3% of the study area is AEY_{mod} higher than 5.0 GWh/yr. In 31% of the study area AEY_{mod} is lower than 3.0 GWh/yr with a tendency towards lower AEY_{mod}-values in the southeast, which is mainly due to low U_{850hPa}-values in this part over the study area. In contrast, generally higher AEY_{mod}-values were calculated in the northeast where U_{850hPa}-values are highest at the landscape level. The spatial AEY_{mod}-pattern indicates that the local wind resource is mainly determined by terrain features and surface roughness.

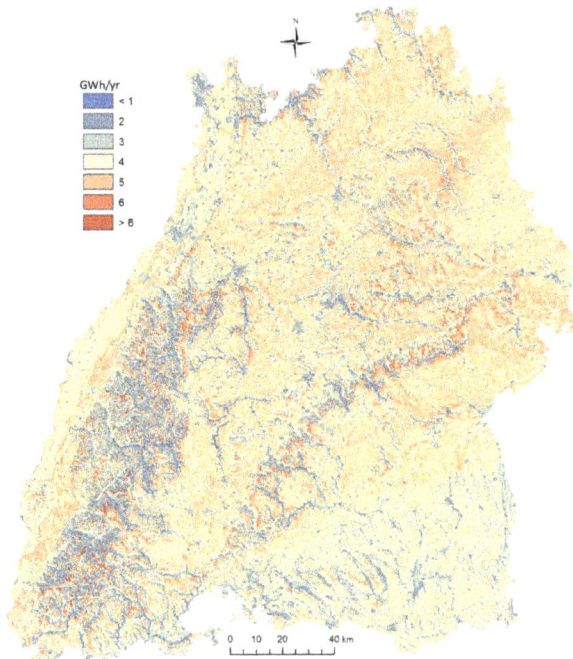

Figure 6. AEY_{mod} in the study area. The legend values indicate highest class values.

This is underlined by the map extract shown in Figure 7. In the topographically structured Black Forest region, it appears that highest and lowest AEY_{mod}-values occur over horizontal distances shorter than 500 m. This finding is in good accordance to a previous study regarding gust speed in the same area [25]. The main wind direction can be inferred from highest AEY_{mod}-values over southwest-facing slopes.

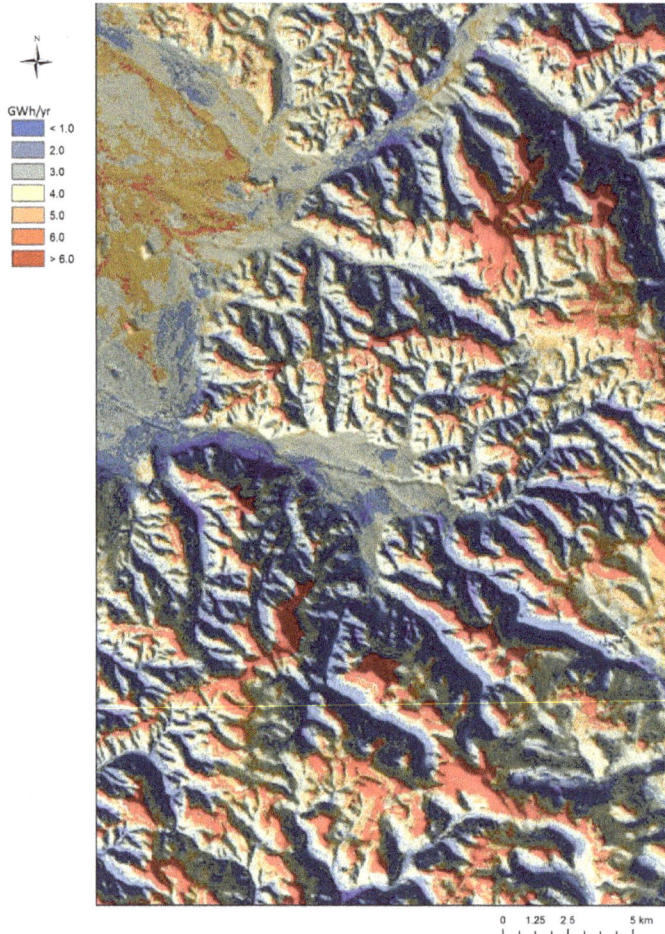

Figure 7. AEY_{mod} in the Southern Black Forest region. The legend values indicate highest class values.

Figure 8 shows r-values which were calculated between AEY_{mod} and various predictor variables. The r-values confirm the results of the PI-evaluation. The highest and lowest r-values are obtained for the most important predictor variables. The highest absolute r-values are ($r = |-0.59|$) for τ_{SW} and ($r = |0.58|$) for Φ_{2500m}. The correlation between ψ and AEY_{mod} is relatively weak ($r = 0.08$). This is due to the fact that some highly elevated Black Forest valleys are sites with the lowest AEY_{mod}-values. The correlation between $U_{850hPa,0.75}$ and AEY_{mod} ($r = 0.08$) is also relatively low since the influence of the large-scale airflow on AEY_{mod} is superimposed by influences of local terrain features and surface roughness. All correlations are highly significant with significance values (p) $p \leqslant 0.0000$.

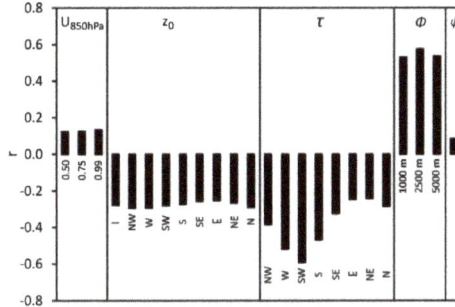

Figure 8. Correlation coefficient (r)-values calculated between AEY_{mod} and various predictor varibles.

The exemplary functional relationships between classes of four important predictor variables and AEY_{mod} are shown in Figure 9. The variability of AEY_{mod}-values as a function of $U_{850hPa,0.75}$ (Figure 9a) is lower than the variability of the other displayed predictor variables. This is interpreted to mean that the variability of $U_{850hPa,0.75}$ is of minor importance for explaining the spatial AEY_{mod}-patterns in the study area. Due to their high roughness, AEY_{mod} is lower over forests and cities (Figure 9b). Areas that are exposed to the southwest ($\tau_{SW} < 2°$) show higher AEY_{mod}-values (median: 3.9 GWh/yr) than sheltered areas ($\tau_{SW} > 18°$) (median: 1.5 GWh/yr) (Figure 9c). The strongest functional relationship is between Φ_{2500m} and AEY_{mod} (Figure 9d). The assigned median AEY_{mod}-values increase from 1.6 GWh/yr at $\Phi_{2500m} < -150$ m to 4.7 GWh/yr at $\Phi_{2500m} > 150$ m.

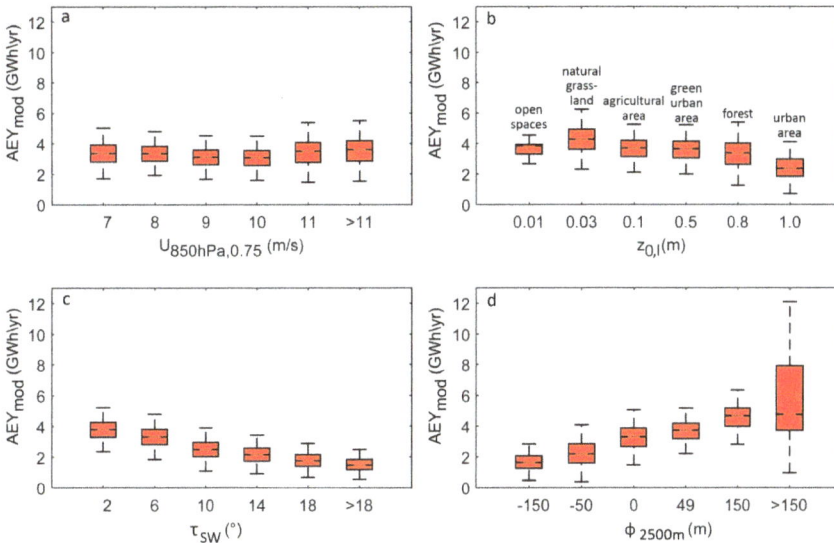

Figure 9. Boxplots of AEY_{mod} as a function of: (**a**) 0.75 percentile of the wind speed at the 850 hPa pressure level ($U_{850hPa,0.75}$); (**b**) local roughness length ($z_{0,l}$); (**c**) topographic exposure in southwest direction (τ_{SW}); and (**d**) relative elevation with outer circle radius of 2500 m (Φ_{2500m}). Boxplot style: red lines indicate medians, boxes indicate interquartile ranges, whiskers indicate 1.5-times interquartile ranges. The legend values indicate highest class values.

3.5. Model Validation

The *MAPE*-values indicate that $U_{100m,mod}$ was simulated accurately (Table 4). They are always below 6% for both DS1 and DS2. The R^2-values are mostly 0.97 for DS1 percentiles and about 0.95 for DS2 percentiles. The largest downward bias is $ME = -0.30$ m/s for $F = 0.99$.

Table 4. Performance measures coefficient of determination (R^2), mean error (*ME*), mean absolute error (*MAE*), mean squared error (*MSE*) and mean absolute percentage error (*MAPE*) calculated from the comparison of empirical and modeled cumulative probabilities (*F*) associated with U_{100m}-time series included in DS1 and DS2.

Data Set	F	R^2	ME (m/s)	MAE (m/s)	MSE (m/s)	MAPE (%)
	0.10	0.97	0.05	0.11	0.02	5.9
	0.20	0.97	0.04	0.11	0.02	5.1
	0.30	0.97	0.01	0.12	0.02	4.6
	0.40	0.97	0.00	0.13	0.03	4.3
DS1	0.50	0.97	0.00	0.14	0.03	4.2
	0.60	0.97	−0.01	0.16	0.04	4.1
	0.70	0.97	−0.01	0.20	0.06	4.2
	0.80	0.97	−0.01	0.25	0.09	4.3
	0.90	0.97	−0.09	0.33	0.17	4.6
	0.99	0.98	0.00	0.38	0.23	3.2

Data Set	F	R^2	ME (m/s)	MAE (m/s)	MSE (m/s)	MAPE (%)
	0.10	0.95	−0.01	0.07	0.01	3.6
	0.20	0.95	−0.03	0.10	0.01	4.5
	0.30	0.96	−0.06	0.10	0.02	4.0
	0.40	0.97	−0.07	0.12	0.02	4.3
DS2	0.50	0.97	−0.07	0.14	0.02	4.4
	0.60	0.96	−0.08	0.17	0.04	4.7
	0.70	0.95	−0.10	0.21	0.07	5.0
	0.80	0.95	−0.12	0.25	0.10	5.1
	0.90	0.95	−0.24	0.36	0.17	5.5
	0.99	0.94	−0.30	0.46	0.42	4.0

The model performance for DS2 is only marginally worse than for DS1. This indicates the portability of LSBM to other data sets.

Performance measures from the comparison of modeled cumulative distribution functions (CDF$_{mod}$) with CDF$_{emp}$ associated with U_{100m}-time series included in DS2 are shown in Table 5.

Table 5. Performance measures from the comparison of modeled cumulative distribution function (CDF$_{mod}$) with empirical cumulative distribution function (CDF$_{emp}$) associated with U_{100m} time series included in DS2.

Station	D	R^2	Station	D	R^2
Bad Dürrheim	0.05	0.9977	Söllingen	0.06	0.9973
Bad Herrenalb	0.05	0.9973	Stockach-Espasingen	0.08	0.9918
Müllheim	0.05	0.9982	Stuttgart-Stadt	0.09	0.9924
Neuhausen ob Eck	0.02	1.0000	Todtmoos	0.07	0.9968
Pforzheim-Ispringen	0.04	0.9991	Weilheim-Bierbronnen	0.06	0.9978

It appears that the GoF measures for modeled WK5 parameters are better than for many statistical distributions that were directly fitted to CDF$_{emp}$ (compare Table 2).

In Figure 10, AEY_{emp} is plotted against AEY_{mod}. Related performance measures for DS1 (Figure 10a) and DS2 (Figure 10b) are $R^2 = \{0.98, 0.97\}$, $ME = \{-0.16 \text{ GWh/yr}, -0.23 \text{ GWh/yr}\}$, $MAE = \{0.32 \text{ GWh/yr}, 0.31 \text{ GWh/yr}\}$, $MSE = \{0.16 \text{ GWh/yr}, 0.13 \text{ GWh/yr}\}$ and $MAPE = \{10.0\%, 17.1\%\}$. Thus, it can be concluded that the calculated AEY_{emp}-values were modeled with sufficient accuracy.

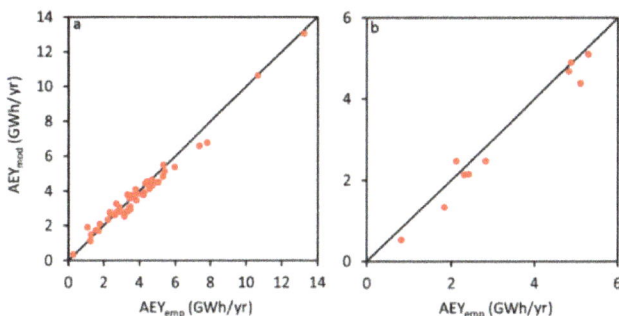

Figure 10. AEY_{emp} plotted against AEY_{mod} for: (**a**) DS1; and (**b**) DS2.

4. Conclusions

A methodology is presented that allows assessing the statistical AEY on a high spatial resolution (50 m × 50 m) grid in an area with mosaic-like land cover pattern and complex topography. It was found that highest and lowest AEY occurs in highly textured terrain within very small distances (<500 m). The results of this study therefore emphasize the need to assess AEY at very small spatial scales. This is demonstrated in particular by the great importance of the predictor variables relative elevation and topographic exposure in the main wind direction.

Since the methodology allows for the calculation of all WK5 parameters, the AEY for any manufacturer power curve can be estimated. The methodology is easily portable to other heights above ground level as well as to other study areas. The only requirements for the portability are the availability of the following: (i) near-surface wind speed time series as measured in meteorological networks; (ii) a DTM; (iii) a land cover data set; and (iv) wind speed data not influenced by local topography or land use.

The proposed modeling approach is a useful first step in the exploration of the most appropriate wind turbine sites based on the local wind resource. The produced model outputs and maps are valuable starting points for further in-depth wind turbine site assessment.

Conflicts of Interest: The author declares no conflict of interest sponsors had no role in the design of the study; in the collection, analyses, or interpretation of data; in the writing of the manuscript, and in the decision to publish the results.

Nomenclature

AEY	Annual wind energy yield
AEY_{emp}	Empirical annual wind energy yield
AEY_{mod}	Modeled annual wind energy yield
B	Regression tree
B_m	Regression tree m
D	Kolmogorov-Smirnov statistic
E	Hellmann exponent
f	Probability density
F	Cumulative probability

h_{850hPa}	Height above ground level of the 850 hPa pressure level
h_s	Measurement height of U_S
M	Total number of weak learners
MAE	Mean absolute error
$MAPE$	Mean absolute percentage error
ME	Mean error
MSE	Mean squared error
N	Number of days in the investigation period
p	Significance value
P	Electrical power output
$\overline{P}$	Average electrical power output
PI	Relative predictor importance
p_m	Weight for model m
r	Correlation coefficient
R^2	Coefficient of determination
U	Wind speed
U_{100m}	Wind speed in 100 m AGL
$U_{100m,distr}$	Wind speed in 100 m AGL fitted to a WK5 distribution
$U_{100m,emp}$	Empirical wind speed in 100 m AGL
$U_{100m,mod}$	Modeled wind speed in 100 m AGL
$\tilde{U}_{100m,mod}$	Median of modeled wind speed in 100 m AGL
U_{850hPa}	Wind speed at the 850 hPa pressure level
$U_{850hPa,0.01}$	1.st percentile of the wind speed at the 850 hPa pressure level
$U_{850hPa,0.30}$	30.th percentile of the wind speed at the 850 hPa pressure level
$U_{850hPa,0.50}$	50.th percentile of the wind speed at the 850 hPa pressure level
$U_{850hPa,0.75}$	75.th percentile of the wind speed at the 850 hPa pressure level
$U_{850hPa,0.99}$	99.th percentile of the wind speed at the 850 hPa pressure level
U_S	Wind speed near the surface
$\tilde{U}_s$	Median wind speed near the surface
$U_{S,distr}$	Wind speed near the surface fitted to a WK5 distribution
v	Learning rate
X	Predictor variables
Y	Target variables
$\tilde{Y}$	Median of target variables
Y_L	Left-hand side of WK5
Y_{pred}	Aggregated prediction of predictor variables
Y_R	Right-hand side of WK5
Y_{R1}	30.th percentile of WK5
Y_{R2}	50.th percentile of WK5
Y_{R3}	75.th percentile of WK5
Y_{R4}	99.th percentile of WK5
z_0	Roughness length
z_{0eff}	Effective roughness length
$z_{0eff,E}$	Effective roughness length in east direction
$z_{0,l}$	Local roughness length
$z_{0eff,N}$	Effective roughness length in north direction
$z_{0eff,NE}$	Effective roughness length in northeast direction
$z_{0eff,NW}$	Effective roughness length in northwest direction
$z_{0eff,S}$	Effective roughness length in south direction
$z_{0,effSE}$	Effective roughness length in southeast direction
$z_{0eff,SW}$	Effective roughness length in southwest direction
$z_{0eff,W}$	Effective roughness length in west direction

Z_1	Number of years in the investigation period
Z_2	Number of days per year
α	Parameter of WK5
β	Parameter of WK5
γ	Parameter of WK5
δ	Parameter of WK5
ΔAEY_{emp}	Empirical annual wind energy yield per wind speed class
ΔAEY_{mod}	Modeled annual wind energy yield per wind speed class
ε	Location parameter of WK5
τ	Topographic exposure
τ_E	Topographic exposure in east direction
τ_N	Topographic exposure in north direction
τ_{NE}	Topographic exposure in northeast direction
τ_{NW}	Topographic exposure in northwest direction
τ_S	Topographic exposure in south direction
τ_{SE}	Topographic exposure in southeast direction
τ_{SW}	Topographic exposure in southwest direction
τ_W	Topographic exposure in west direction
Φ	Relative elevation
Φ_{1000m}	Relative elevation with outer circle radius of 1000 m
Φ_{2500m}	Relative elevation with outer circle radius of 2500 m
Φ_{5000m}	Relative elevation with outer circle radius of 5000 m
ψ	Absolute elevation

Abbreviations

AGL	Above ground level
CDF	Theoretical cumulative distribution function
CDF_{emp}	Empirical cumulative distribution function
CDF_{mod}	Modeled cumulative distribution function
CLC	CORINE Land Cover
CORINE	Coordination of Information on the Environment
DS1	Parameterization dataset
DS2	Validation dataset
DTM	Digital terrain model
DWD	German Weather Service
ERA	European Centre for Medium-Range Weather Forecasts re-analysis
GoF	Goodness of fit
ID	Identification number
LSBM	Least squares boosting model
LSBoost	Least squares boosting
NP	Number of parameters
RK	Rank of distribution according to D-evaluation
WK5	Wakeby distribution
$WK5_{mod}$	Modeled Wakeby distribution

References

1. Kaldellis, J.K.; Zafirakis, D. The wind energy (r) evolution: A short review of a long history. *Renew. Energy* **2011**, *36*, 1887–1901. [CrossRef]
2. Leung, D.Y.; Yang, Y. Wind energy development and its environmental impact: A review. *Renew. Sustain. Energy Rev.* **2012**, *16*, 1031–1039. [CrossRef]
3. Lew, D.J. Alternatives to coal and candles: Wind power in China. *Energy Policy* **2000**, *28*, 271–286. [CrossRef]

4. Büsgen, U.; Dürrschmidt, W. The expansion of electricity generation from renewable energies in Germany: A review based on the Renewable Energy Sources Act Progress Report 2007 and the new German feed-in legislation. *Energy Policy* **2009**, *37*, 2536–2545. [CrossRef]

5. Jacobson, M.Z.; Masters, G.M. Exploiting wind *versus* coal. *Science* **2001**, *293*, 1438. [CrossRef] [PubMed]

6. Menyah, K.; Wolde-Rufael, Y. CO$_2$ emissions, nuclear energy, renewable energy and economic growth in the US. *Energy Policy* **2010**, *38*, 2911–2915. [CrossRef]

7. Christodouleas, J.P.; Forrest, R.D.; Ainsley, C.G.; Tochner, Z.; Hahn, S.M.; Glatstein, E. Short-term and long-term health risks of nuclear-power-plant accidents. *New Engl. J. Med.* **2011**, *364*, 2334–2341. [CrossRef] [PubMed]

8. Lu, X.; McElroy, M.B.; Kiviluoma, J. Global potential for wind-generated electricity. *Proc. Natl. Acad. Sci. USA* **2009**, *106*, 10933–10938. [CrossRef] [PubMed]

9. EUR-Lex. Available online: http://eur-lex.europa.eu/legal-content/EN/ALL/?uri=CELEX%3A32009L0028 (accessed on 10 February 2016).

10. Windenergie. Available online: http://www4.lubw.baden-wuerttemberg.de/servlet/is/224533/ (accessed on 27 January 2016).

11. Newman, J.F.; Klein, P.M. The impacts of atmospheric stability on the accuracy of wind speed extrapolation methods. *Resources* **2014**, *3*, 81–105. [CrossRef]

12. Wieringa, J. Roughness-dependent geographical interpolation of surface wind speed averages. *Q. J. R. Meteorol. Soc.* **1986**, *112*, 867–889. [CrossRef]

13. Kalthoff, N.; Bischoff-Gauß, I.; Fiedler, F. Regional effects of large-scale extreme wind events over orographically structured terrain. *Theor. Appl. Climatol.* **2003**, *74*, 53–67. [CrossRef]

14. Garcia, A.; Torres, J.L.; Prieto, E.; de Franciso, A. Fitting wind speed distributions: A case study. *Sol. Energy* **1998**, *2*, 139–144. [CrossRef]

15. Celik, A.N. Assessing the suitability of wind speed probability distribution functions based on wind power density. *Renew. Energy* **2003**, *28*, 1563–1574. [CrossRef]

16. Van Ackere, S.; Van Eetvelde, G.; Schillebeeckx, D.; Papa, E.; Van Wyngene, K.; Vandevelde, L. Wind resource mapping using landscape roughness and spatial interpolation methods. *Energies* **2015**, *8*, 8682–8703. [CrossRef]

17. Intergovernmental Panel on Climate Change (IPCC). *Renewable Energy Sources and Climate Change Mitigation*; Special Report of the Intergovernmental Panel on Climate Change; Cambridge University Press: New York, NY, USA, 2012.

18. Carta, J.A.; Ramírez, P.; Velazquez, S. A review of wind speed probability distributions used in wind energy analysis. Case studies in the Canary Islands. *Renew. Sustain. Energy Rev.* **2009**, *13*, 933–955. [CrossRef]

19. Morgan, E.C.; Lackner, M.; Vogel, R.M.; Baise, L.G. Probability distributions for offshore wind speeds. *Energy Convers. Manag.* **2011**, *52*, 15–26. [CrossRef]

20. Jung, C.; Schindler, D. Statistical modeling of near-surface wind speed: A case study from Baden-Wuerttemberg (Southwest Germany). *Austin J. Earth Sci.* **2015**, *2*, 1006.

21. Jung, C.; Schindler, D.; Albrecht, A.T.; Buchholz, A. The role of highly-resolved gust speed in simulations of storm damage in forests at the landscape scale: A case study from southwest Germany. *Atmosphere* **2016**, *7*. [CrossRef]

22. Touma, J.S. Dependence of the wind profile power law on stability for various locations. *J. Air Pollut. Control Assoc.* **1977**, *27*, 863–866. [CrossRef]

23. Gualtieri, G.; Secci, S. Methods to extrapolate wind resource to the turbine hub height based on power law: A 1-h wind speed *vs.* Weibull distribution extrapolation comparison. *Renew. Energy* **2012**, *43*, 183–200. [CrossRef]

24. Gualtieri, G.; Secci, S. Extrapolating wind speed time series *vs.* Weibull distribution to assess wind resource to the turbine hub height: A case study on coastal location in Southern Italy. *Renew. Energy* **2014**, *62*, 164–176. [CrossRef]

25. Jung, C.; Schindler, D. Modeling monthly near-surface maximum daily gust speed distributions in Southwest Germany. *Int. J. Climatol.* **2016**. [CrossRef]

26. Dee, D.P.; Uppala, S.M.; Simmons, A.J.; Berrisford, P.; Polia, P.; Kobayashi, S.; Andrae, U.; Balmaseda, M.A.; Balsamo, G.; Bauer, P.; *et al.* The ERA-Interim reanalysis: configuration and performance of the data assimilation system. *Q. J. R. Meteorol. Soc.* **2011**, *137*, 553–597. [CrossRef]

27. Soukissian, T. Use of multi-parameter distributions for offshore wind speed modeling: The Johnson SB distribution. *Appl. Energy* **2013**, *111*, 982–1000. [CrossRef]

28. Poje, D.; Cividini, B. Assessment of wind energy potential in Croatia. *Sol. Energy* **1988**, *41*, 543–554. [CrossRef]

29. Dorvlo, A.S.S. Estimating wind speed distributions. *Energy Convers. Manage.* **2002**, *43*, 2311–2318. [CrossRef]

30. Sulaiman, M.Y.; Akaak, A.M.; Wahab, M.A.; Zakaria, A.; Sulaiman, Z.A.; Suradi, J. Wind characteristics of Oman. *Energy* **2002**, *27*, 35–46. [CrossRef]

31. Houghton, J.C. Birth of a parent: the Wakeby distribution for modeling flood flows. *Water Resour. Res.* **1978**, *14*, 1105–1109. [CrossRef]

32. Öztekin, T. Estimation of the parameters of Wakeby distribution by a numerical least squares method and applying it to the annual peak flows of Turkish Rivers. *Water Resour. Manag.* **2011**, *25*, 1299–1313. [CrossRef]

33. Singh, K.P. Comment on 'Birth of a parent: The Wakeby distribution for modeling flood flows' by John C. Houghton. *Water Resour. Res.* **1979**, *15*, 1285–1287. [CrossRef]

34. UBA. Umweltbundesamt, CORINE Land Cover. 2009. Available online: http://www.umweltbundesamt.de/themen/boden-landwirtschaft/flaechensparen-bodenlandschaften-erhalten/corine-land-cover-clc (accessed on 27 January 2016).

35. Wilson, J.D. Determining a topex score. *Scott. For.* **1984**, *38*, 251–256.

36. Quine, C.P.; White, I.M.S. The potential of distance-limited topex in the prediction of site windiness. *Forestry* **1998**, *71*, 325–332. [CrossRef]

37. Pryor, S.C.; Barthelmie, R.J. Long-term trends in near-surface flow over the Baltic. *Int. J. Climatol.* **2003**, *23*, 271–289. [CrossRef]

38. Hosking, J.R.M.; Wallis, J.R. *Regional Frequency Analysis*; Cambridge University Press: New York, NY, USA, 1997; p. 224.

39. Hosking, J.R.M. *Fortran Routines for Use with the Method of L-Moment*; IBM Research Division: New York, NY, USA, 2000; p. 33.

40. Friedman, J. Greedy function approximation: A gradient boosting machine. *Ann. Stat.* **2001**, *29*, 1189–1232. [CrossRef]

41. Van Heijst, D.; Potharst, R.; Van Wezel, M. A support system for predicting eBay end prices. *Decis. Support Syst.* **2008**, *44*, 970–982. [CrossRef]

42. Willmott, C.J.; Matsuura, K. Advantages of the mean absolute error (MAE) over the root mean square error (RMSE) in assessing average model performance. *Clim. Res.* **2005**, *30*, 79–82. [CrossRef]

43. Hyndman, R.J.; Kohler, A.B. Another look at measures of forecast accuracy. *Int. J. Forecast.* **2006**, *22*, 679–688. [CrossRef]

44. Celik, A.N.; Kolhe, M. Generalized feed-forward based method for wind energy production. *Appl. Energy* **2013**, *101*, 582–588. [CrossRef]

45. Belsley, D.A.; Kuh, E.; Welsh, R.E. *Regression Diagnostics*; John Wiley & Sons: New York, NY, USA, 2001.

46. Lydia, M.; Kumar, S.S.; Selvakumar, A.I.; Kumar, G.E.P. A comprehensive review on wind turbine power curve modeling techniques. *Renew. Sustain. Energy Rev.* **2014**, *30*, 452–460. [CrossRef]

47. Carrillo, C.; Montaño, A.O.; Cidrás, J.; Díaz-Dorado, E. Review of power curve modelling for wind turbines. *Renew. Sustain. Energy Rev.* **2013**, *21*, 572–581. [CrossRef]

48. GE Power & Water Renewable Energy. Available online: http://www.ge-renewable-energy.com/de/wind/produkte/produktuebersicht/25-275-285-32-34/ (accessed on 27 January 2016).

49. Lydia, M.; Kumar, S.S.; Selvakumar, A.I.; Kumar, G.E.P. Wind resource estimation using wind speed and power curve models. *Renew. Energy* **2015**, *83*, 425–434. [CrossRef]

50. Jowder, F.A.L. Wind power analysis and site matching of wind turbine generators in Kingdom of Bahrain. *Appl. Energy* **2009**, *86*, 538–545. [CrossRef]

51. Rehman, S.; Halawan, T.O.; Husain, T. Weibull parameters for wind speed distribution in Saudi Arabia. *Sol. Energy* **1994**, *53*, 473–479. [CrossRef]

52. Lun, I.Y.F.; Lan, J.C. A study of Weibull parameters using long-term wind observations. *Renew. Energy* **2000**, *20*, 145–153. [CrossRef]

53. Seguro, J.V.; Lambert, T.W. Modern estimation of the parameters of the Weibull wind speed distribution for wind energy analysis. *J. Wind. Eng. Ind. Aerodyn.* **2000**, *85*, 75–84. [CrossRef]

54. Ramírez, P.; Carta, J.A. Influence of the data sampling interval in the estimation of the parameters of the Weibull wind speed probability density distribution: A case study. *Energy Convers. Manage.* **2005**, *46*, 2419–2438. [CrossRef]
55. Toure, S. Investigations on the Eigen-coordinates method for the 2-parameter Weibull distribution of wind speed. *Renew. Energy* **2005**, *30*, 511–521. [CrossRef]
56. Cellura, M.; Cirricione, G.; Marvuglia, A.; Miraoui, A. Wind speed spatial estimation for energy planning in Sicily: Introduction and statistical analysis. *Renew. Energy* **2008**, *33*, 1237–1250. [CrossRef]
57. Tar, K. Some statistical characteristics of monthly average wind speed at various heights. *Renew. Sustain. Energy Rev.* **2008**, *12*, 1712–1724. [CrossRef]
58. Zaharim, A.; Razali, A.M.; Abidin, R.Z.; Sopian, K. Fitting of statistical distributions to wind speed data in Malaysia. *Eur. J. Sci. Res.* **2009**, *26*, 6–12.
59. Jaramillo, O.A.; Borja, M.A. Wind speed analysis in La Ventosa, Mexico: a bimodal probability distribution case. *Renew. Energy* **2004**, *29*, 1613–1630. [CrossRef]
60. Kiss, P.; Janosi, I.M. Comprehensive empirical analysis of ERA-40 surface wind speed distribution over Europe. *Energy Convers. Manage.* **2008**, *49*, 2142–2151. [CrossRef]
61. Zhou, J.; Erdem, E.; Li, G.; Shi, J. Comprehensive evaluation of wind speed distribution models: A case study for North Dakota sites. *Energy Convers. Manage.* **2010**, *51*, 1449–1458. [CrossRef]
62. Alodat, M.T.; Anagreh, Y.N. Durations distribution of Rayleigh process with application to wind turbines. *J. Wind Eng. Ind. Aerodyn.* **2011**, *99*, 651–657. [CrossRef]
63. Ahmmad, M.R. Statistical analysis of the wind resources at the importance for energy production in Bangladesh. *Int. J. U E Serv. Sci. Technol.* **2014**, *7*, 127–136. [CrossRef]
64. Tuller, S.E.; Brett, A.C. The characteristics of wind velocity that favor the fitting of a Weibull distribution in wind speed analysis. *J. Clim. Appl. Meteor.* **1984**, *23*, 124–134. [CrossRef]

energies

MDPI

Article

Analyses of the Extensible Blade in Improving Wind Energy Production at Sites with Low-Class Wind Resource

Jiale Li [1] and Xiong (Bill) Yu [1,2,*]

[1] Department of Civil Engineering, Case Western Reserve University, 10900 Euclid Avenue, Bingham Building, Cleveland, OH 44106-7201, USA; jxl780@case.edu
[2] Department of Electrical Engineering and Computer Science, Case Western Reserve University, Cleveland, OH 44106, USA
* Correspondence: xxy21@case.edu; Tel.: +1-216-368-6247

Received: 24 July 2017; Accepted: 27 August 2017; Published: 30 August 2017

Abstract: This paper describes the feasibility analysis of an innovative, extensible blade technology. The blade aims to significantly improve the energy production of a wind turbine, particularly at locations with unfavorable wind conditions. The innovative 'smart' blade will be extended at low wind speed to harvest more wind energy; on the other hand, it will be retracted to its original shape when the wind speed is above the rated wind speed to protect the blade from damages by high wind loads. An established aerodynamic model is implemented in this paper to evaluate and compare the power output of extensible blades versus a baseline conventional blade. The model was first validated with a monitored power production curve based on the wind energy production data of a conventional turbine blade, which is subsequently used to estimate the power production curve of extended blades. The load-on-blade structures are incorporated as the mechanical criteria to design the extension strategies. Wind speed monitoring data at three different onshore and offshore sites around Lake Erie are used to predict the annual wind energy output with different blades. The effects of extension on the dynamic characteristics of blade are analyzed. The results show that the extensive blade significantly increases the annual wind energy production (up to 20% to 30%) with different blade extension strategies. It, therefore, has the potential to significantly boost wind energy production for utility-scale wind turbines located at sites with low-class wind resource.

Keywords: wind turbine blade; extensible blade; smart blade; distributed energy resources; low-class wind resource

---

1. Introduction

Wind turbines have been used by human beings for more than 3000 years [1]. Its roles have evolved from performing mechanical work such as pumping, grinding and cutting to renewable energy production [2]. Modern wind turbines are typically horizontal axis turbine with two or three blades, which are results of optimal design from both efficiency and cost considerations.

Increasing both the wind power output and efficiency have been consistent goals for the wind energy industry. The proper siting of wind turbines is important to achieve such a goal. It is recommended that wind turbines be constructed at sites with high-quality wind resources. According to TC88-MC 2005 [3], wind resources are classified into four levels depending upon the characteristics of the average wind speed. Many investigations have been conducted into optimizing wind turbine locations for a particular wind farm [4,5]. These include constructing the aerodynamic model to account for the variations of wind flow over hills, ridges, valleys, offshore, and other types of complex topography. However, sites with high-quality wind resources are typically located in remote areas far away from cities [6,7]. However, more and more wind turbines named 'distributed energy resources' are installed at locations with lower-class wind resources [8–10] to take advantages of their close proximity to the existing electrical grid or manufacturing infrastructures. It helps to reduce the development and transportation cost, which offsets to a certain extent the disadvantages of the low-quality wind resource. In contrast to wind farms, which typically contain hundreds of wind turbines, distributed generators are mostly small-scale power generators located close to the service loads. There are a significant number of wind turbines built as distributed energy resources. According to the U.S. Department of Energy's Distributed Wind Turbine Market Report, 934 MW of distributed wind capacity was installed between 2013 and 2015, representing nearly 75,000 units across 36 states, Puerto Rico, and the U.S. Virgin Islands. Effective utilizing wind resources at sites with low and medium wind speed helps to make wind energy production to be more geographically dispersed; this also helps to reduce the inherent variabilities of wind energy production [11,12].

Improving the energy production at sites with low-class wind bears an important practice value. One potential method is to increase the wind turbine hub height [4,13–16], which utilizes the benefit that the near-ground wind speed increases with elevation. There are, however, significant cost factors associated with manufacturing, logistic transportation, and the construction of components for the higher supporting tower. An alternative method is to develop innovative wind turbine blades technologies that achieve both improvements in production and resiliency. Another alternative method is to develop innovative wind turbine blades that increase both production and resiliency. Significant progress has been made in this aspect. A new design of a dual-rotor wind turbine (DRWT), which includes rotors in both upwind and downwind directions, has been studied by [17]; the authors used the blade element momentum theory to calculate the aerodynamic forces and the torques generated from each of the rotor blades. This dual-rotor wind turbine is considered to have better performance in extracting energy than a conventional single-rotor wind turbine. Huang [18] studied a novel designed wind turbine blade with sinusoidal protuberances with different amplitudes at the leading edge, which was inspired by the structure observed in humpback whale flippers [19]. They used the wind tunnel to test the performance of both the smooth leading edge blade and the comparative models with leading edge protuberances. The results indicated that this new blade has a better performance at the stall region. In Huang and Wu's study [20], a balloon-type airfoil whose shape changes with the pressure distribution has been introduced. The blade is full of air and is able to change its shape according to the pressure distribution. The authors used the numerical simulation to simulate an NACA0012 airfoil blade and came out with the result that this innovative blade can achieve better aerodynamic performance than the conventional blade. Bhuyan and Biswas [21] described an unsymmetrical cambered airfoil blade for a vertical axis wind turbine (VAWT) which achieves improved performance in self-starting and a high power coefficient. Bottasso [22] investigated a novel passive control concept to mitigate loads and suppress vibrations of wind turbines via a flap or a pitching blade tip that moves passively in response to blade vibrations.

These previous efforts primarily look at dynamically changing the cross-sectional shape of the airfoil in response to wind directions. Meanwhile, the diameter of the rotor is another major factor determining the maximum energy output. Longer blades feature larger sweep areas, and hence capture more kinetic energy. This leads to a lower cost per kilowatt-hour of energy produced, which has been validated by numerous studies [23–27]. According to Jureczko et al. [28], the manufacturing cost of a wind turbine blade is about 15–20% of the total wind turbine production cost. Improving the total power output of a wind turbine via optimizing the wind turbine blades presents an important opportunity to increase the turbine's cost efficiency.

Besides this, as with most mechanical system, the capacity of the blade should be matched to the wind conditions at a particular site to achieve the best performance. Due to variations in the wind conditions across different sites, it is difficult for a fixed length blade to match the varying characteristics of installation sites. In fact, commercial wind turbine manufacturers supply wind turbines of similarly rated power outputs with different blade lengths for sites with different wind conditions. A new concept of variable length blade or telescope blade is proposed recently to increase the power output and the annual energy production of the wind turbine [29–31]. The smart blade has an extensible length that will adjust itself according to the incoming wind speed. The blade will extend at low wind speeds to harvest more wind energy, and it will retract to its original shape when the wind speed is above the 'rated wind speed' to ensure structural safety. Therefore, it will produce more energy while protecting the blade from possible damage under high wind speeds. Although this variable length blade has been proposed for several years, there is very limited information on the aerodynamic performance characteristics of this blade. In addition, the blade concepts in previous studies are only extended at the blade tip.

This paper analyzes the concept of the smart blade with the extensible length adjusted according to the incoming wind speed. The blade will extend at low wind speeds to harvest more wind energy and it will retract to its original shape when the wind speed is above the 'rated wind speed' to ensure structural safety. Therefore, it will provide more wind energy outputs while protecting the blade from possible damage under high wind speed. The performance of this extensible blade was analyzed using blade element momentum (BEM) theory, which is an accepted method by the wind industry for wind turbine blade aerodynamic calculation and therefore provides practice feasible conclusions. The BEM model is firstly validated with the field-monitored energy production data of regular wind turbines. The performance of the extensible blade is then analyzed using field-monitored wind speed data at a few onshore and offshore sites around Lake Erie. The results show the promise of the extensible blade to significantly improve energy production at sites with a low class of wind resources.

2. Extensible Blade Concept

The theoretical power output of a wind turbine is described in Equation (1) [32,33].

$$P = \frac{1}{2} C_p \rho A U_{\text{tot}}{}^3 \tag{1}$$

where ρ is the density of air, C_p is the power coefficient, A is the rotor swept area, and U_{tot} is the inflow wind speed.

The equation shows that, at a certain inflow wind speed and air density (which are primarily decided by the climate condition and the topology of a particular site), the power output of a wind turbine is dependent upon its power coefficient and the rotor swept area. The power coefficient is decided by the mechanical structure of the rotor, with the theoretical maximum given by the Benz limit. The rotor swept area is decided by the length of the blade. The blade length is typically controlled from safety consideration to prevent the structural failure at a critical high wind speed which is rarely exceeded in the turbine service life. In this sense, the fixed length blade is not optimized to work under low wind speed conditions. The low wind speed allows the blade length to be increased to improve the wind turbine production while posing no threat to its structure safety.

The basic idea of the extensible blade is to increase the blade length at lower wind speed to produce more energy; the blade will turn to its original length when the inflow wind speed exceeds the rated wind speed. Therefore, an improved power output curve will be achieved for all the working conditions while mitigating safety risk. To study the technical feasibility without losing generality, two types of blade extension scenery are analyzed to assess the benefits of the extensible blade in wind energy production; i.e., (1) extension at the middle of the blade and (2) extension at the tip of the blade (Figure 1). In these analyses, the extensible part of the blades is assumed to have the same foil size as the connection parts. The parameters of the prototype turbine blade are first determined. Aerodynamics analyses are conducted on the extensible blade at different extension conditions. The performance of the extensible blade in energy produced is compared with regular blade using the wind speed data at three different sites in Lake Erie area, Cleveland, LA, USA.

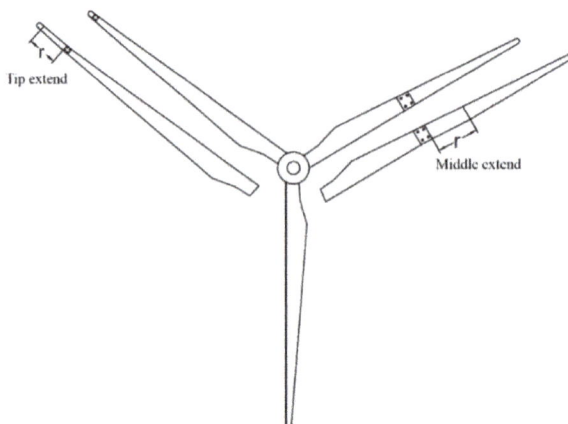

Figure 1. Schematic of the extensible blade concept (with an illustration of extension at the tip and middle of the blade).

3. Specifications of Fixed Length Wind Turbine Blade

The baseline turbine model is built based on a 100 kW utility-scale wind turbine (Northern Power® 100, Northern Power Systems, Barre, VT, USA), installed on the campus of Case Western Reserve University. The key parameters of the turbine are shown in Table 1. The manufacturer power curve is plotted in Figure 2. The turbine was installed in November 2010 with financial support from the Ohio Third Frontier Program. The primary role of the turbine is to serve as a research test-bed for wind energy research [34]. A Campbell-Scientific data acquisition system (DAQ) is installed in the wind turbine tower to monitor the operation data, i.e., wind speed, wind direction, output power, etc. continuously.

Table 1. Prototype wind turbine parameter.

Configuration	Description
Model	Northern Power® 100
Design Class	IEC IIA
Design Life	20 years
Hub Heights	37 m
Power Regulation	Variable speed, stall control
Rotor Diameter	21 m
Rated Wind Speed	14.5 m/s
Rated Electrical Power	100 kW, 3 phase, 480 VAC, 60/50 Hz
Cut-In Wind Speed	3.5 m/s
Cut-Out Wind Speed	25 m/s

Wind energy is produced due to the lift force on the blade produced by the incoming air flow, which drives the rotor. The airfoil shape characteristics are the essential factors determining the lift force. The model used in this research is based upon the airfoil DU-00-W-401 from the well-known NREL 5-MW prototype wind turbine. Because the blade profile data is unavailable for the 100 kW prototype wind turbine, the model used in this research is based upon the airfoil DU-00-W-401 from the well-known NREL 5-MW prototype wind turbine [35] and scaled down to the length of a 100 kW turbine. The detailed profile data of the 5-MW prototype wind turbine is available for research purposes. The lift and the drag coefficient of DU-00-W-401 are plotted in Figure 3. As a simplification, the airfoil is assumed to have the same shape from root to tip of the blade, with a decreasing chord length. The chord length of a blade is defined as the width of the wind turbine blade at a given distance along the length of the blade (Figure 4). In this study, the rotor shape of the NREL 5-MW reference wind turbine is scaled down to the 21 m diameter blade with the corresponding chord lengths scaled and shown in Table 2.

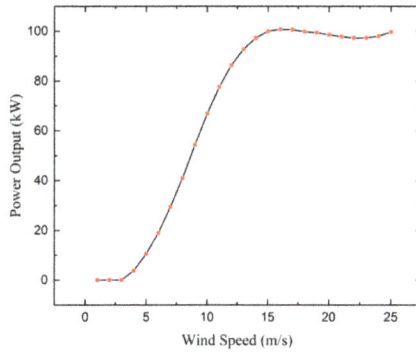

Figure 2. Power curve of the 100 kW wind turbine.

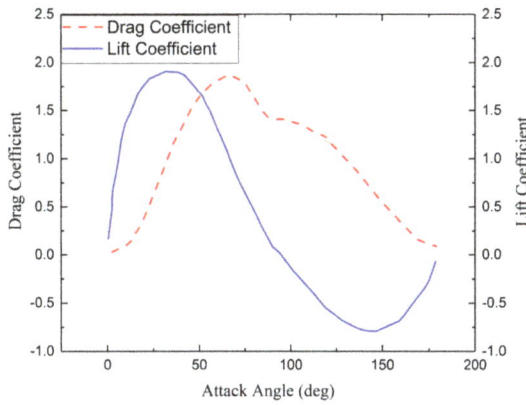

Figure 3. DU-00-W-401 airfoil lift and drag coefficients [35].

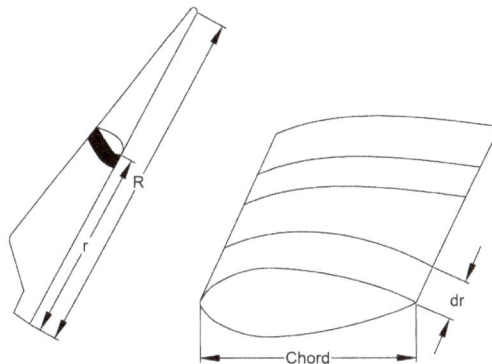

Figure 4. Schematic of the blade with an example airfoil blade element (r is the distance from blade's root to airfoil blade element, R is the blade radius, and the chord length is that of the straight line joining the leading and trailing edges of an airfoil).

Table 2. Parameters for each section along the blade based on 5 MW prototype wind turbine scaling to blade length of 10.52 m [35].

Radius (m)	Twist (Deg)	Chord (m)	Airfoil Shape
0.48	13.31	0.59	Cylinder
0.93	13.31	0.64	Cylinder
1.39	13.31	0.69	Cylinder
1.96	13.31	0.76	DU-00-W-401 [a]
2.65	11.48	0.78	DU-00-W-350
3.33	10.16	0.74	DU-00-W-350
4.02	9.01	0.71	DU-97-W-300
4.70	7.79	0.67	DU-91-W2-250
5.39	6.54	0.63	DU-91-W2-250
6.07	5.36	0.58	DU-93-W-210
6.76	4.19	0.54	DU-93-W-210
7.44	3.13	0.50	NACA64618 [b]
8.13	2.32	0.46	NACA64618
8.81	1.52	0.42	NACA64618
9.38	0.86	0.38	NACA64618
9.84	0.37	0.35	NACA64618
10.29	0.11	0.24	NACA64618
10.52	0	0.15	NACA64618

[a] DU stands for Delft University; [b] NACA stands for National Advisory Committee for Aeronautics.

4. Model and Analyses of the Original Length Blade and Extensible Blade

The aerodynamic analyses are conducted on the original fixed blade as well as the extensible blades using the blade element momentum (BEM) theory. BEM theory is a classical analysis method of wind turbines [36], which has been widely accepted for blade performance analyses; an established model such as BEM is selected for these analyses so that the results provide a practical assessment of the new blade technology. BEM is composed of two different theories; i.e., blade element theory and momentum theory [37]. Blade element theory assumes that blades can be divided into small elements that act independently of the surrounding elements and operate aerodynamically as two-dimensional airfoils as shown in Figure A1, in which α is the attack angle. The characteristics of blade responses (drag and life on each element) are determined by the angle of attack of incoming wind, which is the angle between the center reference line of the geometry and the relative incoming flow W (Figure A1). The momentum theory assumes that the loss of air pressure or the generation of turning momentum in the airfoil blade element is caused by the work done by the incoming airflow [38]. The BEM theory couples these two theories together and calculates the total lift and momentum via an iterative process [39]. The model is subsequently used to determine the power output at a given wind speed.

The BEM theory is implemented via customized code developed with MATHCAD® (MATHCAD 15.0, Parametric Technology Corporation, Needham, MA, USA). Details of the implementation procedures for the BEM theory are provided in the Appendix A as they are not the focus of this paper.

Validation of BEM Model in Blade Power Output Prediction

For implementing the BEM model analyses, the 10.5 m prototype blade is divided into 30 sections each with a width of 0.35 m. The number of section and the section width are determined based on the results of a sensitivity study, which achieves computational efficiencies while ensures the accuracy.

The performance of the developed BEM model in power output prediction is firstly validated by utilizing the monitored power production data from the 100 kW utility-scale wind turbine described in Section 3. The data is collected at 10 min time interval between September 2014 and August 2015, which includes the air density, the rotational speed of the blade, the wind speed at hub height, and the power output. Measured wind speed, blade rotational speed, and air density are used to calculate the power output using the described BEM model. The monitored power outputs at different wind speeds are compared with those predicted by the BEM model in Figure 5. Also shown in this figure are the curve fitting of the measured or BEM model predicted power output. In general, the predicted power output performance matches well with the measured data. The monitored total energy output during the one-year period is 388.87 MWh, while the energy production predicted by the BEM model is 397.52 MWh. For wind speeds under 6 m/s, the curve fitted turbine power curve from the BEM model prediction is slightly beneath that from the monitoring power production data; the trend reverses for wind speeds larger than 12 m/s. One of the causes is the limited amount of data available at high wind speed range. Overall, the maximum error between the BEM model's predicted output and monitored data is within 2.2%. The comparison with the monitored wind turbine power output data validated that the BEM model is accurate in predicting the wind energy output. Subsequently, the validated BEM model was utilized to analyze the performance of the proposed extensible blade in the subsequent section.

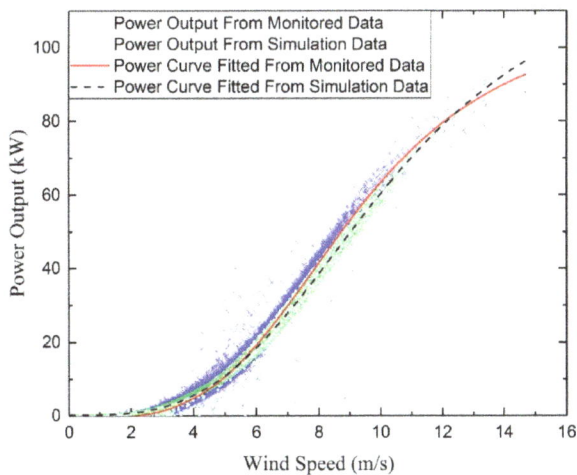

Figure 5. Comparison of the blade element momentum (BEM) model's predicted power output and the monitoring power output of the 100 kW wind turbine.

5. Analyses of Extensible Blade Performance with the BEM Model

5.1. Wind Characteristics at Studied Sites

A few utility scale wind turbines have been erected as part of the efforts of the State of Ohio in promoting renewable wind energy both onshore and offshore [6,40–42]. These wind turbines serve as the case studies in this research. According to the National Oceanic and Atmospheric Administration (NOAA), the monthly average wind speed in Cleveland is 4.69 m/s, or Class 4 according to TC88-MC 2005. Winter is the windiest season in Cleveland with average wind speeds reaching 5.36 m/s, or on average 28% higher than wind speeds in the other seasons. On the other hand, summer features the lowest average wind speed of 3.93 m/s. This pattern of seasonal wind speed is consistent with the seasonality pressure gradients across Cleveland and the Great Lakes region [43]. Three instrumented locations with different typical wind resources are selected to evaluate the potential performance of the extensible blade, including two locations onshore and one location offshore (Figure 6).

Figure 6. Location of selected sites with wind condition measurement.

For all these sites, the data of wind conditions have been monitored over the years (2006–2015). The data set include wind data at 10 min time interval at the three locations as shown in Figure 6, i.e., site A (on the campus of Case Western Reserve Univeristy (CWRU)); site B (along an interstate highway and adjacent to manufacturing facility); and site C (offshore Lake Erie). The data for location A is provided by the data acquisition system (DAQ) installed in the 100 kW wind turbine on CWRU campus. The data for location B is from Lidar measurement of the wind speed. Data from location C is from the met mast, which is installed on a water intake crib 5 miles offshore of Lake Erie.

5.2. Weibull Distribution

Weibull distribution is the most widely used probability distribution to present wind data. The general form of the Weibull density function is a two parameter function, which is given as [44]:

$$f(U_{tot}) = \frac{k}{c}\left(\frac{U_{tot}}{c}\right)^{k-1} e^{-(U_{tot}/c)k} \tag{2}$$

where $f(U_{tot})$ is the probability density function, also referred to as PDF; U_{tot} is the wind speed (m/s); c is the scale factor (m/s), and k is the shape factor. The maximum likelihood method (MLM) is used in this research to calculate the Weibull scale and shape factors according to our 10 min time intervals of data availability [45]. The shape factor and scale factor could be calculated as follows [46,47]:

$$k = \left(\frac{\sum_{i=1}^{N} U_i^k \ln(U_i)}{\sum_{i=1}^{N} U_i^k} - \frac{\sum_{i=1}^{N} \ln(U_i)}{N}\right)^{-1} \tag{3}$$

$$c = \left(\frac{1}{N}\sum_{i=1}^{N} U_i^k\right)^{1/k} \tag{4}$$

where U_i is the average wind speed in time step i and N is the total number of nonzero wind speed data points.

5.3. Adjustment of Wind Speeds with Elevation

The wind speed data collected by Lidar system at site B and met mast at site C are both measured at the height of 30 m above the ground, and the prototype wind turbine has a hub height of 37.5 m. Therefore, the wind speed is adjusted to the hub height of the prototype wind turbine. The most common method to adjust the wind speed over different height is the power law model, where the wind speed at any height above the ground can be determined using the following expression [48]:

$$U_z(z) = U_{ref} \left(\frac{z}{z_{ref}} \right)^{\alpha_0} \tag{5}$$

where z is the target height, z_{ref} is the reference height above the ground [49], and the U_{ref} is the reference wind velocity measured at reference height. The exponent, α_0, will change with the terrain roughness and the surrounding building height range, which also refers as the wind shear coefficient (WSC). The WSC in the above equation depends on the terrain type from very flat terrain to dense urban, and its value at different terrain types can be referred to previous studies [50]. The WSC value for site B is chosen as 1/4, which is the suggested value for the rural area; the WSC for site C is chosen as 1/9, which is the suggested value for the water surface. The measured wind speed data at site B and site C are adjusted from 30 m to 37.5 m from Equation (5). The wind speed distribution at 37.5 m height in a typical year is shown in Figure 7 for each of the three sites. It can be concluded from Figure 7 that the wind speed distribution at the same location has slightly bias in different years, but the overall trend is similar. The wind speed at site A is more concentrated with the highest frequency at 4 m/s, and the wind speed at site C is more distributed, varying from 0 m/s to 25 m/s. The statistical characteristics of the wind speed data for these three locations are summarized in Table 3. Overall, site A has the lowest mean wind speed since the site is surrounded by a few buildings with heights of up to 20 m. The wind speed of site B is slightly higher because the site is located in a rural area that most of the surrounding buildings are under 10 m in height. The offshore site C has the strongest wind speed since the terrain is flatter at the offshore location. Both Sites A and B are classified as Class 4 while Site C barely qualifies as Class 3 wind site according to IEC standard [3].

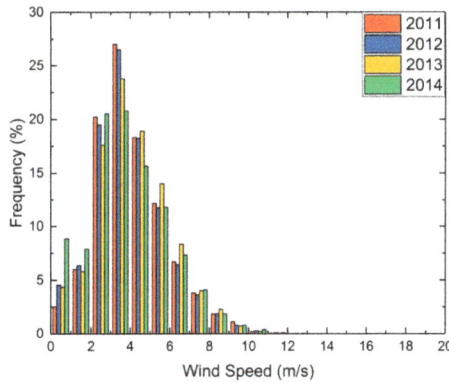

(a)

Figure 7. *Cont.*

(b)

(c)

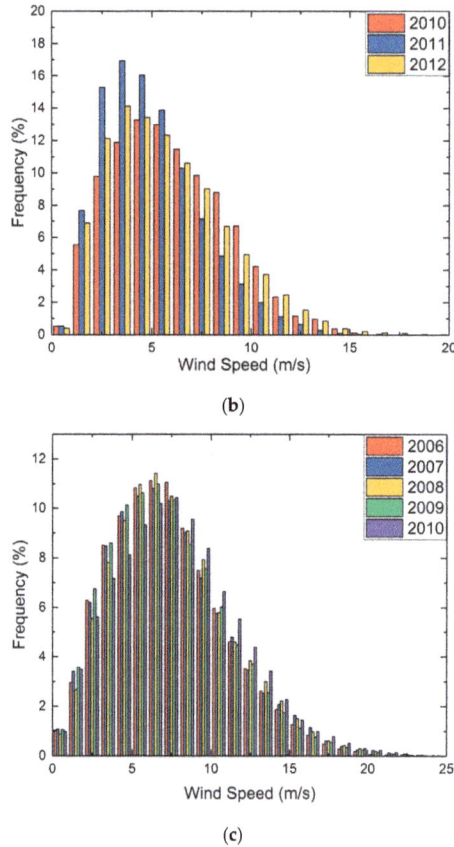

Figure 7. Weibull distribution of wind speed data at 10 min intervals at (**a**) Site A; (**b**) Site B; (**c**) Site C.

Table 3. Locations and mean wind speed characteristics.

Location	Year	Mean Wind Speed (m/s)	S.D. (m/s)	Weibull Shape Factor k	Weibull Scale Factor c (m/s)	Note
Site A	2011	4.01	SD = 1.797	2.26	4.53	NA
41°30′08.6″ N	2012	3.96	SD = 1.05	2.00	4.37	NA
81°36′19.9″ W	2013	3.95	SD = 0.69	2.30	4.65	NA
(IEC Class 4)	2014	3.80	SD = 1.52	1.48	4.07	NA
Site B	2010	5.99	SD = 2.84	2.25	6.78	October to December
41°36′07.8″ N						
81°29′48.7″ W	2011	4.98	SD = 2.49	2.13	5.64	April to December
(IEC Class 4)	2012	5.71	SD = 2.96	2.05	6.46	January to April
Site C	2006	7.35	SD = 3.64	2.13	8.31	NA
41°32′53.7″ N	2007	7.46	SD = 3.85	2.04	8.43	NA
81°44′58.7″ W	2008	7.62	SD = 3.78	2.12	8.61	NA
(IEC Class 3)	2009	7.29	SD = 3.73	2.06	8.24	NA
	2010	7.82	SD = 3.87	2.13	8.83	NA

5.4. Analyze the Energy Production Performance of the Extensive Blade

5.4.1. Determination of the Working Range of Wind Speed for Extensible Blades

The extended blade is subjected to a higher wind load. Therefore, determining the range of working wind speed is firstly conducted to ensure the safety of the blade. Since the focus of this study is to assess the feasibility of the extensible blade for improving energy production, simplified mechanical analyses are conducted rather than sophisticated evaluations. The maximum allowed working wind speed is determined based on the corresponding bending moment on the extended blade, whose value should not exceed the bending moment of the original length blade at the cut-out wind speed [51].

With these criteria, the ranges of working wind speeds for two types of extensible blades are analyzed; i.e., (1) different extent of extension at blade tip; and (2) the different extent of extension in the middle of the blade. By using the BEM model, maximal in-plane and out-of-plane bending moments in the original blade and extended blades at different wind speeds are shown in Figure 8, and the intersection points are limits that determine the range of operational wind speed for the extensible blade. Ranges of safe working wind speeds corresponding to the different extension of the blade are determined, which are summarized in Table 4. As a note, from an operation perspective, the scheme of extension is designed to be simple (i.e., extension at steps of 25%, 20%, 0%) so that blade extension is not too frequent to conserve the energy needed for blade actuation. More sophisticated extension schema can be designed based on further analyses of wind characteristics.

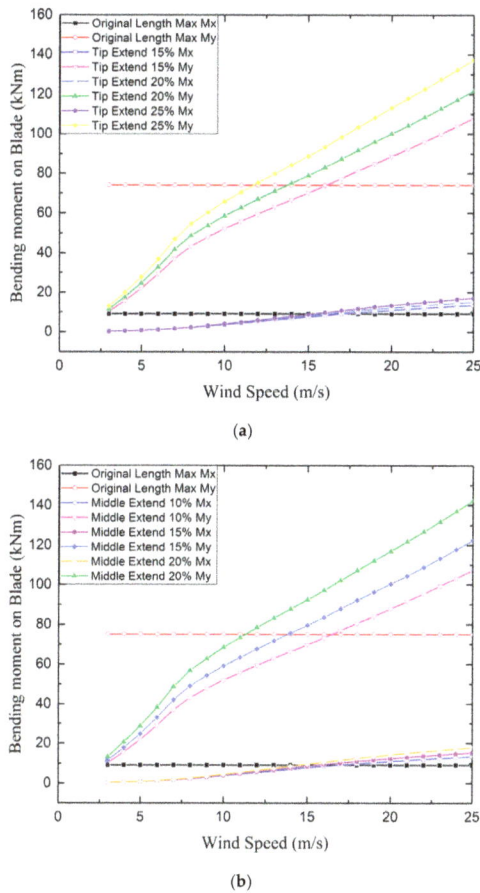

(a)

(b)

Figure 8. (**a**) Determination of the wind speed range for (**a**) blade extension at tip and (**b**) blade extension in the middle.

Table 4. Blade extends type in the research.

Extension Method	Wind Speed Range (m/s)	Extension (%)
Extensible blade with tip extension	3–10	25
	10–14	20
	14+	0
Extensible blade with middle extension	3–10	20
	10–14	10
	14+	0

5.4.2. Modal Analysis

A modal analysis determines the vibration characteristics (natural frequencies and mode shapes) of a structure. The natural frequencies and mode shapes are important parameters affecting the response and design of a structure for dynamic loading conditions. A good design for reducing vibration is to separate the natural frequencies of the structure from the harmonics of rotor speed [52]. The modal analysis of the extensible blade helps us understand how the natural frequencies change, thus avoid resonance when the large amplitudes of vibration could damage the wind turbine.

The FEM software COMSOL® (COMSOL 5.0, COMSOL, Inc., Burlington, MA, USA) is used to calculate the un-damped modal characteristics of the turbine. The wind turbine blade is considered as a cantilever beam with blade root fixed. The program solves the following eigenvalue problem [53] utilizing the model's stiffness and mass matrices.

$$[K - \omega^2 M]\{\phi\} = \{0\} \tag{6}$$

Equation (6) is a typical real eigenvalue problem; therefore, ϕ has a non-zero solution if the value of its determinant coefficient is zero.

Typically, only the first few natural modes are of interest for structural engineering design as they typically contain most of the modal mass and have natural frequencies close to the excitation frequency of the wind. In this research, only the first four natural frequencies are considered; as the finite element model considered here is a simplification of the structure intended to capture global structural dynamic demands, the higher mode results will likely be less accurate. For comparison purposes, both the tip-extend and middle-extend strategy are extensions of 20% of its length.

Table 5 presents the results of modal analysis with the first four modes. Overall, increasing the length of the blades reduces its natural frequencies. From the results of the modal analysis, the dominant vibration mode for the horizontal across wind direction has a natural frequency of 1.356 Hz for original length blade, 0.8982 Hz for middle extend 20% blade and 1.104 Hz for tip extend 20%. The natural vibration mode shapes are shown in Figure 9.

Table 5. Modal Frequency.

Model Shape	Original Length (Hz)	Middle Extend 20% (Hz)	Tip Extend 20% (Hz)
1	1.3562	0.8982	1.104
2	5.2671	4.0438	2.6484
3	6.4427	4.2213	5.3358
4	12.147	8.8918	6.0251

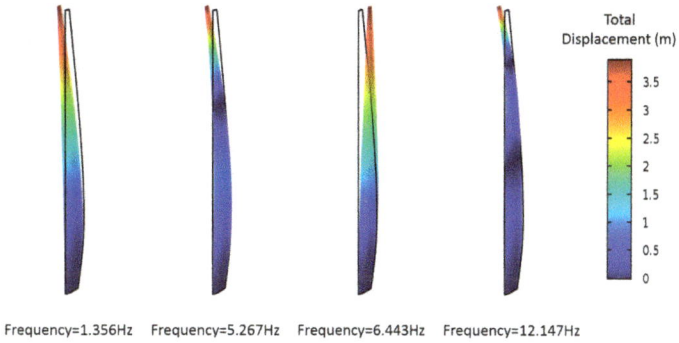

Frequency=1.356Hz   Frequency=5.267Hz   Frequency=6.443Hz   Frequency=12.147Hz

(**a**)

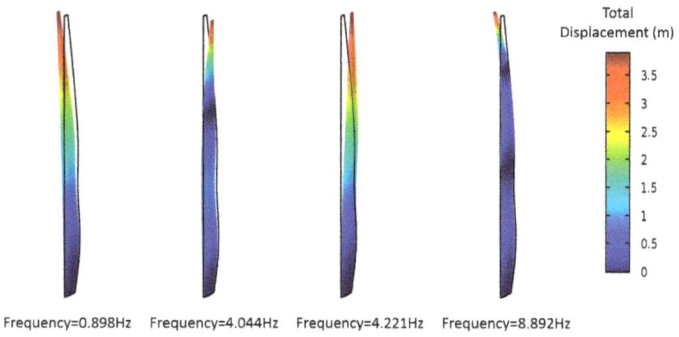

Frequency=0.898Hz   Frequency=4.044Hz   Frequency=4.221Hz   Frequency=8.892Hz

(**b**)

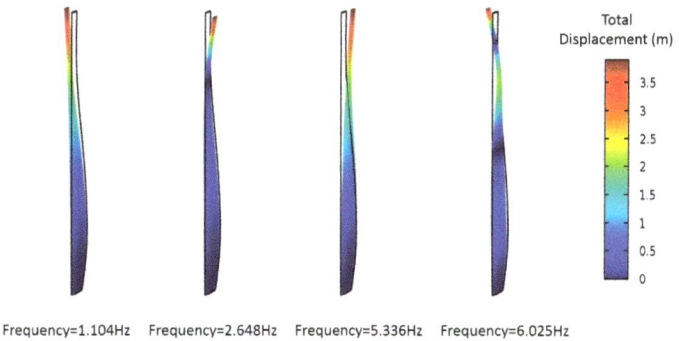

Frequency=1.104Hz   Frequency=2.648Hz   Frequency=5.336Hz   Frequency=6.025Hz

(**c**)

Figure 9. The shapes of first four modes for (**a**) original length blade; (**b**) blade-extension of 20% at the tip; (**c**) blade extension of 20% in the middle.

5.4.3. Performance of Extensible Blade in Wind Energy Production

With the extension strategy defined by base structural safety considerations, which are summarized in Table 3, Figure 10 compares the corresponding power output curves of the extensible blades with that of the original blade. Both the manufacturer's power curve and the power curve from a curve fitting of the monitoring data are plotted for comparison purposes. It is noted that the power curve from the monitoring data does not cover a high wind speed range. The predicted power curves of the extensible blade by the BEM model with two different extension strategies are also plotted. The comparison clearly shows that the extended blade has a much higher power output at wind speeds lower than the blade's rated wind speed of 14 m/s. There are different extents of shift in the power production curves at wind speed of 10 m/s is due to the proposed blade extension strategy that changes the extent of blade extension at 10 m/s (Table 4). It is assumed that the maximum output is limited to 100 kW to match the capacity of the generator (modified power curve shown in Figure 10). It can be seen from the figure that the power curve from the BEM model and the power curve from the monitored data are closer to each other but different from the manufacturer's power curve. This is because the manufacturer's power curve is measured under certain meteorology conditions which are different to the real conditions [54]. The blade production curves of extensible blades are similar to the original blade as they are completely retracted to the original length. Since the wind speed at the three selected sites was under 14 m/s for most of the time (Figure 7), it is expected that a turbine with extensible blades will consistently produce more energy than a regular turbine for the majority of the year. The annual energy output of the extensible and original length blade can be calculated using the corresponding power curves and the wind speed data at the three test sites.

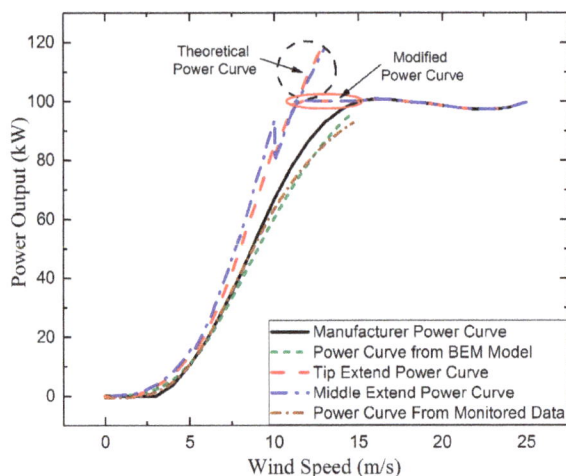

Figure 10. Comparison of the power curves for the original length blade, tip-extended blade and middle-extended blade.

The monitored yearly wind speed data at the three locations with the low-class wind (A and B: Class 4 and C: Class 3) are utilized to estimate the total wind energy outputs, following the validated procedures described in the earlier context. Figure 11 shows the histogram of the predicted average power output in 10 min intervals for each site for different years with the baseline blade and extensible blades. Overall, the comparison shows that the original baseline blade has a higher occurrence of low-energy output periods than the extended blades. In another word, the extensible blades shift the wind energy production to higher energy output than an original blade for these sites with a low class of wind resource.

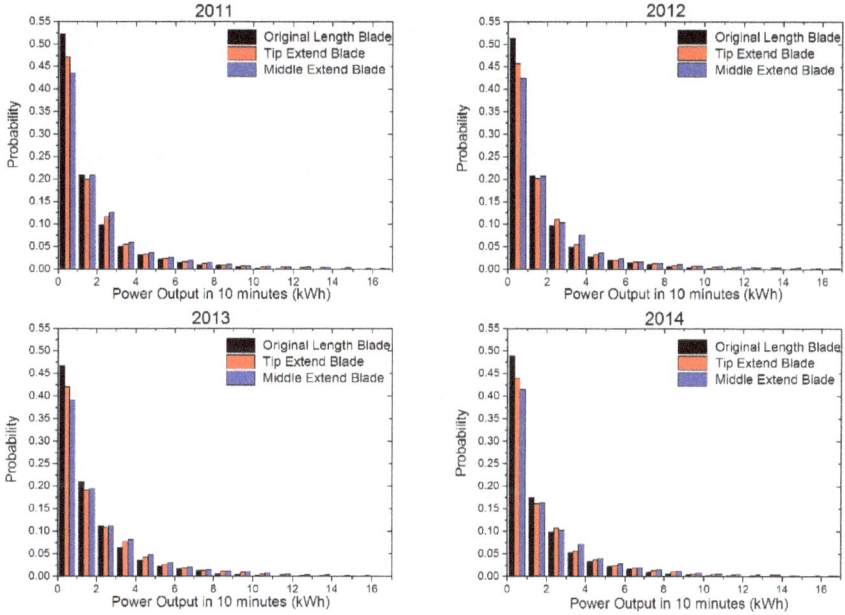

(**a**)

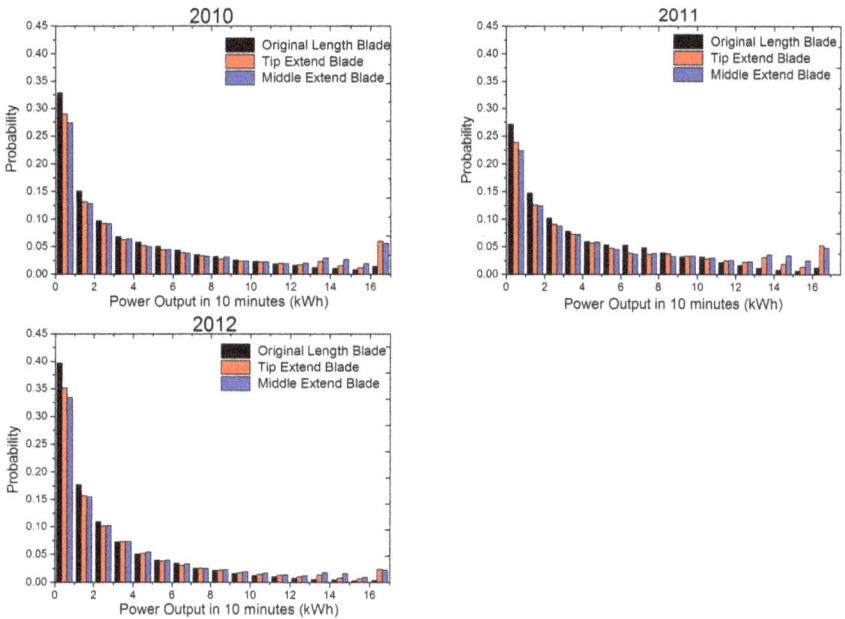

(**b**)

Figure 11. *Cont.*

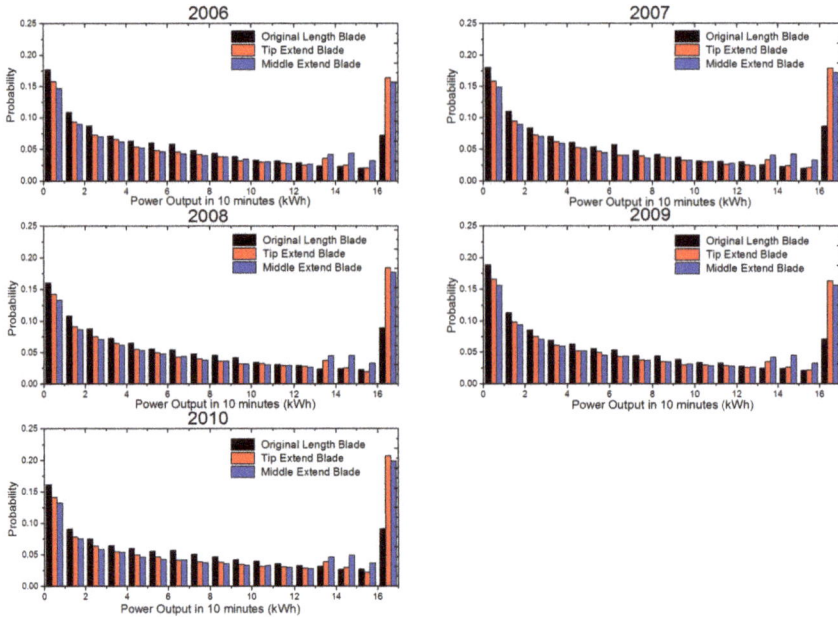

Figure 11. Statistical distribution of 10 min of energy output for the original blade and extensible blades at (**a**) Site A; (**b**) Site B; (**c**) Site C.

Table 6 summarizes the predicted total annual energy production by different types of blades (original versus extensible blades). The results show that the innovative extensible wind turbine blades will potentially increase the total annual wind energy production for all sites with a low class of wind. For Site A, the extensible blade that extends at the tip will increase the power output by around 19%; the extensible blade that extends in the middle will increase the power output by 32%. For Site B, the corresponding increases in total energy production are 22% and 31% by the two types of extensible blades. For Site C, the amount of increase in the annual energy production for tip extension and middle extension blades are around 19% and 25% respectively. The extensible blade that extends in the middle provides a larger increase in the energy output than that extends at the tip due to the larger wind carry areas; besides this, the percentage increase in energy production is more significant at site with a low class of wind (i.e., Sites A and B) than site with high class of the wind (i.e., Site C). These are a clear demonstration of the benefits of the extensible blade to boost energy production for a site with low classes of wind. In the meantime, the extension scheme is designed so that the extensible blade is protected with a similar structural safety to a regular blade.

Table 6. Comparison of total annual energy production by the original blade versus extensible blades at different sites.

Year	Energy Output by Original Blade (kWh)	Energy Output by Tip Extended Blade (kWh)	Increase Percentage (%)	Energy Output by Middle Extended Blade (kWh)	Increase Percentage (%)
		Site A (Class 4, Onshore)			
2011	80,316.25	95,692.79	19.14	106,387.10	32.46
2012	73,393.83	87,562.00	19.30	97,316.17	32.59
2013	78,277.17	92,783.17	18.53	103,071.30	31.67
2014	75,296.00	89,820.17	19.29	99,600.17	32.28
		Site B (Class 4, Onshore)			
2010	44,819.83	55,296.00	23.37	59,269.00	32.24
2011	70,515.67	85,525.50	21.29	92,656.17	31.40
2012	93,464.50	114,019.3	21.99	121,786.50	30.30
		Site C (Class 3, Offshore)			
2006	308,698.30	368,768.30	19.46	387,289.30	25.46
2007	314,257.70	372,854.80	18.65	390,657.80	24.31
2008	322,774.70	382,870.50	18.62	401,353.30	24.34
2009	305,493.20	364,615.70	19.35	382,645.30	25.25
2010	239,212.20	284,880.20	19.09	297,576.30	24.40

6. Conclusions

Wind farms are ideally located at locations with high-class wind. However, there are a large number of distributed wind turbines constructed at sites close to communities, with non-ideal wind conditions. This paper describes the analyses of an innovative, extensible blade technology that aims to utilize wind energy in areas with low-class wind resources. The extensible blade functions by adjusting its length depending on the wind conditions (i.e., it will extend at low wind speed and retract at high wind speed). Based on the principle that the larger the sweep area, the higher the turbine energy output, dynamically adjusting the blade length helps to increase the energy output under low wind speed while mitigating safety risks under high wind speed. The computational model is developed based on the blade element momentum (BEM) theory, which determines the aerodynamic load and power output of the blade at different wind conditions. The model is firstly validated with monitored energy output data of in-service wind turbine. The validated model is subsequently used to estimate the annual energy production by the extensible blades and regular blade at three locations inland and offshore of the Lake Erie area, where yearly wind data are continuously monitored. Two types of extensible blade scheme are analyzed; i.e., extension in the middle of the blade versus extension at the tip of the blade. The extension and contraction scheme of these extensible blades are determined based on a limiting of the maximum bending moment acting on the blade, which helps ensure their structural safety. The influence of blade extension on the dynamic characteristics of blade structure is analyzed. The results show that the extensible blade will potentially increase annual energy output up to 20% to 30% for the sites analyzed. Besides this, the lower the wind speed, the more effective the extensible blade in increasing energy production. Overall, the results of this paper point to the promise of this innovative, extensible blade in improving the wind energy production.

Acknowledgments: This research is partially supported by the US National Science Foundation CMMI-1300149.

Author Contributions: Xiong (Bill) Yu conceived the concept of extensible blade and control criteria for blade extension. Jiale Li conducted the detailed modeling and analyses under the guidance to quantify the performance of extensible blade in improving energy production. Both authors contributed to the write up and refinement of this paper.

Conflicts of Interest: The authors declare no conflict of interest. The option raised by this paper does not represent the opinion of the sponsor.

Appendix A. Implementation Procedures of Beam Element Momentum (BEM) Theory for Turbine Energy Output Prediction

BEM is composed of two different theories, i.e., blade element theory and momentum theory [37]. Blade element theory assumes that blades can be divided into small elements that act independently of the surrounding elements and operate aerodynamically as two-dimensional airfoils as shown in Figure A1, in which α is the attack angle. The characteristics of blade responses (drag and life on each element) are determined by the angle of attack of incoming wind, which is the angle between the center reference line of the geometry and the relative incoming flow W (Figure A1). The momentum theory assumes that the loss of air pressure or generation of turning momentum in the airfoil blade element is caused by the work done by the incoming airflow [38]. The BEM theory couples these two theories together and calculates the total lift and momentum via an iterative process [39].

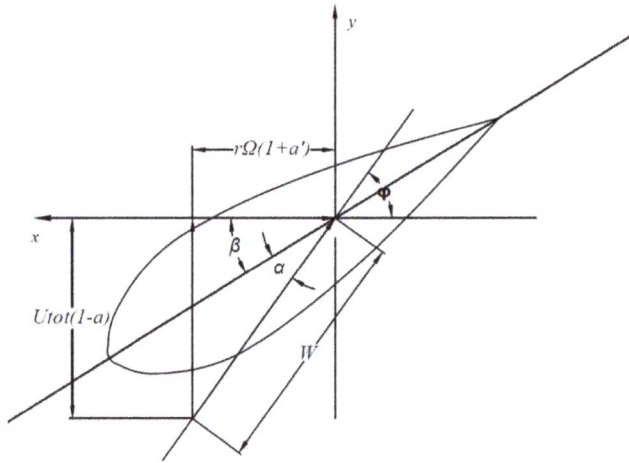

Figure A1. Blade Element velocity components.

Appendix A.1. Blade Element Momentum (BEM) Theory

The actual wind flow acting on the turbine rotor is rather complex and can be simplified by the use of the blade element theory. The velocity components in the radial positions of the blade can be expressed regarding the wind speed, the axial induction factor (a), tangential flow induction factors (a') and the rotational speed of the rotor (Ω). The axial flow induction factor (a) and the tangential flow induction factor (a') are critical parameters in the BEM theory. Figure A1 illustrates the conceptual model to calculate the lift and draft forces on each airfoil blade element. The airfoil is assumed to have a blade pitch angle of β, and the wind acts on the airfoil with an attack angle α. The pitch angle is the angle between the blade chord and normal direction of the rotor plane, which is an important parameter for maximizing blade lift and determines the load acting on the blade. The component of wind velocity in the direction of the blade is ignored as it does not contribute to the torque on the blade rotation. Therefore, the inflow angle ϕ which is the intersection angle between the inflow wind velocity and the rotation plane of the blades, satisfying the following relationship:

$$\varphi = \alpha + \beta \tag{A1}$$

BEM theory does not include the effects of tip losses, hub losses, skewed values, dynamic stall, and tower shadow. The lift and drag forces generated by the airfoil along the blade and the momentum equations are used to produce the induction factors. The calculation step was then organized into a series of equations that can be solved iteratively, which is further elaborated in the following sections.

Appendix A.2. The Calculation of Relative Inflow Wind Velocity

An important assumption of the blade element momentum (BEM) theory is that the lift and drag forces acting on a blade element are solely responsible for the momentum which caused by the air passing through the blade swept annulus [55]. Lift and draft forces are determined by the relative wind velocity act on the airfoil. The wind velocity perpendicular to the rotor plane is the inflow wind velocity $U_{tot}(t)$ reduced by the amount of $a \times U_{tot}(t)$ due to axial interference (i.e., $(1-a) \times U_{tot}(t)$). Assuming the rotor rotates with angular speed Ω, the blade element at a distance r from the rotor axis will be moving with a tangential speed Ωr [56]. When the wind passes through the rotor plane and interacts with the moving rotor, a tangential slipstream (or wake rotation) of wind velocity $a'\Omega r$ is introduced. The resultant inflow wind velocity about the rotor blade W is shown in Figure A1 and can be calculated via the procedures are shown in the following:

$$W = \sqrt{U_{tot}^2(1-a)^2 + [\Omega r(1+a\prime)]^2} \tag{A2}$$

And the inflow angle φ could also express using the velocity:

$$\varphi = \arctan[\frac{U_{tot}(1-a)}{\Omega r(1+a\prime)}] \tag{A3}$$

To calculate the relative incoming wind speed, W at each position r along the length of the blade and for each total wind speed U_{tot}, the axial flow induction factor a and tangential flow induction factor a' need to be calculated first. Typically, this is done via an iterative numerical procedure, with the basic steps as follows [2,57,58]:

a. Assume an initial choice of a and a'. (for example $a = a' = 0$ as an initial guess). Calculate the inflow angle via $\varphi = \arctan[\frac{U_{tot}(1-a)}{\Omega r(1+a\prime)}]$, where Ω is the rotor angular speed.

b. Calculate $\alpha = \varphi - \beta$;

c. Read C_l and C_d from the lift and drag coefficient curves shown in Figure 3 with the result of α from step b. Calculate the coefficient of sectional blade element force normal to the rotor plane C_x and coefficient of sectional blade element force parallel to the rotor plane C_y:

$$C_x = C_l \times \cos\varphi + C_d \times \sin\varphi$$

$$C_y = C_l \times \sin\varphi + C_d \times \cos\varphi$$

d. Substitute C_x *and* C_y into the following expressions to calculate new values for a and a'

$$\frac{a}{1-a} = \frac{\sigma_r}{4 \times \sin^2\varphi}(C_x - \frac{\sigma_r}{4\sin^2\varphi}C_y{}^2)$$

$$\frac{a\prime}{1+a\prime} = \frac{\sigma_r C_y}{4 \times \sin\varphi\cos\varphi}$$

$$\sigma_r = 3 \times \frac{C(r)}{2\pi r}$$

e. Evaluate convergence of the solution by comparing the calculated a and a' from step e with the assumed a and a' from step a.

f.   If the differences between values are smaller than designated threshold, the process stops. Otherwise, update a and a' values and continue the iteration between (b) and (e) until the results converge.

g.   Take the result of a and a' into Equation (A2) to calculate the relative wind speed W.

The procedure shown above applies to different types of turbine blades. It also needs to note that as the lift and drag coefficients vary with attack angle, variable pitch wind turbine modulates the wind attack angle by dynamically adjust the pitch angle of blades.

Appendix A.3. Blade Lift and Drag Force Calculation

The relative wind velocity gives rise to aerodynamic lift and drag forces acting on each segment of the blade, which can be calculated as follows:

$$F_L(r) = \frac{1}{2}C_l\rho C(r)W^2 r \tag{A4}$$

$$F_D(r) = \frac{1}{2}C_d\rho C(r)W^2 r \tag{A5}$$

where $C(r)$ is blade chord length; r stands for the distance from the hub of a section of the blade; C_l is the lift coefficient, C_d is the drag coefficient.

The differential torque act on a blade section is

$$dT = rdF_T = F_L \sin \varphi - F_D \cos \varphi = \frac{1}{2}\rho W^2 C(r) C_y \tag{A6}$$

The shaft power is calculated via total torque and rotor angular speed

$$P_m = T\Omega = \int_0^r dT\Omega \tag{A7}$$

$$P_w = C_p P_m \tag{A8}$$

where P_w is wind turbine production power, Ω is rotor speed; P_m is shaft power; C_p is the power coefficient.

In summary, the driving force on a wind turbine is generated by lift force when the wind flows past the airfoils. The lift force increases with attack angle, which is also accompanied by increases in undesirable drag force. While the tangential component of lift force supports blade rotation, drag force opposes it at the same time. Therefore, a wind turbine will achieve the best performance when the ratio of lift force to drag force is maximum, or at its optimum attack angle. Airfoil cross sections are aligned in a way to operate at close to optimum attack angle. The torque is dependent on the blade section chord length (C), and the relative inflow wind velocity W, which varies along the blade length. They are also dependent on the air density. The power output can be calculated by multiplying the rotational speed and the torque acting on blades. The procedure is also illustrated in Figure A2.

Start

Input: Ω, U_{top}, ρ, R, r, $C(r)$, $\beta(r)$, a, a_1, a'_1

If $a-a1 \geq 0.01$, $a'-a \geq 0.01$

N

Y

$a_1 = a$, $a'_1 = a'$

Calculate angle of relative wind from rotor plane

$$\varphi = \arctan\left[\frac{U_{tot}(1-a)}{\Omega r(1+a')}\right]$$

Calculate attack angle

$$\alpha = \varphi - \beta$$

Read C_l and C_d from the lift and drag curves shown in the figure with the calculated attack angle

Calculate coefficient of sectional blade element force normal and parallel to the rotor plane

$$C_x = C_l \times \cos\varphi + C_d \times \sin\varphi$$
$$C_y = C_l \times \sin\varphi + C_d \times \cos\varphi$$

Substitute C_x and C_y into the following equations to calculate new values of a and a'

$$\frac{a}{1-a} = \frac{\sigma_r}{4 \times \sin^2\varphi}\left(C_x - \frac{\sigma_r}{4\sin^2\varphi}C_y^2\right)$$

$$\frac{a'}{1+a'} = \frac{\sigma_r C_y}{4 \times \sin\varphi\cos\varphi}$$

$$\sigma_r = 3 \times \frac{C(r)}{2\pi r}$$

If $R \geq r$

N

End

Y

Calculate relative wind velocity

$$W = \sqrt{U_{tot}^2(1-a)^2 + [\Omega r(1+a')]^2}$$

Calculate blade Torque

$$dT = rdF_T = F_L \sin\varphi - F_D \cos\varphi = \frac{1}{2}\rho W^2 C(r)C_y$$

Calculate the Power

$$P_m = T\Omega$$

r=r+0.35

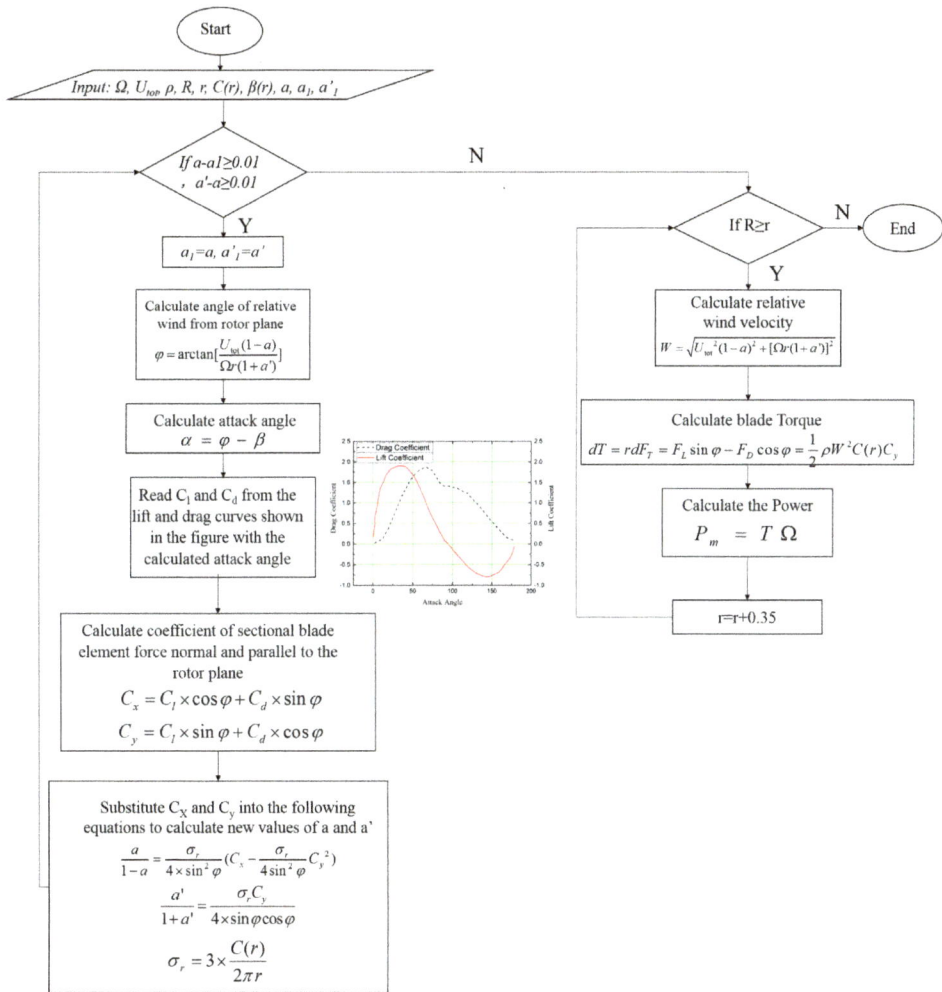

Figure A2. Flowchart for calculating blade production power using BEM theory.

References

1. Carlin, P.W.; Laxson, A.S.; Muljadi, E. The history and state of the art of variable-speed wind turbine technology. *Wind Energy* **2003**, *6*, 129–159. [CrossRef]
2. Burton, T.; Sharpe, D.; Jenkins, N.; Bossanyi, E. *Wind Energy Handbook*; John Wiley & Sons: Hoboken, NJ, USA, 2001.
3. *IEC 61400-3: Wind Turbines—Part 1: Design Requirements*; TC88-MT; International Electrotechnical Commission: Geneva, Switzerland, 2005.
4. Chen, K.; Song, M.; Zhang, X. The iteration method for tower height matching in wind farm design. *J. Wind Eng. Ind. Aerodyn.* **2014**, *132*, 37–48. [CrossRef]
5. Li, J.; Yu, X.B. LiDAR technology for wind energy potential assessment: Demonstration and validation at a site around Lake Erie. *Energy Convers. Manag.* **2017**, *144*, 252–261. [CrossRef]
6. Wang, X.; Yang, X.; Zeng, X. Centrifuge modeling of lateral bearing behavior of offshore wind turbine with suction bucket foundation in sand. *Ocean Eng.* **2017**, *139*, 140–151. [CrossRef]

7. Wang, X.; Yang, X.; Zeng, X. Seismic Centrifuge Modelling of Suction Bucket Foundation for Offshore Wind Turbine. *Renew. Energy* **2017**, in press. [CrossRef]

8. Calvert, S.; Thresher, R.; Hock, S.; Laxson, A.; Smith, B. US department of energy wind energy research program for low wind speed technology of the future. *J. Sol. Energy Eng.* **2002**, *124*, 455–458. [CrossRef]

9. Robinson, M.; Veers, P. US national laboratory research supporting low wind speed technology. *Trans. Am. Soc. Mech. Eng. J. Sol. Energy Eng.* **2002**, *124*, 458–459. [CrossRef]

10. Zhang, J.; Zhou, Z.; Lei, Y. Design and Research of High-Performance Low-Speed Wind Turbine Blades. In Proceedings of the World Non-Grid-Connected Wind Power and Energy Conference (WNWEC 2009), Nanjing, China, 24–26 September 2009; pp. 1–5.

11. Barnes, R.; Morozov, E.; Shankar, K. Improved methodology for design of low wind speed specific wind turbine blades. *Compos. Struct.* **2015**, *119*, 677–684. [CrossRef]

12. Wichser, C.; Klink, K. Low wind speed turbines and wind power potential in Minnesota, USA. *Renew. Energy* **2008**, *33*, 1749–1758. [CrossRef]

13. Alam, M.M.; Rehman, S.; Meyer, J.P.; Al-Hadhrami, L.M. Review of 600–2500 kW sized wind turbines and optimization of hub height for maximum wind energy yield realization. *Renew. Sustain. Energy Rev.* **2011**, *15*, 3839–3849. [CrossRef]

14. Lee, J.; Kim, D.R.; Lee, K.-S. Optimum hub height of a wind turbine for maximizing annual net profit. *Energy Convers. Manag.* **2015**, *100*, 90–96. [CrossRef]

15. Rehman, S.; Al-Hadhrami, L.M.; Alam, M.M.; Meyer, J.P. Empirical correlation between hub height and local wind shear exponent for different sizes of wind turbines. *Sustain. Energy Technol. Assess.* **2013**, *4*, 45–51. [CrossRef]

16. Pellegrino, A.; Meskell, C. Vortex shedding from a wind turbine blade section at high angles of attack. *J. Wind Eng. Ind. Aerodyn.* **2013**, *121*, 131–137. [CrossRef]

17. No, T.; Kim, J.-E.; Moon, J.; Kim, S. Modeling, control, and simulation of dual rotor wind turbine generator system. *Renew. Energy* **2009**, *34*, 2124–2132. [CrossRef]

18. Huang, G.-Y.; Shiah, Y.; Bai, C.-J.; Chong, W. Experimental study of the protuberance effect on the blade performance of a small horizontal axis wind turbine. *J. Wind Eng. Ind. Aerodyn.* **2015**, *147*, 202–211. [CrossRef]

19. Miklosovic, D.; Murray, M.; Howle, L.; Fish, F. Leading-edge tubercles delay stall on humpback whale (Megaptera novaeangliae) flippers. *Phys. Fluids* **2004**, *16*, L39–L42. [CrossRef]

20. Huang, D.; Wu, G. Preliminary study on the aerodynamic characteristics of an adaptive reconfigurable airfoil. *Aerosp. Sci. Technol.* **2013**, *27*, 44–48. [CrossRef]

21. Bhuyan, S.; Biswas, A. Investigations on self-starting and performance characteristics of simple H and hybrid H-Savonius vertical axis wind rotors. *Energy Convers. Manag.* **2014**, *87*, 859–867. [CrossRef]

22. Bottasso, C.L.; Croce, A.; Gualdoni, F.; Montinari, P. Load mitigation for wind turbines by a passive aeroelastic device. *J. Wind Eng. Ind. Aerodyn.* **2016**, *148*, 57–69. [CrossRef]

23. Veers, P.S.; Ashwill, T.D.; Sutherland, H.J.; Laird, D.L.; Lobitz, D.W.; Griffin, D.A.; Mandell, J.F.; Musial, W.D.; Jackson, K.; Zuteck, M. Trends in the design, manufacture and evaluation of wind turbine blades. *Wind Energy* **2003**, *6*, 245–259. [CrossRef]

24. Bir, G.S. Computerized method for preliminary structural design of composite wind turbine blades. *J. Sol. Energy Eng.* **2001**, *123*, 372–381. [CrossRef]

25. Jackson, K.; Zuteck, M.; Van Dam, C.; Standish, K.; Berry, D. Innovative design approaches for large wind turbine blades. *Wind Energy* **2005**, *8*, 141–171. [CrossRef]

26. Griffin, D.A. *Blade System Design Studies Volume I: Composite Technologies for Large Wind Turbine Blades*; Paper No. SAND-1879 2002; Sandia National Laboratories: Livermore, CA, USA, 2002.

27. Griffin, D.A. *WindPACT Turbine Design Scaling Studies Technical Area 1—Composite Blades for 80- to 120-m Rotor*; Technical Report; National Renewable Energy Laboratory: Livermore, CA, USA, 2001.

28. Jureczko, M.; Pawlak, M.; Mężyk, A. Optimisation of wind turbine blades. *J. Mater. Process. Technol.* **2005**, *167*, 463–471. [CrossRef]

29. Sharma, R.; Madawala, U. The concept of a smart wind turbine system. *Renew. Energy* **2012**, *39*, 403–410. [CrossRef]

30. Imraan, M.; Sharma, R.N.; Flay, R.G. Wind tunnel testing of a wind turbine with telescopic blades: The influence of blade extension. *Energy* **2013**, *53*, 22–32. [CrossRef]
31. McCoy, T.J.; Griffin, D.A. Control of rotor geometry and aerodynamics: Retractable blades and advanced concepts. *Wind Eng.* **2008**, *32*, 13–26. [CrossRef]
32. Singh, M.; Santoso, S. *Dynamic Models for Wind Turbines and Wind Power Plants*; National Renewable Energy Laboratory: Livermore, CA, USA, 2011.
33. Dai, J.; Liu, D.; Wen, L.; Long, X. Research on power coefficient of wind turbines based on SCADA data. *Renew. Energy* **2016**, *86*, 206–215. [CrossRef]
34. Li, J.; Yu, X. Model and Procedures for Reliable Near Term Wind Energy Production Forecast. *Wind Eng.* **2015**, *39*, 595–608. [CrossRef]
35. Jonkman, J.M.; Butterfield, S.; Musial, W.; Scott, G. *Definition of a 5-MW Reference Wind Turbine for Offshore System Development*; National Renewable Energy Laboratory: Livermore, CA, USA, 2009.
36. Glauert, H. Airplane propellers. In *Aerodynamic Theory*; Springer: New York, NY, USA, 1935; pp. 169–360.
37. Leishman, J. *Principles of Helicopter Aerodynamics*; CambridgeUniversity Press: New York, NY, USA, 2000.
38. Moriarty, P.J.; Hansen, A.C. *AeroDyn Theory Manual*; National Renewable Energy Laboratory: Golden, CO, USA, 2005.
39. Kulunk, E. Aerodynamics of wind turbines. In *Fundamental and Advanced Topics in Wind Powe*; InTech: Rijeka, Croatia, 1970.
40. Gorsevski, P.V.; Cathcart, S.C.; Mirzaei, G.; Jamali, M.M.; Ye, X.; Gomezdelcampo, E. A group-based spatial decision support system for wind farm site selection in Northwest Ohio. *Energy Policy* **2013**, *55*, 374–385. [CrossRef]
41. Mekonnen, A.D.; Gorsevski, P.V. A web-based participatory GIS (PGIS) for offshore wind farm suitability within Lake Erie, Ohio. *Renew. Sustain. Energy Rev.* **2015**, *41*, 162–177. [CrossRef]
42. Wang, X.; Yang, X.; Zeng, X. Lateral capacity assessment of offshore wind suction bucket foundation in clay via centrifuge modelling. *J. Renew. Sustain. Energy* **2017**, *9*, 033308. [CrossRef]
43. Klink, K. Atmospheric circulation effects on wind speed variability at turbine height. *J. Appl. Meteorol. Climatol.* **2007**, *46*, 445–456. [CrossRef]
44. Soler-Bientz, R. Preliminary results from a network of stations for wind resource assessment at North of Yucatan Peninsula. *Energy* **2011**, *36*, 538–548. [CrossRef]
45. Saleh, H.; Aly, A.A.E.-A.; Abdel-Hady, S. Assessment of different methods used to estimate Weibull distribution parameters for wind speed in Zafarana wind farm, Suez Gulf, Egypt. *Energy* **2012**, *44*, 710–719. [CrossRef]
46. Seguro, J.; Lambert, T. Modern estimation of the parameters of the Weibull wind speed distribution for wind energy analysis. *J. Wind Eng. Ind. Aerodyn.* **2000**, *85*, 75–84. [CrossRef]
47. Carta, J.A.; Ramirez, P.; Velazquez, S. A review of wind speed probability distributions used in wind energy analysis: Case studies in the Canary Islands. *Renew. Sustain. Energy Rev.* **2009**, *13*, 933–955. [CrossRef]
48. Foley, C.; Fournelle, R.; Ginal, S.J.; Peronto, J.L. *Structural Analysis of Sign Bridge Structures and Luminaire Supports*; Northwestern University: Evanston, IL, USA, 2004.
49. Durst, C. Wind speeds over short periods of time. *Meteorol. Mag.* **1960**, *89*, 181–186.
50. Holmes, J.D. *Wind Loading of Structures*; CRC Press: Boca Raton, FL, USA, 2015.
51. Pourazarm, P.; Caracoglia, L.; Lackner, M.; Modarres-Sadeghi, Y. Stochastic analysis of flow-induced dynamic instabilities of wind turbine blades. *J. Wind Eng. Ind. Aerodyn.* **2015**, *137*, 37–45. [CrossRef]
52. Chaudhari, N. Dynamic Characteristics of Wind Turbine Blade. *Int. J. Eng. Res. Technol.* **2014**, *8*.
53. Yanbin, C.; Lei, S.; Feng, Z. Modal Analysis of Wind Turbine Blade Made of Composite laminated plates. In Proceedings of the 2010 Asia-Pacific Power and Energy Engineering Conference (APPEEC), Chengdu, China, 28–31 March 2010; pp. 1–4.
54. Taslimi-Renani, E.; Modiri-Delshad, M.; Elias, M.F.M.; Rahim, N.A. Development of an enhanced parametric model for wind turbine power curve. *Appl. Energy* **2016**, *177*, 544–552. [CrossRef]

55. Det Norske Veritas and Riso National Laboratory. *Guidelines for Design of Wind Turbines*, 2nd ed.; DNV: Oslo, Norway, 2002.

56. Sedaghat, A.; Assad, M.E.H.; Gaith, M. Aerodynamics performance of continuously variable speed horizontal axis wind turbine with optimal blades. *Energy* **2014**, *77*, 752–759. [CrossRef]

57. Tenguria, N.; Mittal, N.; Ahmed, S. Investigation of blade performance of horizontal axis wind turbine based on blade element momentum theory (BEMT) using NACA airfoils. *Int. J. Eng. Sci. Technol.* **2010**, *2*, 25–35. [CrossRef]

58. Kulunk, E. *Aerodynamics of Wind Turbines, Fundamental and Advanced Topics in Wind Power*; Rupp, C., Ed.; InTech: Rijeka, Croatia, 2011; ISBN 978-953-307-508-2.

energies

MDPI

Article

A New Fault Location Approach for Acoustic Emission Techniques in Wind Turbines

Carlos Quiterio Gómez Muñoz * and Fausto Pedro García Márquez

Ingenium Research Group, Department of Business Management, University of Castilla-La Mancha, Ciudad Real 13071, Spain; FaustoPedro.Garcia@uclm.es
* Correspondence: carlosquiterio.gomez@uclm.es; Tel.: +34-650-603-707

Academic Editor: Frede Blaabjerg
Received: 30 October 2015; Accepted: 5 January 2016; Published: 12 January 2016

Abstract: The renewable energy industry is undergoing continuous improvement and development worldwide, wind energy being one of the most relevant renewable energies. This industry requires high levels of reliability, availability, maintainability and safety (RAMS) for wind turbines. The blades are critical components in wind turbines. The objective of this research work is focused on the fault detection and diagnosis (FDD) of the wind turbine blades. The FDD approach is composed of a robust condition monitoring system (CMS) and a novel signal processing method. CMS collects and analyses the data from different non-destructive tests based on acoustic emission. The acoustic emission signals are collected applying macro-fiber composite (MFC) sensors to detect and locate cracks on the surface of the blades. Three MFC sensors are set in a section of a wind turbine blade. The acoustic emission signals are generated by breaking a pencil lead in the blade surface. This method is used to simulate the acoustic emission due to a breakdown of the composite fibers. The breakdown generates a set of mechanical waves that are collected by the MFC sensors. A graphical method is employed to obtain a system of non-linear equations that will be used for locating the emission source. This work demonstrates that a fiber breakage in the wind turbine blade can be detected and located by using only three low cost sensors. It allows the detection of potential failures at an early stages, and it can also reduce corrective maintenance tasks and downtimes and increase the RAMS of the wind turbine.

Keywords: acoustic emission; wind turbine; fault detection and diagnosis; macro-fiber composite; non-destructive testing

1. Introduction

The renewable energy industry is undergoing continuous improvement to cover the current demands of electricity, wind energy being one of the most important. The new technologies, communication systems and advances in mathematical models for signal processing aid in achieving that goal [1]. The complexity of these devices causes a reduction of the reliability, availability, maintainability and safety of the system (RAMS) and increases the maintenance costs due to the occurrence of non-monitored failures [2–4].

Nowadays, fault detection and diagnosis (FDD) by non-destructive testing (NDT) is employed in maintenance management [5–7], for example in structural health monitoring (SHM) [8]. SHM enables identifying and diagnosing the fault and its location by detecting changes in the static and dynamic features of the structure [9,10]. SHM can be remotely managed, reducing the costs of manual inspections and the time between the fault occurrences, and this has been noted [11,12]. This will lead to an increase in the productivity, reducing the potential downtimes for the wind farms and increasing the RAMS of the wind turbine [13–15].

The purpose of this paper is to design an FDD model for the SHM of a wind turbine blade [16–18]. The case study proposes a novel localization method using signals from macro-fiber composite (MFC) sensors. Three MFC sensors are strategically located along a blade section to detect incipient breakages in the structure [19,20]. The case study involves some considerations, e.g., the appearance of the scattering phenomena, the orientation of the sensors when the excitation is received, *etc.* However, it will be demonstrated that the proposed method can set the location with high accuracy. The analysis identifies a single point obtained from a graphical method that is analytically set by nonlinear equations.

The accuracy of this method depends on the transducer sensitivity, the type of composite material, irregularities in the material, the environmental noise, *etc.* The localization precision of the emission source will be affected by the type of composite material, the sensitivity of the materials, environmental noise, false positives due to impacts on the piece, *etc.* Moreover, in real working conditions, considering environmental conditions, e.g., rain or hail, or impacts on the blade, it can cause false alarms.

In working conditions, it would be possible to distinguish between the frequencies associated with the vibration of the blade (low frequencies) and the frequencies associated with the acoustic emission of the fiber breakage (frequencies within the audible range and the ultrasonic range) [21]. It is possible to filter the frequencies associated with the vibration from the collected signal. The authors demonstrated this in [22].

2. Experiments

The experiments are done in a section of the wind turbine blade. The fragment, shown in Figure 1, is made of glass fiber-reinforced polymer (GFRP), with dimensions of 100×79.5 cm. The section is composed of a honeycomb central layer embedded between two fiberglass layers made of polyester resin. This type of material has good structural properties, resistance to fatigue and other advantages. The attenuation of the acoustic emission in the blade is high, and it depends on the material, wave frequency and travelling distance between the failure source and the sensor location [23].

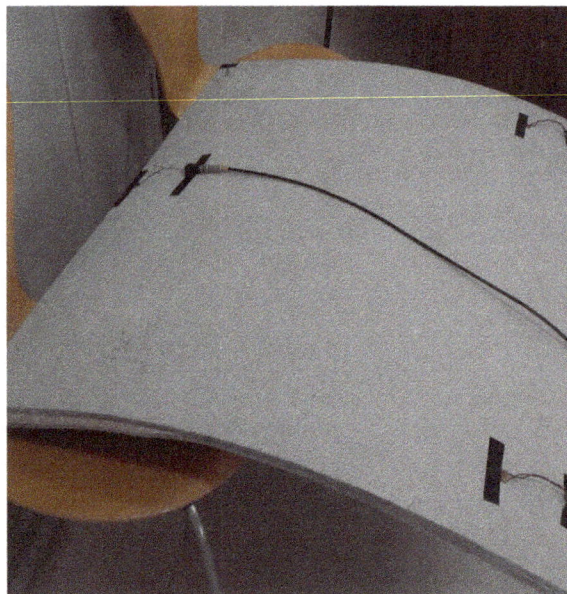

Figure 1. Wind turbine section with sensors for acoustic emission location.

The waves with the same velocity form a circular wave front when they propagate through an isotropic material. The velocity generally does not depend on the direction of propagation, but in anisotropic materials, e.g., the composite materials of the wind turbines, the velocity depends on the direction of propagation. A slowness factor could be introduced in order to consider the propagation direction, e.g., it has been observed that the configuration of layers (+45/−45) for a composite has a strong dependency of the direction of propagation. However, it has been demonstrated that the direction of propagation does not affect the velocity in the blade section studied in this paper. Therefore, the slowness factor has not been introduced in these experiments.

3. Location of the Fiber Breakage by the Triangulation Methodology

The SHM on wind turbine blades is employed to detect the defect online and to locate it with accuracy [24]. The wind turbine blades are becoming larger and more complex, and this requires setting the exact location of a fiber breakage to reduce the maintenance cost and the productivity.

The glass fiber breakages of a wind turbine blade have been simulated in the laboratory on a real blade. A novel location method by triangulation has been developed. The aim of the paper is to locate the acoustic emission source in four different points on the blade section. The acoustic emission produced by the division of the glass fibers is simulated by breaking the tip of the lead from a mechanical pencil [25–27]. Three MFC transducers (A, B and C) were used to detect the acoustic emissions. The three transducers are used as sensors that collect the wave front of the mechanical wave produced by the acoustic emission. These signals received by the sensors present a low amplitude, and therefore, they need to be pre-amplified before being acquired by the oscilloscope [28].

In working conditions, there are many factors that could influence the configuration of the arrangement of the sensors on a blade, for example the length of the blades, the intensity of the acoustic emission, the accuracy of the sensors, the background noise, attenuation, *etc.* Depending on these factors, many groups of three sensors would be established, as they are required to cover the entire blade.

The propagation velocity of the acoustic emission (see Figure 2) has been experimentally calculated by breaking a pencil lead and measuring the delays in the excitement of the sensors S1 and S2 (Figure 3).

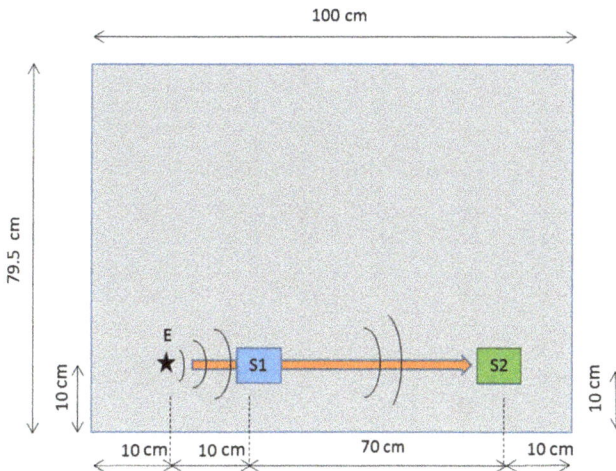

Figure 2. Measuring the experimental propagation velocity in the composite material.

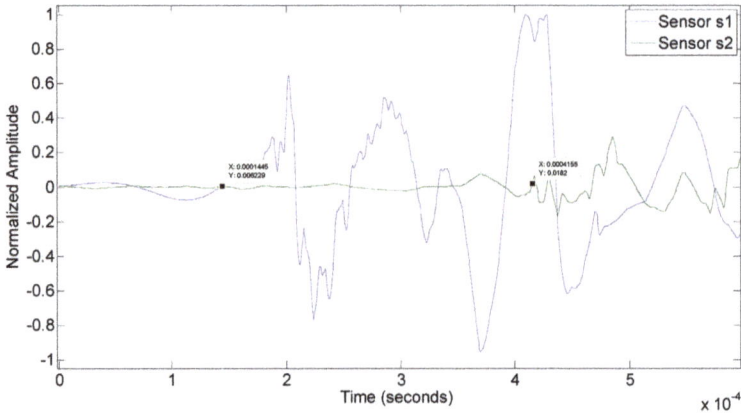

Figure 3. Peak detection of the acoustical emission collected by Sensor 1 (**blue**) and Sensor 2 (**green**) to obtain the experimental propagation velocity in the composite material.

The delay between Signals 1 and 2 is 271 μs (542 samples), and the propagation velocity for the composite material is 2583 m/s.

Four experiments have been conducted at four different locations of the acoustic emission. Twelve tests have been done applying the same force, angle of inclination and length (1 mm approximately). The main objective is to get similar signals for all of the case studies. The data are also filtered for the signal processing, where undesired frequencies are filtered [29]. The peak detection algorithm identifies the wave front of each signal. This process is complex because the waves are compounded by a large number of frequencies. Moreover, there are multiple elements in the blade that could affect the scattering of the acoustic signal, such as the edges of the geometry, the junction with the beam, adhesives, *etc.*

The signal processing consists of a pass band filter that eliminates low and high frequencies and carries out a comparison of the peaks of the wave front in the same frequency range of Signals A, B and C, generated by the above-mentioned MFC sensors, A, B and C (Figure 4).

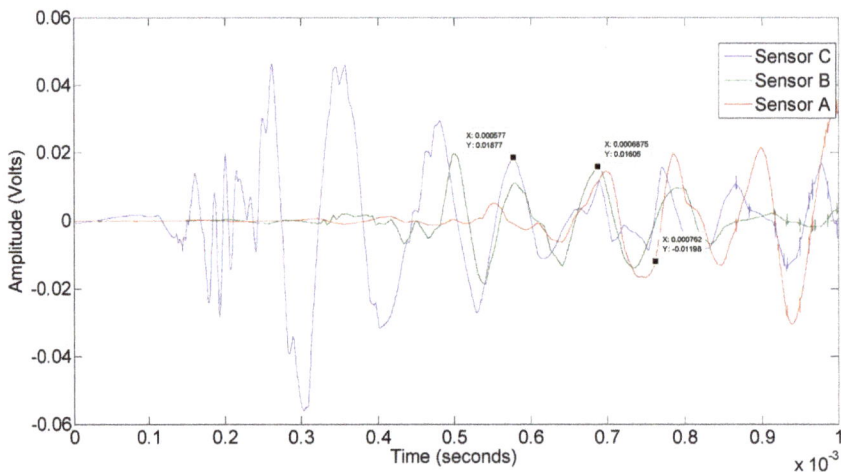

Figure 4. Pre-processing of the signal. Wave front collected by Sensors C (**blue**), B (**green**) and A (**red**).

The MFC Sensors A, B and C are placed as an equilateral triangle (see Figure 5). D is the location of the emission source by the breaking, D (D is known in this experiment).

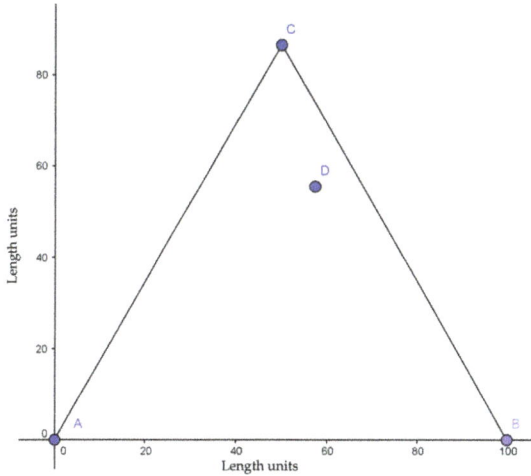

Figure 5. Location of Vertices A, B and C and the, defect D.

The nearest Sensor C is the first to be excited due to the wave front coming from the acoustic emission (Figure 6).

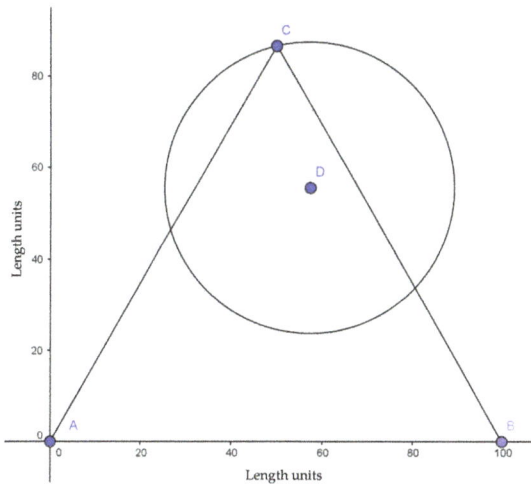

Figure 6. Wave front of the acoustic emission collected by the nearest Sensor C.

The delay between the excitation of the first Sensor C and the second closest Sensor B to the defect, D, is given by the distance from E to B (Figure 7). The delay time and the speed of the wave propagation on the blade is calculated by Equation (1):

$$D_{EB} = v \times t_{CB} \tag{1}$$

where D_{EB} is the distance between E and B, v is the propagation velocity of the wave (obtained experimentally) and t_{CB} is the time delay between the excitation of Sensors C and B.

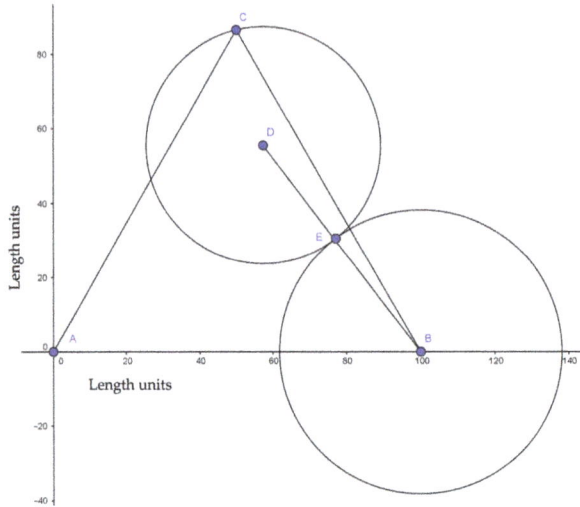

Figure 7. Location of Point E, set by the delay between the excitation time in Sensors C and B.

The delay between Sensor C and Sensor A, the farthest one from Defect D, is given by the distance from F to A D_{FA} in Equation (2).

$$D_{FA} = v \times t_{CA} \tag{2}$$

where t_{CA} is the time delay between the excitation of Sensor C and Sensor A. Figure 8 shows the scheme of the triangulation approach, the delay being represented by a circle.

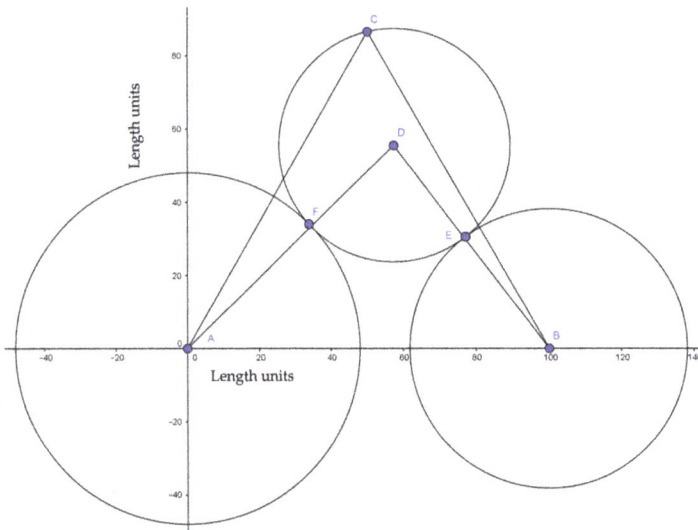

Figure 8. Scheme of the acoustic emission delays for locating the source.

In a real case study, Point D is unknown regarding the time and location, and the delays between the different sensors can be calculated. This condition is shown in Figure 9, where the circumferences represent the delays of the signal that comes to each sensor with respect to the first sensor (C).

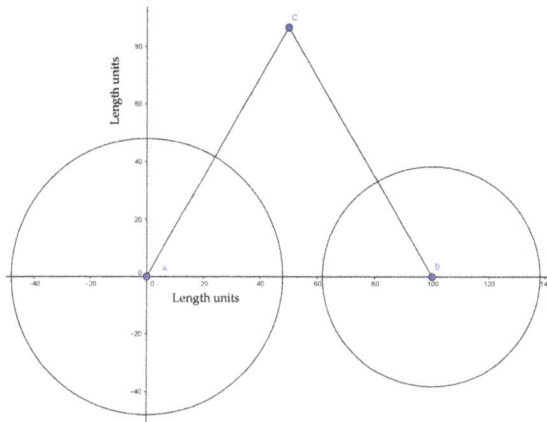

Figure 9. Initial conditions to locate the source of the acoustic emission.

The objective is to find the source of the acoustic emission D mentioned above. This point is the center of a circle that is tangential to two given circles and passes through Point C (see Figure 8). The solution is obtained in this paper employing a graphical method and an analytical method using a system of seven nonlinear equations.

4. Triangulation Equations System

The seven nonlinear equations to solve this problem are given by Equations (3) to (9), considering the scheme shown in Figure 4, where the MFC sensors are located at Points A, B and C, and the defect is at Point D. The coordinates and radius are:

- x_c: x-coordinate at the top of the triangle.
- y_c: y-coordinate at the top of the triangle.
- x_a: x-coordinate at the left lower corner of the triangle.
- y_a: y-coordinate at the left lower corner of the triangle.
- x_b: x-coordinate at the right lower corner of the triangle.
- y_b: y-coordinate at the right lower corner of the triangle.
- r_a: radius of the circle originated from A (delay of Sensor A).
- r_b: radius of the circle originated from B (delay of Sensor B).

The data mentioned above are known. The unknown variables are x_1, x_2, x_3, x_4, x_5, x_6 and x_7, being:

- x_1 and x_2 the coordinates of the emission Source D.
- x_3 and x_4 the coordinates of the tangency of Point F.
- x_5 and x_6 the coordinates of the tangency of Point E.
- x_7 is the radius of the circumference with the center D.

The following equations define the method analytically.
Equation (3) considers a circle with the center at D and passing through C:

$$F(1) = (x_c - x_1)^2 + (y_c - x_2)^2 - (x_7)^2 \qquad (3)$$

123

Equation (4) represents a circle with the center at D and passing through F:

$$F(2) = (x_3 - x_1)^2 + (x_4 - x_2)^2 - (x_7)^2 \qquad (4)$$

Equation (5) sets a circle with the center at D and passing through E:

$$F(3) = (x_5 - x_1)^2 + (x_6 - x_2)^2 - (x_7)^2 \qquad (5)$$

Equation (6) represents a circle with the center at A and passing through F:

$$F(4) = (x_1 - x_a)^2 + (x_4 - y_a)^2 - r_a^2 \qquad (6)$$

Equation (7) considers a circle with the center at B and passing through E:

$$F(5) = (x_5 - x_b)^2 + (x_6 - y_b)^2 - r_b^2 \qquad (7)$$

Equation (8) provides the straight line passing through Points A and F:

$$F(6) = \frac{(x_4 - y_a)}{(x_3 - x_a)} \times x_1 + \left(y_a - \frac{(x_4 - y_a)}{(x_3 - x_a)} \times x_a\right) - x_2 \qquad (8)$$

Equation (9) sets the straight line passing through Points B and E:

$$F(7) = \frac{(y_b - x_6)}{(x_b - x_5)} \times x_1 + \left((y_b - \frac{(y_b - x_6)}{(x_b - x_5)})\right) \times x_b) - x_2 \qquad (9)$$

5. Experimental Procedure and Results

The time of flight and distances are set in this section for Sensors B and A regarding C, C being the first sensor to receive the acoustic signal of the breakage. The experiments are repeated four times to take into account the deviations of the results. The algorithm gives the exact location of the defect, as well as a graphic outline, knowing the radius of the circles with centers at B and C. The dimensions of the blade section, the distribution of the sensors and the emission source (star) in the wind turbine blade are shown in Figure 10. The mathematical results obtained with the algorithm are given in Tables 1–8.

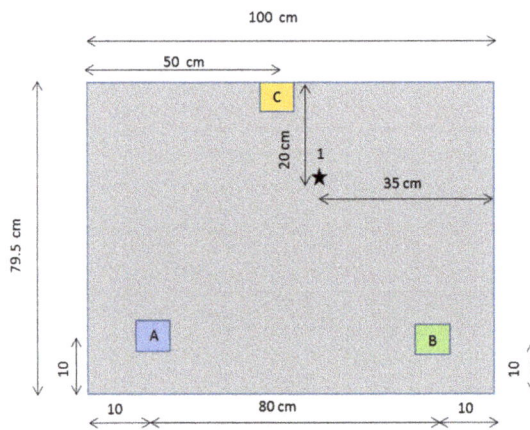

Figure 10. First experiment. Case Study 1.

Table 1. First case study: detection time; delay with C; delay; theoretical distance; experimental distance.

Sensors	Detection Time (Samples)	Delay with C (Samples)	Delay (s)	Theoretical Distance (m)	Experimental Distance (m)
C	1152	-	-	-	-
B	1381	229	1.15×10^{-4}	0.30	0.30
A	1528	376	1.88×10^{-4}	0.49	0.49

Table 2. Initial data of the first case study.

Locations	x-Coordinate (m)	y-Coordinate (m)	Radius (m)
A	0	0	0.49
B	0.8	0	0.30
C	0.4	0.69	-
1	0.55	0.495	-

Table 3. Second case study: detection time; delay with C; delay; theoretical distance; experimental distance.

Sensors	Detection Time (Samples)	Delay with C (Samples)	Delay (s)	Theoretical Distance (m)	Experimental Distance (m)
C	912	-	-	-	-
B	1063	151	7.55×10^{-5}	0.20	0.20
A	1296	384	1.92×10^{-4}	0.50	0.50

Table 4. Initial data of the second case study.

Locations	x-Coordinate (m)	y-Coordinate (m)	Radius (m)
A	0	0	0.50
B	0.8	0	0.20
C	0.4	0.69	-
2	0.65	0.495	-

Table 5. Third case study: detection time; delay with C; delay; theoretical distance; experimental distance.

Sensors	Detection Time (Samples)	Delay with C (Samples)	Delay (s)	Theoretical Distance (m)	Experimental Distance (m)
C	962	-	-	-	-
B	1298	336	1.68×10^{-4}	0.43	0.43
A	1087	125	6.25×10^{-5}	0.17	0.16

Table 6. Initial data of the third case study.

Locations	x-Coordinate (m)	y-Coordinate (m)	Radius (m)
A	0	0	0.16
B	0.8	0	0.43
C	0.4	0.69	-
3	0.2	0.445	-

Table 7. Fourth case study: detection time; delay with C; delay; theoretical distance; experimental distance.

Sensors	Detection Time (Samples)	Delay with C (Samples)	Delay (s)	Theoretical Distance (m)	Experimental Distance (m)
C	1155	-	-	-	-
B	1650	495	2.48×10^{-4}	0.64	0.64
A	1385	230	1.15×10^{-4}	0.29	0.30

Table 8. Initial data of the fourth case study.

Locations	x-Coordinate (m)	y-Coordinate (m)	Radius (m)
A	0	0	0.30
B	0.8	0	0.64
C	0.4	0.69	/
4	0.05	0.645	/

5.1. Case Study 1

The breaking of the lead is made in the following coordinates from Sensor A at Point 1 (star); see Figure 10. Sensor A is the coordinate origin.

- Coordinate x: 0.55.
- Coordinate y: 0.495.

The location of the source employing the algorithm is: Point 1: (x: 0.5533, y: 0.4920). The error in the location is: coordinate x: 3.3 mm; coordinate y: 30 mm.

5.2. Case Study 2

In this case, the emission source was generated at Point 2 (star), shown in Figure 11:

- Coordinate x: 0.65.
- Coordinate y: 0.495.

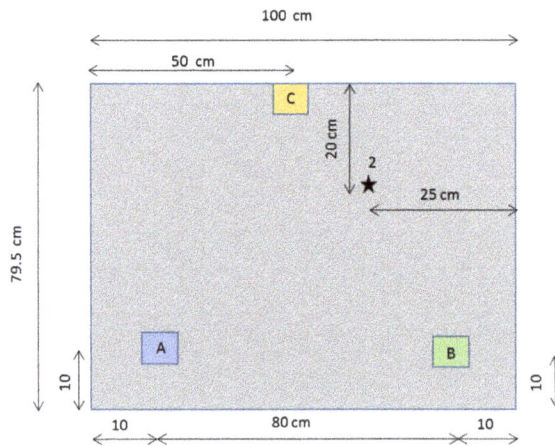

Figure 11. Scheme for Case Study 2.

The location of the source employing the algorithm is: Point 2: (x: 0.6502, y: 0.4950). The errors in the location are: coordinate x: 0.2 mm; coordinate y: 0.00 mm.

5.3. Case Study 3

In this case, the emission source was generated at Point 3 (star); see Figure 12:

- Coordinate x: 0.20.
- Coordinate y: 0.445.

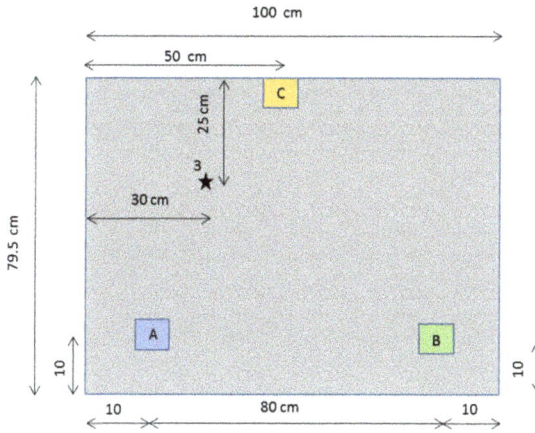

Figure 12. Scheme for Case Study 3.

The location of the source employing the algorithm is: Point 3 (x: 0.1914, y: 0.4434). The errors in the location are: coordinate x: 8.6 mm; coordinate y: 1.6 mm.

5.4. Case Study 4

In this case, the emission source was generated at Point 4 (star), and it is shown in Figure 13:

- Coordinate x: 0.05.
- Coordinate y: 0.645.

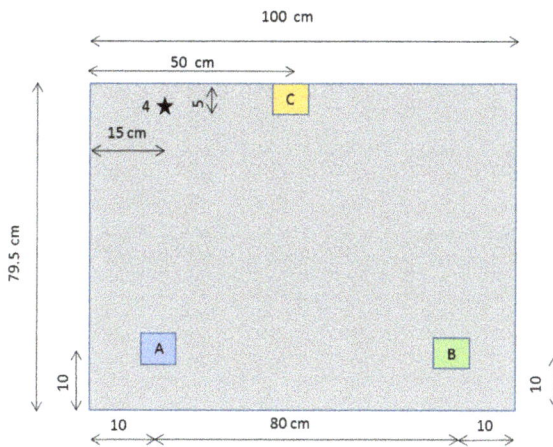

Figure 13. Scheme for Case Study 4.

The location of the source employing the algorithm is: Point 4: (x: 0.050, y: 0.6495). The errors in the location are: coordinate x: 0 mm; coordinate y: 4.5 mm.

Different waves with different speeds appear as a result of the scattering phenomena when a large number of frequencies are excited by the breakage. This makes the identification of peaks to measure the delays of the signals complicated. The orientation of the sensors, when they receive the excitation, can affect the shape of the signal collected.

It is observed that the algorithm provides correct and coherent results. It detects the location of the acoustic emission with an accuracy of two decimals (millimeters). The maximum error registered was 9 mm.

Finally, the algorithm shows the position of the acoustic emission point with the real dimensions of the blade. Figure 14 shows the location of the acoustic emission for the first case study.

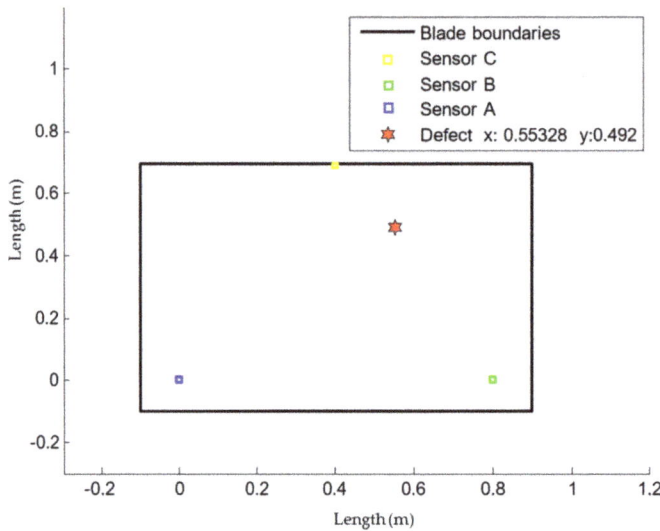

Figure 14. Scheme of the location of the acoustic emission for the first case study.

6. Conclusions

The development of a localization approach presented in this paper is set using macro-fiber composites to detect cracks in blades in an SHM system. This approach, based on NDT, automatically identifies and locates an acoustic emission source coming from a fiber's breakage in a wind turbine blade section by a novel signal processing method. It can be extrapolated to other similar structures, e.g., airplane wings.

Three sensors are strategically located in the blade. It is demonstrated that the approach is able to detect the location of the simulated defect accurately employing acoustic emissions signals. The signal processing is based on a graphical method of triangulation and seven nonlinear equations. The signals are previously filtered. Different experiments are performed to demonstrate the effectiveness of the proposed method.

The approach detects the location of the acoustic emission with high accuracy, 9 mm being the maximum error registered.

There are conditions that affect the accuracy of the emission source location, e.g., the type of composite material, the sensitivity of the transducers, environmental noise, false positives due to impacts on the piece, *etc.* The method shows the position of the acoustic emission point with the real dimensions of the blade.

Acknowledgments: This project is partly funded by the EC under the FP7 framework program (Ref. 322430), OPTIMUS and the MINECO project WindSeaEnergy (Ref. DPI2012-31579), where OPTIMUS is supporting to the new condition Monitoring System, and WindSeaEnergy the novel signal processing approach presented in this research paper.

Author Contributions: Carlos Quiterio Gómez Muñoz and Fausto Pedro García Marquez conceived and designed the experiments; Carlos Quiterio Gómez Muñoz and Fausto Pedro García Marquez performed the experiments; Carlos Quiterio Gómez Muñoz and Fausto Pedro García Marquez analyzed the data; Carlos Quiterio Gómez Muñoz and Fausto Pedro García Marquez contributed reagents/materials/analysis tools; Carlos Quiterio Gómez Muñoz and Fausto Pedro García Marquez wrote the paper.

Conflicts of Interest: The authors declare no conflict of interest.

References

1. Dai, D.; He, Q. Structure damage localization with ultrasonic guided waves based on a time–frequency method. *Signal Process.* **2014**, *96*, 21–28. [CrossRef]
2. Spinato, F.; Tavner, P.J.; van Bussel, G.; Koutoulakos, E. Reliability of wind turbine subassemblies. *IET Renew. Power Gen.* **2009**, *3*, 387–401. [CrossRef]
3. Ajayi, O.O.; Fagbenle, R.O.; Katende, J.; Ndambuki, J.M.; Omole, D.O.; Badejo, A.A. Wind energy study and energy cost of wind electricity generation in nigeria: Past and recent results and a case study for south west nigeria. *Energies* **2014**, *7*, 8508–8534. [CrossRef]
4. Marugán, A.P.; Márquez, F.P.G. A novel approach to diagnostic and prognostic evaluations applied to railways: A real case study. *Proc. Inst. Mech. Eng. F J. Rail Rapid Transit* **2015**. [CrossRef]
5. Chen, X.; Zhao, W.; Zhao, X.L.; Xu, J.Z. Failure test and finite element simulation of a large wind turbine composite blade under static loading. *Energies* **2014**, *7*, 2274–2297. [CrossRef]
6. Márquez, F.G.; Roberts, C.; Tobias, A.M. Railway point mechanisms: Condition monitoring and fault detection. *Proc. Inst. Mech. Eng. F J. Rail Rapid Transit* **2010**, *224*, 35–44. [CrossRef]
7. Pliego Marugán, A.; García Márquez, F.P.; Lorente, J. Decision making process via binary decision diagram. *Int. J. Manag. Sci. Eng. Manag.* **2015**, *10*, 3–8. [CrossRef]
8. Marquez, F.P.G. An approach to remote condition monitoring systems management. In Proceedings of the Institution of Engineering and Technology International Conference on Railway Condition Monitoring, Birmingham, UK, 29–30 Novemmber 2006; pp. 156–160.
9. Light-Marquez, A.; Sobin, A.; Park, G.; Farinholt, K. Structural damage identification in wind turbine blades using piezoelectric active sensing. In *Structural Dynamics and Renewable Energy*; Springer: New York, NY, USA, 2011; pp. 55–65.
10. García, F.P.; Pinar, J.M.; Papaelias, M.; Ruiz de la Hermosa, R. Wind turbines maintenance management based on FTA and BDD. *Renew. Energy Power Qual. J.* **2012**. Available online: http://icrepq.com/icrepq'12/699-garcia.pdf (accessed on 7 January 2016).
11. Pedregal, D.J.; García, F.P.; Roberts, C. An algorithmic approach for maintenance management based on advanced state space systems and harmonic regressions. *Ann. Oper. Res.* **2009**, *166*, 109–124. [CrossRef]
12. Yang, H.-H.; Huang, M.-L.; Yang, S.-W. Integrating auto-associative neural networks with hotelling T^2 control charts for wind turbine fault detection. *Energies* **2015**, *8*, 12100–12115. [CrossRef]
13. Chen, X.; Qin, Z.W.; Zhao, X.L.; Xu, J.Z. Structural performance of a glass/polyester composite wind turbine blade with flatback and thick airfoils. In Proceedings of the American Society of Mechanical Engineers (ASME) 2014 International Mechanical Engineering Congress and Exposition, Montreal, QC, Canada, 14–20 November 2014.
14. De la Hermosa González, R.R.; Márquez, F.P.G.; Dimlaye, V. Maintenance management of wind turbines structures via mfcs and wavelet transforms. *Renew. Sustain. Energy Rev.* **2015**, *48*, 472–482. [CrossRef]
15. Márquez, F.P.G.; Pedregal, D.J.; Roberts, C. New methods for the condition monitoring of level crossings. *Int. J. Syst. Sci.* **2015**, *46*, 878–884. [CrossRef]
16. García, F.P.; Pedregal, D.J.; Roberts, C. Time series methods applied to failure prediction and detection. *Reliab. Eng. Syst. Saf.* **2010**, *95*, 698–703. [CrossRef]
17. García Márquez, F.P.; García-Pardo, I.P. Principal component analysis applied to filtered signals for maintenance management. *Qual. Reliab. Eng. Int.* **2010**, *26*, 523–527. [CrossRef]

18. Márquez, F.P.G.; Muñoz, J.M.C. A pattern recognition and data analysis method for maintenance management. *Int. J. Syst. Sci.* **2012**, *43*, 1014–1028. [CrossRef]
19. Michaels, J.E. Detection, localization and characterization of damage in plates with an *in situ* array of spatially distributed ultrasonic sensors. *Smart Mater. Struct.* **2008**, *17*, 035035. [CrossRef]
20. Chen, H.; Yan, Y.; Chen, W.; Jiang, J.; Yu, L.; Wu, Z. Early damage detection in composite wingbox structures using hilbert-huang transform and genetic algorithm. *Struct. Health Monit.* **2007**, *6*, 281–297. [CrossRef]
21. Eftekharnejad, B.; Carrasco, M.; Charnley, B.; Mba, D. The application of spectral kurtosis on acoustic emission and vibrations from a defective bearing. *Mech. Syst. Signal Process.* **2011**, *25*, 266–284. [CrossRef]
22. Gómez, C.Q.; Villegas, M.A.; García, F.P.; Pedregal, D.J. Big data and web intelligence for condition monitoring: A case study on wind turbines. In *Handbook of Research on Trends and Future Directions in Big Data and Web Intelligence*; Information Science Reference, IGI Global: Hershey, PA, USA, 2015.
23. Bohse, J. Acoustic emission characteristics of micro-failure processes in polymer blends and composites. *Compos. Sci. Technol.* **2000**, *60*, 1213–1226. [CrossRef]
24. Márquez, F.P.G.; Pérez, J.M.P.; Marugán, A.P.; Papaelias, M. Identification of critical components of wind turbines using FTA over the time. *Renew. Energy* **2015**, *87*, 869–883. [CrossRef]
25. Gorman, M.R. Plate wave acoustic emission. *J. Acoust. Soc. Am.* **1991**, *90*, 358–364. [CrossRef]
26. Ruiz de la Hermosa, R.; García Márquez, F.P.; Dimlaye, V.; Ruiz-Hernández, D. Pattern recognition by wavelet transforms using macro fibre composites transducers. *Mechan. Syst. Signal Process.* **2014**, *48*, 339–350. [CrossRef]
27. Betz, D.C.; Staszewski, W.J.; Thursby, G.; Culshaw, B. Structural damage identification using multifunctional bragg grating sensors: II. Damage detection results and analysis. *Smart Mater. Struct.* **2006**, *15*, 1313–1322. [CrossRef]
28. Coverley, P.; Staszewski, W. Impact damage location in composite structures using optimized sensor triangulation procedure. *Smart Mater. Struct.* **2003**, *12*, 795–803. [CrossRef]
29. Gómez, C.Q.; Ruiz de la Hermosa, R.; Trapero, J.R.; Garcia, F.P. A novel approach to fault detection and diagnosis on wind turbines. *Glob. Nest J.* **2014**, *16*, 1029–1037.

![energies logo] *energies*

MDPI

Article

Wind Turbine Fault Detection through Principal Component Analysis and Statistical Hypothesis Testing

Francesc Pozo * and Yolanda Vidal

Control, Dynamics and Applications (CoDAlab), Departament de Matemàtiques, Escola Universitària d'Enginyeria Tècnica Industrial de Barcelona (EUETIB), Universitat Politècnica de Catalunya (UPC), Comte d'Urgell, 187, Barcelona 08036, Spain; yolanda.vidal@upc.edu
* Correspondence: francesc.pozo@upc.edu; Tel.: +34-934-137-316; Fax: +34-934-137-401

Academic Editor: Frede Blaabjerg
Received: 17 November 2015; Accepted: 14 December 2015; Published: 23 December 2015

Abstract: This paper addresses the problem of online fault detection of an advanced wind turbine benchmark under actuators (pitch and torque) and sensors (pitch angle measurement) faults of different type: fixed value, gain factor, offset and changed dynamics. The fault detection scheme starts by computing the baseline principal component analysis (PCA) model from the healthy or undamaged wind turbine. Subsequently, when the structure is inspected or supervised, new measurements are obtained are projected into the baseline PCA model. When both sets of data—the baseline and the data from the current wind turbine—are compared, a statistical hypothesis testing is used to make a decision on whether or not the wind turbine presents some damage, fault or misbehavior. The effectiveness of the proposed fault-detection scheme is illustrated by numerical simulations on a well-known large offshore wind turbine in the presence of wind turbulence and realistic fault scenarios. The obtained results demonstrate that the proposed strategy provides and early fault identification, thereby giving the operators sufficient time to make more informed decisions regarding the maintenance of their machines.

Keywords: wind turbine; fault detection; principal component analysis; statistical hypothesis testing; FAST (Fatigue, Aerodynamics, Structures and Turbulence)

---

1. Introduction

Wind energy is currently the fastest growing source of renewable energy in the world. As wind turbines (WT) increase in size, and their operating conditions become more extreme, a number of current and future challenges exist. A major issue with wind turbines, specially those located offshore, is the relatively high cost of maintenance [1]. Since the replacement of main components of a wind turbine is a difficult and costly affair, improved maintenance procedures can lead to essential cost reductions. Autonomous online fault detection algorithms allow early warnings of defects to prevent major component failures. Furthermore, side effects on other components can be reduced significantly. Many faults can be detected while the defective component is still operational. Thus necessary repair actions can be planned in time and need not to be taken immediately and this fact is specially important for off-shore turbines where bad weather conditions can prevent any repair actions. Therefore the implementation of fault detection (FD) systems is crucial.

The past few years have seen a rapid growth in interest in wind turbine fault detection [2] through the use of condition monitoring and structural health monitoring (SHM) [3,4]. The SHM techniques are based on the idea that the change in mechanical properties of the structure will be captured by a change in its dynamic characteristics [5]. Existing techniques for fault detection can be broadly classified into two major categories: model-based methods and signal processing-based methods. For model-based fault detection, the system model could be mathematical—or knowledge-based [6]. Faults are detected based on the residual generated by state variable or model parameter estimation [7–11]. For signal processing-based fault detection, mathematical or statistical operations are performed on the measurements (see, for example, [12,13]), or artificial intelligence techniques are applied to the measurements to extract the information about the faults (see [14,15]).

With respect to signal-processing-based fault detection, principal component analysis (PCA) is used in this framework as a way to condense and extract information from the collected signals. Following this structure, this paper is focused on the development of a wind turbine fault detection strategy that combines a data driven baseline model—reference pattern obtained from the healthy structure—based on PCA and hypothesis testing. A different approach in the frequency domain can be found in [16], where a Karhunen-Loeve basis is used.

Most industrial wind turbines are manufactured with an integrated system that can monitor various turbine parameters. These monitored data are collated and stored via a supervisory control and data acquisition (SCADA) system that archives the information in a convenient manner. These data quickly accumulates to create large and unmanageable volumes that can hinder attempts to deduce the health of a turbine's components. It would prove beneficial if the data could be analyzed and interpreted automatically (online) to support the operators in planning cost-effective maintenance activities [17–19]. This paper describes a technique that can be used to identify incipient faults in the main components of a turbine through the analysis of this SCADA data. The SCADA data sets are already generated by the integrated monitoring system, and therefore, no new installation of specific sensors or diagnostic equipment is required. The strategy developed is based on principal component analysis and statistical hypothesis testing. The final section of the paper shows the performance of the proposed techniques using an enhanced benchmark challenge for wind turbine fault detection, see [2]. This benchmark proposes a set of realistic fault scenarios considered in an aeroelastic computer-aided engineering tool for horizontal axes wind turbines called FAST, see [20].

The structure of the paper is the following. In Section 2 the wind turbine benchmark is recalled as well as the fault scenarios studied in this work. Section 3 presents the design of the proposed fault detection strategy. The simulation results obtained with the proposed approach applied to the wind turbine benchmark are given in Section 4. Concluding remarks are given in Section 5.

2. Wind Turbine Benchmark Model

A complete description of the wind turbine benchmark model, as well as the used baseline torque and pitch controllers, can be found in [2]. In this benchmark challenge, a more sophisticated wind turbine model—a modern 5 MW turbine implemented in the FAST software—and updated fault scenarios are presented. These updates enhance the realism of the challenge and will therefore lead to solutions that are significantly more useful to the wind industry. Hereafter, a brief review of the reference wind turbine is given and the generator-converter actuator model and the pitch actuator model are recalled, as the studied faults affect those subsystems. A complete description of the tested fault scenarios is given.

2.1. Reference Wind Turbine

The numerical simulations use the onshore version of a large wind turbine that is representative of typical utility-scale land- and sea-based multimegawatt turbines described by [21]. This wind turbine is a conventional three-bladed upwind variable-speed variable blade-pitch-to-feather-controlled turbine of 5 MW. The wind turbine characteristics are given in Table 1. In this work we deal with the full load region of operation (also called region 3). That is, the proposed controllers main objective is that the electric power follows the rated power.

Table 1. Gross properties of the wind turbine.

Reference Wind Turbine	Magnitude
Rated power	5 MW
Number of blades	3
Rotor/Hub diameter	126 m, 3 m
Hub Height	90 m
Cut-In, Rated, Cut-Out Wind Speed	3 m/s, 11.4 m/s, 25 m/s
Rated generator speed (ω_{ng})	1173.7 rpm
Gearbox ratio	97

In the simulations, new wind data sets with turbulence intensity set to 10% are generated with TurbSim [22]. It can be seen from Figure 1 that the wind speed covers the full load region, as its values range from 12.91 m/s up to 22.57 m/s.

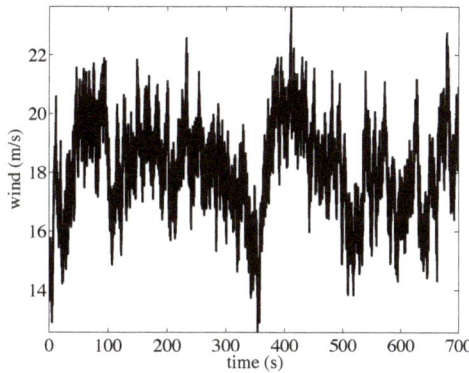

Figure 1. Wind speed signal with turbulence intensity set to 10%.

2.2. Generator-Converter Model

The generator-converter system can be approximated by a first-order ordinary differential equation, see [2], which is given by:

$$\dot{\tau}_r(t) + \alpha_{gc}\tau_r(t) = \alpha_{gc}\tau_c(t) \tag{1}$$

where τ_r and τ_c are the real generator torque and its reference (given by the controller), respectively. In the numerical simulations, $\alpha_{gc} = 50$, see [21]. Moreover, the power produced by the generator, $P_e(t)$, is given by (see [2]):

$$P_e(t) = \eta_g\omega_g(t)\tau_r(t) \tag{2}$$

where η_g is the efficiency of the generator and ω_g is the generator speed. In the numerical experiments, $\eta_g = 0.98$ is used, see [2].

2.3. Pitch Actuator Model

The hydraulic pitch system consists of three identical pitch actuators, which are modeled as a linear differential equation with time-dependent variables, pitch angle $\beta(t)$ and its reference $\beta_r(t)$. In principle, it is a piston servo-system, which can be expressed as a second-order ordinary differential equation [2]:

$$\ddot{\beta}(t) + 2\zeta\omega_n\dot{\beta}(t) + \omega_n^2\beta(t) = \omega_n^2\beta_r(t) \tag{3}$$

where ω_n and ζ are the natural frequency and the damping ratio, respectively. For the fault-free case, the parameters $\zeta = 0.6$ and $\omega_n = 11.11$ rad/s are used, see [2].

2.4. Fault Scenarios

Both actuator and sensor faults are considered. All the described faults originate from actual faults in wind turbines [2]. Table 2 summarizes all the considered fault scenarios.

Table 2. Fault scenarios.

Number	Fault	Type
F1	Pitch actuator	Change in dynamics: air content in oil
F2	Pitch actuator	Change in dynamics: pump wear
F3	Pitch actuator	Change in dynamics: hydraulic leakage
F4	Torque actuator	Offset
F5	Generator speed sensor	Scaling
F6	Pitch angle sensor	Stuck
F7	Pitch angle sensor	Scaling

2.4.1. Actuator Faults

Pitch actuator faults are studied as they are the actuators with highest failure rate in wind turbines. A fault may change the dynamics of the pitch system by varying the damping ratio (ζ) and natural frequencies (ω_n) from their nominal values to their faulty values. The parameters for the pitch system under different conditions are given in Table 3. The normal air content in the hydraulic oil is 7%, whereas the high air content in oil fault (F1) corresponds to 15%. Pump wear (F2) represents the situation of 75% pressure in the pitch system while the parameters stated for hydraulic leakage (F3) correspond to a pressure of only 50%. The three faults are modeled by changing the parameters ω_n and ζ in the relevant pitch actuator model.

Table 3. Change in dynamics pitch actuator faults.

Faults	ω_n (rad/s)	ζ
Fault Free(FF)	11.11	0.6
High air content in oil (F1)	5.73	0.45
Pump wear (F2)	7.27	0.75
Hydraulic leakage (F3)	3.42	0.9

For the test, the change in dynamics faults given in Table 3 are introduced only in the third pitch actuator (thus β_1 and β_2 are always fault-free).

A torque actuator fault (F4) is also studied. This fault is an offset on the generated torque, which can be due to an error in the initialization of the converter controller. This fault can occur since the converter torque is estimated based on the currents in the converter. If this estimate is initialized incorrectly it will result in an offset on the estimated converter torque, which leads to the offset on the generator torque. The offset value is 2000 Nm.

2.4.2. Sensor Faults

The generator speed measurement uses encoders and its elements are subject to electrical and mechanical failures, which can result in a changed gain factor on the measurement. The simulated fault, F5, is a gain factor on ω_g equal to 1.2.

Faults 6 and 7 result in blade 3 having a stuck pitch angle sensor, which holds a constant value of 5° (F6) and 10° (F7), respectively.

Finally, the fault of a gain factor on the measurement of the third pitch angle sensor is studied (F7). The measurement is scaled by a factor of 1.2.

Table 4. Assumed available measurements. These sensors are representative of the types of sensors that are available on a MW-scale commercial wind turbine.

Number	Sensor Type	Symbol	Units
1	Generated electrical power	$P_{e,m}$	kW
2	Rotor speed	$\omega_{r,m}$	rad/s
3	Generator speed	$\omega_{g,m}$	rad/s
4	Generator torque	$\tau_{c,m}$	Nm
5	first pitch angle	$\beta_{1,m}$	deg
6	second pitch angle	$\beta_{2,m}$	deg
7	third pitch angle	$\beta_{3,m}$	deg
8	fore-aft acceleration at tower bottom	$a^b_{fa,m}$	m/s^2
9	side-to-side acceleration at tower bottom	$a^b_{ss,m}$	m/s^2
10	fore-aft acceleration at mid-tower	$a^m_{fa,m}$	m/s^2
11	side-to-side acceleration at mid-tower	$a^m_{ss,m}$	m/s^2
12	fore-aft acceleration at tower top	$a^t_{fa,m}$	m/s^2
13	side-to-side acceleration at tower top	$a^t_{ss,m}$	m/s^2

3. Fault Detection Strategy

The overall fault detection strategy is based on principal component analysis and statistical hypothesis testing. A baseline pattern or PCA model is created with the healthy state of the wind turbine in the presence of wind turbulence. When the current state of the wind turbine has to be diagnosed, the collected data is projected using the PCA model. The final diagnosis is performed using statistical hypothesis testing.

The main paradigm of vibration based structural health monitoring is based on the basic idea that a change in physical properties due to structural changes or damage will cause detectable changes in dynamical responses. This idea is illustrated in Figure 2, where the healthy structure is excited by a signal to create a pattern. Subsequently, the structure to be diagnosed is excited by the same signal and the dynamic response is compared with the pattern. The scheme in Figure 2 is also know as guided waves in structures for structural health monitoring [23].

However, in our application, the only available excitation of the wind turbines is the wind turbulence. Therefore, guided waves in wind turbines for SHM as in Figure 2 cannot be considered as a realistic scenario. In spite of that, the new paradigm described in Figure 3 is based on the fact that, even with a different wind turbulence, the fault detection strategy based on PCA and statistical hypothesis testing will be able to detect some damage, fault or misbehavior. More precisely, the key idea behind the detection strategy is the assumption that a change in the behavior of the overall system, even with a different excitation, has to be detected. The results presented in Section 4 confirm this hypothesis.

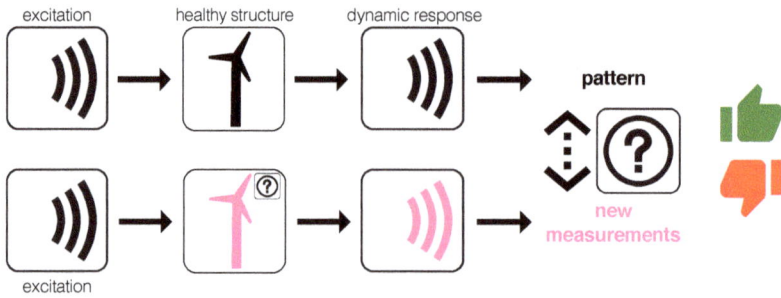

Figure 2. Guided waves in structures for structural health monitoring. The healthy structure is excited by a signal and the dynamic response is measured to create a baseline pattern. Then, the structure to diagnose is excited by the same signal and the dynamic response is also measured and compared with the baseline pattern. A significant difference in the pattern would imply the existence of a fault.

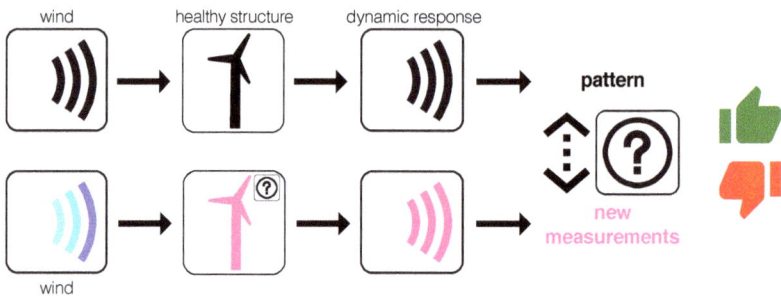

Figure 3. Even with a different wind turbulence, the fault detection strategy is able to detect some damage, fault or misbehavior.

3.1. Data Driven Baseline Modeling Based on PCA

Let us start the PCA modeling by measuring, from a healthy wind turbine, a sensor during $(nL-1)\Delta$ seconds, where Δ is the sampling time and $n, L \in \mathbb{N}$. The discretized measures of the sensor are a real vector

$$\left(\begin{array}{ccccccccccc} x_{11} & x_{12} & \cdots & x_{1L} & x_{21} & x_{22} & \cdots & x_{2L} & \cdots & x_{n1} & x_{n2} & \cdots & x_{nL} \end{array} \right) \in \mathbb{R}^{nL} \tag{4}$$

where the real number x_{ij}, $i = 1, \ldots, n$, $j = 1, \ldots, L$ corresponds to the measure of the sensor at time $((i-1)L + (j-1))\,\Delta$ seconds. This collected data can be arranged in matrix form as follows:

$$\begin{pmatrix} x_{11} & x_{12} & \cdots & x_{1L} \\ \vdots & \vdots & \ddots & \vdots \\ x_{i1} & x_{i2} & \cdots & x_{iL} \\ \vdots & \vdots & \ddots & \vdots \\ x_{n1} & x_{n2} & \cdots & x_{nL} \end{pmatrix} \in \mathcal{M}_{n \times L}(\mathbb{R}) \tag{5}$$

where $\mathcal{M}_{n \times L}(\mathbb{R})$ is the vector space of $n \times L$ matrices over $\mathbb{R}$. When the measures are obtained from $N \in \mathbb{N}$ sensors also during $(nL-1)\Delta$ seconds, the collected data, for each sensor, can be arranged in a matrix as in Equation (5). Finally, all the collected data coming from the N sensors is disposed in a matrix $\mathbf{X} \in \mathcal{M}_{n \times (N \cdot L)}$ as follows:

$$
\mathbf{X} = \begin{pmatrix}
x_{11}^1 & x_{12}^1 & \cdots & x_{1L}^1 & x_{11}^2 & \cdots & x_{1L}^2 & \cdots & x_{11}^N & \cdots & x_{1L}^N \\
\vdots & \vdots & \ddots & \vdots & \vdots & \ddots & \vdots & \ddots & \vdots & \ddots & \vdots \\
x_{i1}^1 & x_{i2}^1 & \cdots & x_{iL}^1 & x_{i1}^2 & \cdots & x_{iL}^2 & \cdots & x_{i1}^N & \cdots & x_{iL}^N \\
\vdots & \vdots & \ddots & \vdots & \vdots & \ddots & \vdots & \ddots & \vdots & \ddots & \vdots \\
x_{n1}^1 & x_{n2}^1 & \cdots & x_{nL}^1 & x_{n1}^2 & \cdots & x_{nL}^2 & \cdots & x_{n1}^N & \cdots & x_{nL}^N
\end{pmatrix} \tag{6}
$$

$$
= \left(\; \mathbf{X}^1 \mid \mathbf{X}^2 \mid \cdots \mid \mathbf{X}^N \; \right)
$$

where the superindex $k = 1, \ldots, N$ of each element x_{ij}^k in the matrix represents the number of sensor.

The objective of the principal component analysis is to find a linear transformation orthogonal matrix $\mathbf{P} \in \mathcal{M}_{(N \cdot L) \times (N \cdot L)}(\mathbb{R})$ that will be used to transform or project the original data matrix $\mathbf{X}$ according to the subsequent matrix product:

$$
\mathbf{T} = \mathbf{XP} \in \mathcal{M}_{n \times (N \cdot L)}(\mathbb{R}) \tag{7}
$$

where $\mathbf{T}$ is a matrix having a diagonal covariance matrix.

Group Scaling

Since the data in matrix $\mathbf{X}$ is affected by diverse wind turbulence, come from several sensors and could have different scales and magnitudes, it is required to apply a preprocessing step to rescale the data using the mean of all measurements of the sensor at the same column and the standard deviation of all measurements of the sensor [24].

More precisely, for $k = 1, 2, \ldots, N$ we define

$$
\mu_j^k = \frac{1}{n} \sum_{i=1}^{n} x_{ij}^k, \; j = 1, \ldots, L, \tag{8}
$$

$$
\mu^k = \frac{1}{nL} \sum_{i=1}^{n} \sum_{j=1}^{L} x_{ij}^k, \tag{9}
$$

$$
\sigma^k = \sqrt{ \frac{1}{nL} \sum_{i=1}^{n} \sum_{j=1}^{L} (x_{ij}^k - \mu^k)^2 } \tag{10}
$$

where μ_j^k is the mean of the measures placed at the same column, that is, the mean of the n measures of sensor k in matrix $\mathbf{X}^k$ at time instants $((i-1)L + (j-1))\,\Delta$ seconds, $i = 1, \ldots, n$; μ^k is the mean of all the elements in matrix $\mathbf{X}^k$, that is, the mean of all the measures of sensor k; and σ^k is the standard deviation of all the measures of sensor k. Therefore, the elements x_{ij}^k of matrix $\mathbf{X}$ are scaled to define a new matrix $\check{\mathbf{X}}$ as

$$
\check{x}_{ij}^k := \frac{x_{ij}^k - \mu_j^k}{\sigma^k}, \; i = 1, \ldots, n, \; j = 1, \ldots, L, \; k = 1, \ldots, N. \tag{11}
$$

When the data are normalized using Equation (11), the scaling procedure is called variable scaling or group scaling [25].

For the sake of clarity, and throughout the rest of the paper, the scaled matrix $\check{\mathbf{X}}$ is renamed as simply $\mathbf{X}$. The mean of each column vector in the scaled matrix $\mathbf{X}$ can be computed as

$$\frac{1}{n}\sum_{i=1}^{n}\check{x}_{ij}^{k} = \frac{1}{n}\sum_{i=1}^{n}\frac{x_{ij}^{k}-\mu_{j}^{k}}{\sigma^{k}} = \frac{1}{n\sigma^{k}}\sum_{i=1}^{n}\left(x_{ij}^{k}-\mu_{j}^{k}\right) \tag{12}$$

$$= \frac{1}{n\sigma^{k}}\left[\left(\sum_{i=1}^{n}x_{ij}^{k}\right)-n\mu_{j}^{k}\right] \tag{13}$$

$$= \frac{1}{n\sigma^{k}}\left(n\mu_{j}^{k}-n\mu_{j}^{k}\right) = 0 \tag{14}$$

Since the scaled matrix $\mathbf{X}$ is a mean-centered matrix, it is possible to calculate its covariance matrix as follows:

$$\mathbf{C_X} = \frac{1}{n-1}\mathbf{X}^{T}\mathbf{X} \in \mathcal{M}_{(N\cdot L)\times(N\cdot L)}(\mathbb{R}) \tag{15}$$

The covariance matrix $\mathbf{C_X}$ is a $(N \cdot L) \times (N \cdot L)$ symmetric matrix that measures the degree of linear relationship within the data set between all possible pairs of columns. At this point it is worth noting that each column can be viewed as a virtual sensor and, therefore, each column vector $\mathbf{X}(:,j) \in \mathbb{R}^{n}$, $j = 1,\ldots,N \cdot L$, represents a set of measurements from one virtual sensor.

The subspaces in PCA are defined by the eigenvectors and eigenvalues of the covariance matrix as follows:

$$\mathbf{C_X}\mathbf{P} = \mathbf{P}\Lambda \tag{16}$$

where the columns of $\mathbf{P} \in \mathcal{M}_{(N\cdot L)\times(N\cdot L)}(\mathbb{R})$ are the eigenvectors of $\mathbf{C_X}$. The diagonal terms of matrix $\Lambda \in \mathcal{M}_{(N\cdot L)\times(N\cdot L)}(\mathbb{R})$ are the eigenvalues λ_i, $i = 1,\ldots,N \cdot L$, of $\mathbf{C_X}$ whereas the off-diagonal terms are zero, that is,

$$\Lambda_{ii} = \lambda_{i},\ i = 1,\ldots,N \cdot L \tag{17}$$

$$\Lambda_{ij} = 0,\ i,j = 1,\ldots,N \cdot L,\ i \neq j \tag{18}$$

The eigenvectors p_j, $j = 1,\ldots,N \cdot L$, representing the columns of the transformation matrix $\mathbf{P}$ are classified according to the eigenvalues in descending order and they are called the principal components or the loading vectors of the data set. The eigenvector with the highest eigenvalue, called the first principal component, represents the most important pattern in the data with the largest quantity of information.

Matrix $\mathbf{P}$ is usually called the principal components of the data set or loading matrix and matrix $\mathbf{T}$ is the transformed or projected matrix to the principal component space, also called score matrix. Using all the $N \cdot L$ principal components, that is, in the full dimensional case, the orthogonality of $\mathbf{P}$ implies $\mathbf{P}\mathbf{P}^{\mathrm{T}} = \mathbf{I}$, where $\mathbf{I}$ is the $(N \cdot L) \times (N \cdot L)$ identity matrix. Therefore, the projection can be inverted to recover the original data as

$$\mathbf{X} = \mathbf{T}\mathbf{P}^{T} \tag{19}$$

However, the objective of PCA is, as said before, to reduce the dimensionality of the data set $\mathbf{X}$ by selecting only a limited number $\ell < N \cdot L$ of principal components, that is, only the eigenvectors related to the ℓ highest eigenvalues. Thus, given the reduced matrix

$$\hat{\mathbf{P}} = (p_1|p_2|\cdots|p_\ell) \in \mathcal{M}_{N\cdot L\times\ell}(\mathbb{R}) \tag{20}$$

matrix $\hat{\mathbf{T}}$ is defined as

$$\hat{\mathbf{T}} = \mathbf{X}\hat{\mathbf{P}} \in \mathcal{M}_{n\times\ell}(\mathbb{R}) \tag{21}$$

Note that opposite to $\mathbf{T}$, $\hat{\mathbf{T}}$ is no longer invertible. Consequently, it is not possible to fully recover $\mathbf{X}$ although $\hat{\mathbf{T}}$ can be projected back onto the original $N \cdot L$−dimensional space to get a data matrix $\hat{\mathbf{X}}$ as follows:

$$\hat{\mathbf{X}} = \hat{\mathbf{T}}\hat{\mathbf{P}}^T \in \mathcal{M}_{n \times (N \cdot L)}(\mathbb{R}) \tag{22}$$

The difference between the original data matrix $\mathbf{X}$ and $\hat{\mathbf{X}}$ is defined as the residual error matrix $\mathbf{E}$ or $\tilde{\mathbf{X}}$ as follows:

$$\mathbf{E} = \mathbf{X} - \hat{\mathbf{X}} \tag{23}$$

or, equivalenty,

$$\mathbf{X} = \hat{\mathbf{X}} + \mathbf{E} = \hat{\mathbf{T}}\hat{\mathbf{P}}^T + \mathbf{E} \tag{24}$$

The residual error matrix $\mathbf{E}$ describes the variability not represented by the data matrix $\hat{\mathbf{X}}$, and can also be expressed as

$$\mathbf{E} = \mathbf{X}(\mathbf{I} - \hat{\mathbf{P}}\hat{\mathbf{P}}^T) \tag{25}$$

Even though the real measures obtained from the sensors as a function of time represent physical magnitudes, when these measures are projected and the scores are obtained, these scores no longer represent any physical magnitude [26]. The key aspect in this approach is that the scores from different experiments can be compared with the reference pattern to try to detect a different behavior.

3.2. Fault Detection Based on Hypothesis Testing

The current wind turbine to diagnose is subjected to a wind turbulence as described in Sections 2 and 3.1. When the measures are obtained from $N \in \mathbb{N}$ sensors during $(\nu L - 1)\Delta$ seconds, a new data matrix $\mathbf{Y}$ is constructed as in Equation (6):

$$\mathbf{Y} = \begin{pmatrix} y_{11}^1 & y_{12}^1 & \cdots & y_{1L}^1 & y_{11}^2 & \cdots & y_{1L}^2 & \cdots & y_{11}^N & \cdots & y_{1L}^N \\ \vdots & \vdots & \ddots & \vdots & \vdots & \ddots & \vdots & \ddots & \vdots & \ddots & \vdots \\ y_{i1}^1 & y_{i2}^1 & \cdots & y_{iL}^1 & y_{i1}^2 & \cdots & y_{iL}^2 & \cdots & y_{i1}^N & \cdots & y_{iL}^N \\ \vdots & \vdots & \ddots & \vdots & \vdots & \ddots & \vdots & \ddots & \vdots & \ddots & \vdots \\ y_{\nu 1}^1 & y_{\nu 2}^1 & \cdots & y_{\nu L}^1 & y_{\nu 1}^2 & \cdots & y_{\nu L}^2 & \cdots & y_{\nu 1}^N & \cdots & y_{\nu L}^N \end{pmatrix} \in \mathcal{M}_{\nu \times (N \cdot L)}(\mathbb{R}) \tag{26}$$

It is worth remarking that the natural number ν (the number of rows of matrix $\mathbf{Y}$) is not necessarily equal to n (the number of rows of $\mathbf{X}$), but the number of columns of $\mathbf{Y}$ must agree with that of $\mathbf{X}$; that is, in both cases the number N of sensors and the number of samples per row must be equal.

Before the collected data arranged in matrix $\mathbf{Y}$ is projected into the new space spanned by the eigenvectors in matrix $\mathbf{P}$ in Equation (16), the matrix has to be scaled to define a new matrix $\check{\mathbf{Y}}$ as in Equation (11):

$$\check{y}_{ij}^k := \frac{y_{ij}^k - \mu_j^k}{\sigma^k}, \ i = 1, \ldots, \nu, \ j = 1, \ldots, L, \ k = 1, \ldots, N, \tag{27}$$

where μ_j^k and σ^k are defined in Equations (8) and (10), respectively.

The projection of each row vector $r^i = \check{\mathbf{Y}}(i,:) \in \mathbb{R}^{N \cdot L}, i = 1, \ldots, \nu$ of matrix $\check{\mathbf{Y}}$ into the space spanned by the eigenvectors in $\hat{\mathbf{P}}$ is performed through the following vector to matrix multiplication:

$$t^i = r^i \cdot \hat{\mathbf{P}} \in \mathbb{R}^\ell \tag{28}$$

For each row vector $r^i, i = 1, \ldots, \nu$, the first component of vector t^i is called the first score or score 1; similarly, the second component of vector t^i is called the second score or score 2, and so on.

In a standard application of the principal component analysis strategy in the field of structural health monitoring, the scores allow a visual grouping or separation [27]. In some other cases, as in [28], two classical indices can be used for damage detection, such as the Q index (also known as SPE, square prediction error) and the Hotelling's T^2 index. The Q index of the ith row y_i^T of matrix $\check{\mathbf{Y}}$ is defined as follows:

$$Q_i = y_i^T (\mathbf{I} - \hat{\mathbf{P}}\hat{\mathbf{P}}^T) y_i \tag{29}$$

The T^2 index of the ith row y_i^T of matrix $\check{\mathbf{Y}}$ is defined as follows:

$$T_i^2 = y_i^T (\hat{\mathbf{P}}\Lambda^{-1}\hat{\mathbf{P}}^T) y_i \tag{30}$$

In this case, however, it can be observed in Figure 4—where the projection onto the two first principal components of samples coming from the healthy and faulty wind turbines are plotted—that a visual grouping, clustering or separation cannot be performed. A similar conclusion is deducted from Figure 5. In this case, the plot of the natural logarithm of indices Q and T^2—defined in Equations (29) and (30)—of samples coming from the healthy and faulty wind turbines does not allow any visual grouping. Therefore, a more powerful and reliable tool is needed to be able to detect a fault in the wind turbine.

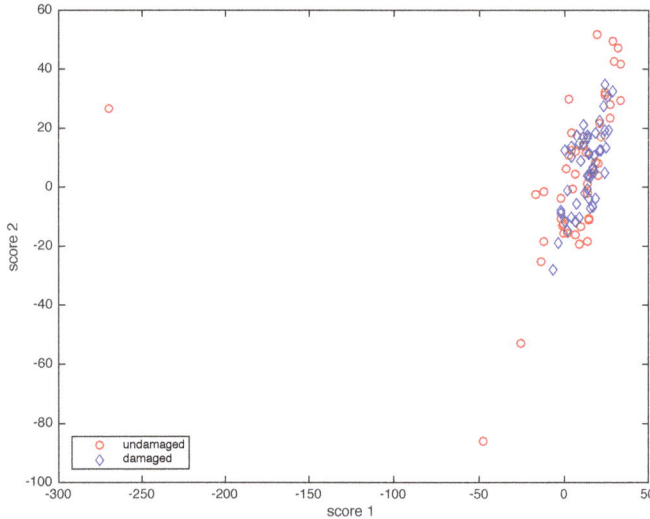

Figure 4. Projection onto the two first principal components of samples coming from the healthy wind turbine (red, circle) and from the faulty wind turbine (blue, diamond).

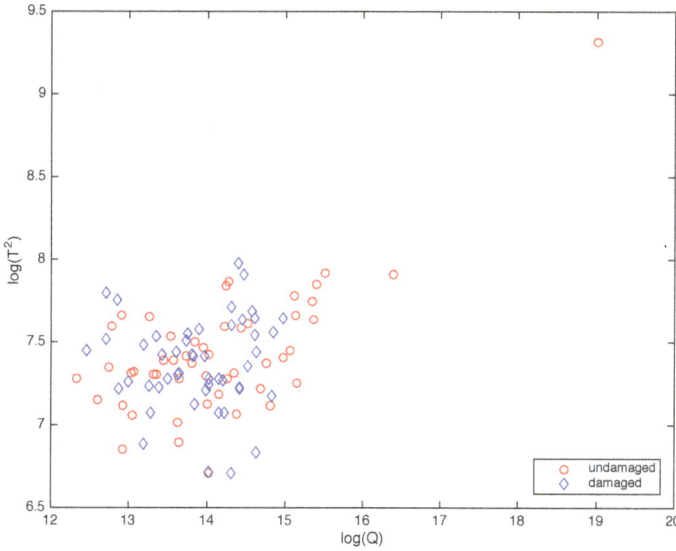

Figure 5. Natural logarithm of indices Q and T^2 of samples coming from the healthy wind turbine (red, circle) and from the faulty wind turbine (blue, diamond).

3.2.1. The Random Nature of the Scores

Since the turbulent wind can be considered as a random process, the dynamic response of the wind turbine can be considered as a stochastic process and the measurements in r^i are also stochastic. Therefore, each component of t^i acquires this stochastic nature and it will be regarded as a random variable to construct the stochastic approach in this paper.

3.2.2. Test for the Equality of Means

The objective of the present work is to examine whether the current wind turbine is healthy or subjected to a fault as those described in Table 2. To achieve this end, we have a PCA model (matrix $\hat{\mathbf{P}}$ in Equation (20)) built as in Section 3.1 with data coming from a wind turbine in a full healthy state. For each principal component $j = 1, \ldots, \ell$, the baseline sample is defined as the set of n real numbers computed as the j–th component of the vector to matrix multiplication $\mathbf{X}(i, :) \cdot \hat{\mathbf{P}}$. Note that n is the number of rows of matrix $\mathbf{X}$ in Equation (6). That is, we define the baseline sample as the set of numbers $\{\tau_j^i\}_{i=1,\ldots,n}$ given by

$$\tau_j^i := (\mathbf{X}(i, :) \cdot \hat{\mathbf{P}})(j) = \mathbf{X}(i, :) \cdot \hat{\mathbf{P}} \cdot \mathbf{e}_j, \ i = 1, \ldots, n, \tag{31}$$

where $\mathbf{e}_j$ is the j–th vector of the canonical basis.

Similarly, and for each principal component $j = 1, \ldots, \ell$, the sample of the current wind turbine to diagnose is defined as the set of v real numbers computed as the j–th component of the vector t^i in Equation (28). Note that v is the number of rows of matrix $\mathbf{Y}$ in Equation (26). That is, we define the sample to diagnose as the set of numbers $\{t_j^i\}_{i=1,\ldots,v}$ given by

$$t_j^i := t^i \cdot \mathbf{e}_j, \ i = 1, \ldots, v. \tag{32}$$

As said before, the goal of this paper is to obtain a fault detection method such that when the distribution of the current sample is related to the distribution of the baseline sample a healthy state is predicted and otherwise a fault is detected. To that end, a test for the equality of means will be performed. Let us consider that, for a given principal component, (a) the baseline sample is a random sample of a random variable having a normal distribution with unknown mean μ_X and unknown standard deviation σ_X; and (b) the random sample of the current wind turbine is also normally distributed with unknown mean μ_Y and unknown standard deviation σ_Y. Let us finally consider that the variances of these two samples are not necessarily equal. As said previously, the problem that we will consider is to determine whether these means are equal, that is, $\mu_X = \mu_Y$, or equivalently, $\mu_X - \mu_Y = 0$. This statement leads immediately to a test of the hypotheses

$$H_0 : \mu_X - \mu_Y = 0 \text{ versus} \tag{33}$$
$$H_1 : \mu_X - \mu_Y \neq 0 \tag{34}$$

that is, the null hypothesis is "the sample of the wind turbine to be diagnosed is distributed as the baseline sample" and the alternative hypothesis is "the sample of the wind turbine to be diagnosed is not distributed as the baseline sample". In other words, if the result of the test is that the null hypothesis is not rejected, the current wind turbine is categorized as healthy. Otherwise, if the null hypothesis is rejected in favor of the alternative, this would indicate the presence of some faults in the wind turbine.

The test is based on the Welch-Satterthwaite method [29], which is outlined below. When random samples of size n and v, respectively, are taken from two normal distributions $N(\mu_X, \sigma_X)$ and $N(\mu_Y, \sigma_Y)$ and the population variances are unknown, the random variable

$$W = \frac{(\bar{X} - \bar{Y}) + (\mu_X - \mu_Y)}{\sqrt{\left(\dfrac{S_X^2}{n} + \dfrac{S_Y^2}{v}\right)}} \tag{35}$$

can be approximated with a t-distribution with ρ degrees of freedom, that is

$$W \hookrightarrow t_\rho \tag{36}$$

where

$$\rho = \left\lfloor \frac{\left(\dfrac{s_X^2}{n} + \dfrac{s_Y^2}{v}\right)^2}{\dfrac{(s_X^2/n)^2}{n-1} + \dfrac{(s_Y^2/v)^2}{v-1}} \right\rfloor \tag{37}$$

and where $\bar{X}, \bar{Y}$ is the sample mean as a random variable; S^2 is the sample variance as a random variable; s^2 is the variance of a sample; and $\lfloor \cdot \rfloor$ is the floor function.

The value of the standardized test statistic using this method is defined as

$$t_{\text{obs}} = \frac{\bar{x} - \bar{y}}{\sqrt{\left(\dfrac{s_X^2}{n} + \dfrac{s_Y^2}{v}\right)}} \tag{38}$$

where $\bar{x}, \bar{y}$ is the mean of a particular sample. The quantity t_{obs} is the fault indicator. We can then construct the following test:

$$|t_{\text{obs}}| \leq t^* \implies \text{Fail to reject } H_0 \tag{39}$$

$$|t_{\text{obs}}| > t^* \implies \text{Reject } H_0 \tag{40}$$

where t^* is such that

$$P\left(t_\rho < t^*\right) = 1 - \frac{\alpha}{2} \tag{41}$$

and α is the chosen risk (significance) level for the test. More precisely, the null hypothesis is rejected if $|t_{\text{obs}}| > t^*$ (this would indicate the existence of a fault in the wind turbine). Otherwise, if $|t_{\text{obs}}| \leq t^*$ there is no statistical evidence to suggest that both samples are normally distributed but with different means, thus indicating that no fault in the wind turbine has been found. This idea is represented in Figure 6.

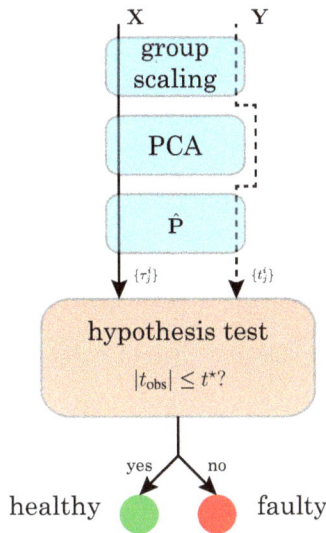

Figure 6. Fault detection will be based on testing for significant changes in the distributions of the baseline sample and the sample coming from the wind turbine to diagnose.

4. Simulation Results

4.1. Type I and Type II errors

To validate the fault detection strategy presented in Section 3, we first consider a total of 24 samples of $\nu = 50$ elements each, according to the following distribution:

- 16 samples of a healthy wind turbine; and
- 8 samples of a faulty wind turbine with respect to each of the eight different fault scenarios described in Table 2.

In the numerical simulations in this Section, each sample of $\nu = 50$ elements is composed by the measures obtained from the $N = 13$ sensors detailed in Table 4 during $(\nu \cdot L - 1)\Delta = 312.4875$ seconds, where $L = 500$ and the sampling time $\Delta = 0.0125$ seconds. The measures of each sample are then arranged in a $v \times (N \cdot L)$ matrix as in Equation (26).

For the first four principal components (score 1 to score 4), these 24 samples plus the baseline sample of $n = 50$ elements are used to test for the equality of means, with a level of significance $\alpha = 0.36$ (the choice of this level of significance will be justified in Section 4.2). Each sample of $v = 50$ elements is categorized as follows: (i) number of samples from the healthy wind turbine (healthy sample) which were classified by the hypothesis test as "healthy" (fail to reject H_0); (ii) faulty sample classified by the test as "faulty" (reject H_0); (iii) samples from the faulty structure (faulty sample) classified as "healthy"; and (iv) faulty sample classified as "faulty". The results for the first four principal components presented in Table 6 are organized according to the scheme in Table 5. It can be stressed from each principal component in Table 6 that the sum of the columns is constant: 16 samples in the first column (healthy wind turbine) and 8 more samples in the second column (faulty wind turbine).

Table 5. Scheme for the presentation of the results in Table 6.

	Undamaged Sample (H_0)	Damaged Sample (H_1)
Fail to reject H_0	Correct decision	Type II error (missing fault)
Reject H_0	Type I error (false alarm)	Correct decision

Table 6. Categorization of the samples with respect to the presence or absence of damage and the result of the test for each of the four scores when the size of the samples to diagnose is $v = 50$.

	score 1		score 2		score 3		score 4	
	H_0	H_1	H_0	H_1	H_0	H_1	H_0	H_1
Fail to reject H_0	16	0	12	1	11	5	9	1
Reject H_0	0	8	4	7	5	3	7	7

In Table 6, it is worth noting that two kinds of misclassification are presented which are denoted as follows:

1. Type I error (false positive or false alarm), when the wind turbine is healthy but the null hypothesis is rejected and therefore classified as faulty. The probability of committing a type I error is α, the level of significance.
2. Type II error (false negative or missing fault), when the structure is faulty but the null hypothesis is not rejected and therefore classified as healthy. The probability of committing a type II error is called γ.

It can be observed from Table 6 that, in the numerical simulations, Type I errors (false alarms) and Type II errors (missing faults) appear only when scores 2, 3 or 4 are considered, while when the first score is used all the decisions are correct. The better performance of the first score is an expected result in the sense that the first principal component is the component that accounts for the largest possible variance.

4.2. Sensitivity and Specificity

Two more statistical measures can be selected here to study the performance of the test: *the sensitivity* and *the specificity*. The sensitivity, also called as the power of the test, is defined, in the context of this work, as the proportion of samples from the faulty wind turbine which are correctly identified as such. Thus, the sensitivity can be computed as $1 - \gamma$. The specificity of the test is defined, also in this context, as the proportion of samples from the healthy structure that are correctly identified and can be expressed as $1 - \alpha$.

The sensitivity and the specificity of the test with respect to the 24 samples and for each of the first four principal components –organized as shown in Table 7– have been included in Table 8.

Table 7. Relationship between type I and type II errors.

	Undamaged Sample (H_0)	Damaged Sample (H_1)
Fail to reject H_0	Specificity $(1 - \alpha)$	False negative rate (γ)
Reject H_0	False positive rate (α)	Sensitivity $(1 - \gamma)$

Table 8. Sensitivity and specificity of the test for each of the four scores when the size of the samples to diagnose is $\nu = 50$.

	score 1 H_0	score 1 H_1	score 2 H_0	score 2 H_1	score 3 H_0	score 3 H_1	score 4 H_0	score 4 H_1
Fail to reject H_0	1.00	0.00	0.75	0.13	0.69	0.62	0.56	0.13
Reject H_0	0.00	1.00	0.25	0.87	0.31	0.38	0.44	0.87

It is worth mentioning that type I errors are frequently considered to be more serious than type II errors. However, in this application, a type II error is related to a *missing fault* whereas a type I error is related to a *false alarm*. In consequence, type II errors should be reduced. Therefore a small level of significance of 1%, 5% or even 10% would lead to a reduced number of *false alarms* but to a higher rate of *missing faults*. That is the reason of the choice of a level of significance of 36% in the hypothesis test.

The results in Table 8 show that the sensitivity of the test $1 - \gamma$ is close to 100%, as desired, with an average value of 78.00%. The sensitivity with respect to the first, second and fourth principal component is increased, in mean, to a 91.33%. The average value of the specificity is 75.00%, which is very close to the expected value of $1 - \alpha = 64\%$.

4.3. Reliability of the Results

Using the scheme in Table 9, the results are computed and given in Table 10. This table is based on the Bayes' theorem [30], where $P(H_1|\text{accept } H_0)$ is the proportion of samples from the faulty wind turbine that have been incorrectly classified as healthy (*true rate of false negatives*) and $P(H_0|\text{accept } H_1)$ is the proportion of samples from the healthy wind turbine that have been incorrectly classified as faulty (*true rate of false positives*).

Table 9. Relationship between the proportion of false negatives and false positives.

	Undamaged Sample (H_0)	Damaged Sample (H_1)
Fail to reject H_0	$P(H_0|\text{accept } H_0)$	true rate of false negatives $P(H_1|\text{accept } H_0)$
Reject H_0	true rate of false positives $P(H_0|\text{accept } H_1)$	$P(H_1|\text{accept } H_1)$

Table 10. True rate of false positives and false negatives for each of the four scores when the size of the samples to diagnose is $\nu = 50$.

	score 1 H_0	score 1 H_1	score 2 H_0	score 2 H_1	score 3 H_0	score 3 H_1	score 4 H_0	score 4 H_1
Fail to reject H_0	1.00	0.00	0.92	0.08	0.69	0.31	0.90	0.10
Reject H_0	0.00	1.00	0.36	0.64	0.62	0.38	0.50	0.50

4.4. The Receiver Operating Curves (ROC)

An additional study has been developed based on the ROC curves to determine the overall accuracy of the proposed method. These curves represent the trade-off between the false positive rate and the sensitivity in Table 10 for different values of the level of significance that is used in the statistical hypothesis testing. Note that the false positive rate is defined as the complementary of the specificity, and therefore these curves can also be used to visualize the close relationship between specificity and sensitivity. It can also be remarked that the sensitivity is also called true positive rate or probability of detection [31]. More precisely, for each principal component and for a given level of significance the pair of numbers

$$(\text{false positive rate}, \text{sensitivity}) \in [0,1] \times [0,1] \subset \mathbb{R}^2 \tag{42}$$

is plotted. We have considered 49 levels of significance within the range $[0.02, 0.98]$ and with a difference of 0.02. Therefore, for each of the first four principal components, 49 connected points are depicted, as can be seen in Figure 7.

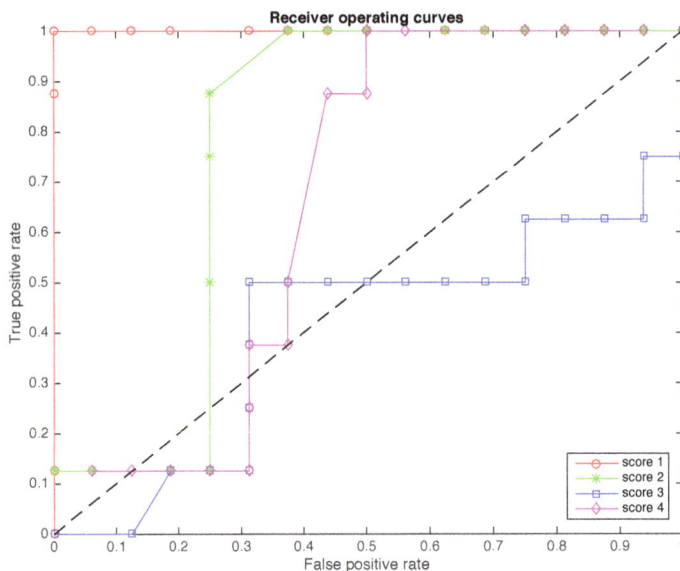

Figure 7. The Receiver Operating Curves (ROCs) for the four scores when the size of the samples to diagnose is $v = 50$.

The placement of these points can be interpreted as follows. Since we are interested in reducing the number of false positives while we increase the number of true positives, these points must be placed in the upper-left corner as much as possible. However, this is not always possible because there is also a relationship between the level of significance and the false positive rate. Therefore, a method can be considered acceptable if those points lie within the upper-left half-plane. In this sense, the results presented in Figure 7, particularly with respect to score 1, are quite remarkable. The overall behavior of scores 2 and 4 are also acceptable, while the results of score 3 cannot be considered, in this case, as satisfactory.

In Figures 8 and 9 a further study is performed. While in Figure 7 we present the ROCs when the size of the samples to diagnose is $v = 50$, in Figure 8 the reliability of the method is analyzed in terms of 48 samples of $v = 25$ elements each and in Figure 9 the reliability of the method is analyzed in terms of 120 samples of $\nu = 10$ elements each. The effect of reducing the number of elements in each sample is the reduction in the total time needed for a diagnostic. More precisely, if we keep $L = 500$, when the size of the samples is reduced from $\nu = 50$ to $\nu = 25$ and $\nu = 10$, the total time needed for a diagnostic is reduced from about 312 s to 156 and 62 s, respectively. Another effect of the reduction in the number of elements in each sample is a slight deterioration of the overall accuracy of the detection method. However, the results of scores 1 and 2 in Figures 8 and 9 are perfectly acceptable.

A very interesting alternative to keep a very good performance of the method without almost no degradation in its accuracy is by reducing L –the number of time instants per row per sensor— instead of reducing the number of elements per sample ν. This way, if we keep $\nu = 50$, when L is reduced from 500 to 50, the total time needed for a diagnostic is reduced from about 312 s to 31 s.

We can finally say that the ROC curves provide a statistical assessment of the efficacy of a method and can be used to visualize and compare the performance of multiple scenarios.

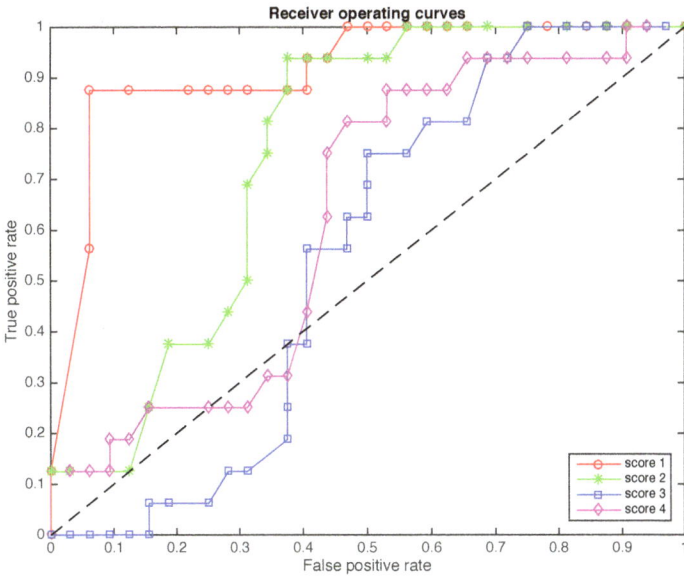

Figure 8. The ROCs for the four scores when the size of the samples to diagnose is $\nu = 25$.

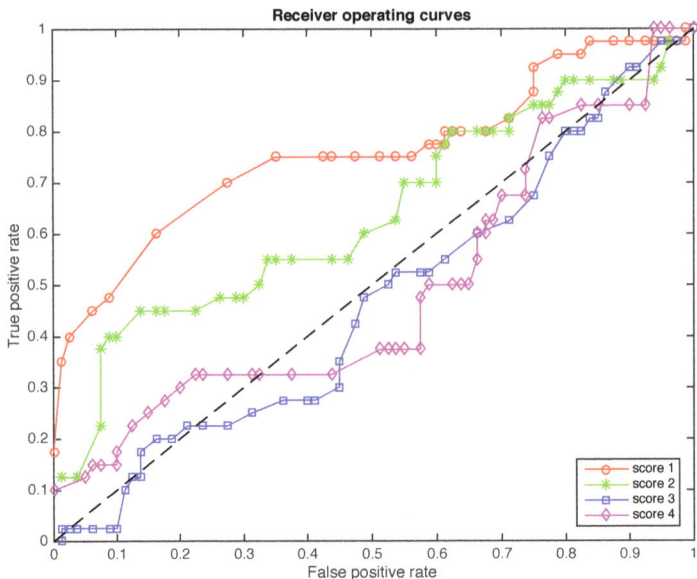

Figure 9. The ROCs for the four scores when the size of the samples to diagnose is $\nu = 10$.

5. Conclusions

The silver bullet for offshore operators is to eliminate unscheduled maintenance. Therefore, the implementation of fault detection systems is crucial. The main challenges of the wind turbine fault detection lie in its nonlinearity, unknown disturbances as well as significant measurement noise. In this work, numerical simulations (with a well-known benchmark wind turbine) show that the proposed PCA plus statistical hypothesis testing is a valuable tool in fault detection for wind turbines. It is noteworthy that, in the simulations, when the first score is used all the decisions are correct (there are no false alarms and no missing faults).

We believe that PCA plus statistical hypothesis testing has tremendous potential in decreasing maintenance costs. Therefore, we view the work described in this paper as only the beginning of a large project. For future work, we plan to develop a complete fault detection, isolation, and reconfiguration method (FDIR). That is, a reconfigurable control strategy in response to faults. In the near future, the next step is to focus our research into efficient fault feature extraction.

Acknowledgments: This work has been partially funded by the Spanish Ministry of Economy and Competitiveness through the research projects DPI2011-28033-C03-01, DPI2014-58427-C2-1-R, and by the Generalitat de Catalunya through the research project 2014 SGR 859.

Author Contributions: All authors contributed equally.

Conflicts of Interest: The authors declare no conflict of interest.

Nomenclature

β	Pitch angle
β_r	Pitch angle reference
η_g	Generator efficiency
ω_g	Generator speed
P_e	Electrical Power
τ_c	Reference generator torque
τ_r	Real generator torque
α	Significance level for the test (probability of committing a type I error)
γ	Probability of committing a type II error
L	Number of time instants per row per sensor
N	Number of sensors
ν	Size of the samples to diagnose
P	Principal components of the data set (loading matrix)
T	Transformed (or projected) matrix to the principal component space (score matrix)
E	Residual error matrix
X	Data matrix (original)
Y	Data matrix to diagnose
$\{\tau_j^i\}_{i=1,...,n}$	Baseline sample
$\{t_j^i\}_{i=1,...,\nu}$	Sample to diagnose

References

1. Sinha, Y.; Steel, J. A progressive study into offshore wind farm maintenance optimisation using risk based failure analysis. *Renew. Sustain. Energy Rev.* **2015**, *42*, 735–742.
2. Odgaard, P.; Johnson, K. Wind turbine fault diagnosis and fault tolerant control—An enhanced benchmark challenge. In Proceedings of the 2013 American Control Conference (ACC), Washington, DC, USA, 17–19 June 2013; pp. 1–6.
3. Soman, R.N.; Malinowski, P.H.; Ostachowicz, W.M. Bi-axial neutral axis tracking for damage detection in wind-turbine towers. *Wind Energy* **2015**, doi:10.1002/we.1856.
4. Adams, D.; White, J.; Rumsey, M.; Farrar, C. Structural health monitoring of wind turbines: Method and application to a HAWT. *Wind Energy* **2011**, *14*, 603–623.
5. Griffith, D.T.; Yoder, N.C.; Resor, B.; White, J.; Paquette, J. Structural health and prognostics management for the enhancement of offshore wind turbine operations and maintenance strategies. *Wind Energy* **2014**, *17*, 1737–1751.
6. Ding, S.X. *Model-Based Fault Diagnosis Techniques: Design Schemes, Algorithms, and Tools*; Springer Science & Business Media: London, UK, 2008.
7. Odgaard, P.F.; Stoustrup, J.; Nielsen, R.; Damgaard, C. Observer based detection of sensor faults in wind turbines. In Proceedings of the European Wind Energy Conference, Marseille, France, 16–19 March 2009; pp. 4421–4430.
8. Zhang, X.; Zhang, Q.; Zhao, S.; Ferrari, R.M.; Polycarpou, M.M.; Parisini, T. Fault detection and isolation of the wind turbine benchmark: An estimation-based approach. In Proceedings of the International Federation of Automatic Control (IFAC) World Congress, Milano, Italy, 28 August–2 September 2011; Volume 2, pp. 8295–8300.
9. Shaker, M.S.; Patton, R.J. Active sensor fault tolerant output feedback tracking control for wind turbine systems via T-S model. *Eng. Appl. Artif. Intell.* **2014**, *34*, 1–12.
10. Shi, F.; Patton, R. An active fault tolerant control approach to an offshore wind turbine model. *Renew. Energy* **2015**, *75*, 788–798.
11. Vidal, Y.; Tutiven, C.; Rodellar, J.; Acho, L. Fault diagnosis and fault-tolerant control of wind turbines via a discrete time controller with a disturbance compensator. *Energies* **2015**, *8*, 4300–4316.
12. Dong, J.; Verhaegen, M. Data driven fault detection and isolation of a wind turbine benchmark. In Proceedings of the International Federation of Automatic Control (IFAC) World Congress, Milano, Italy, 28 August–2 September 2011; Volume 2, pp. 7086–7091.

13. Simani, S.; Castaldi, P.; Tilli, A. Data-driven approach for wind turbine actuator and sensor fault detection and isolation. In Proceedings of the International Federation of Automatic Control (IFAC) World Congress, Milano, Italy, 28 August–2 September 2011; pp. 8301–8306.

14. Laouti, N.; Sheibat-Othman, N.; Othman, S. Support vector machines for fault detection in wind turbines. In Proceedings of International Federation of Automatic Control (IFAC) World Congress, Milano, Italy, 28 August–2 September 2011; Volume 2, pp. 7067–7072.

15. Stoican, F.; Raduinea, C.F.; Olaru, S.; others. Adaptation of set theoretic methods to the fault detection of wind turbine benchmark. In Proceedings of the International Federation of Automatic Control (IFAC) World Congress, Milano, Italy, 28 August–2 September 2011; pp. 8322–8327.

16. Odgaard, P.F.; Stoustrup, J. Gear-box fault detection using time-frequency based methods. *Annu. Rev. Control* **2015**, *40*, 50–58.

17. Pang-Ning, T.; Steinbach, M.; Kumar, V. *Introduction to Data Mining*; Pearson Addison Wesley: Boston, MA, USA, 2005.

18. Zaher, A.; McArthur, S.; Infield, D.; Patel, Y. Online wind turbine fault detection through automated SCADA data analysis. *Wind Energy* **2009**, *12*, 574, doi:10.1002/we.319.

19. Kusiak, A.; Li, W.; Song, Z. Dynamic control of wind turbines. *Renew. Energy* **2010**, *35*, 456–463.

20. Jonkman, J. NWTC Information Portal (FAST). Available online: https://nwtc.nrel.gov/FAST (accessed on 18 December 2015).

21. Jonkman, J.M.; Butterfield, S.; Musial, W.; Scott, G. *Definition of a 5-MW Reference Wind Turbine for Offshore System Development*. National Renewable Energy Laboratory, Golden, CO, USA, 2009.

22. Kelley, N.; Jonkman, B. NWTC Information Portal (Turbsim). Available online: https://nwtc.nrel.gov/TurbSim (accessed on 18 December 2015).

23. Ostachowicz, W.; Kudela, P.; Krawczuk, M.; Zak, A. *Guided Waves in Structures for SHM: The Time-Domain Spectral Element Method*; John Wiley & Sons, Ltd: Hoboken, NJ, USA, 2012.

24. Anaya, M.; Tibaduiza, D.; Pozo, F. A bioinspired methodology based on an artificial immune system for damage detection in structural health monitoring. *Shock Vibration* **2015**, *2015*, 1–15.

25. Anaya, M.; Tibaduiza, D.; Pozo, F. Detection and classification of structural changes using artificial immune systems and fuzzy clustering. *Int. J. Bio-Inspired Comput.* in press.

26. Mujica, L.E.; Ruiz, M.; Pozo, F.; Rodellar, J.; Güemes, A. A structural damage detection indicator based on principal component analysis and statistical hypothesis testing. *Smart Mater. Struct.* **2014**, *23*, 1–12.

27. Mujica, L.E.; Rodellar, J.; Fernández, A.; Güemes, A. Q-statistic and T^2-statistic PCA-based measures for damage assessment in structures. *Struct. Health Monit.* **2011**, *10*, 539–553.

28. Odgaard, P.F.; Lin, B.; Jorgensen, S.B. Observer and data-driven-model-based fault detection in power plant coal mills. *IEEE Trans. Energy Convers.* **2008**, *23*, 659–668.

29. Ugarte, M.D.; Militino, A.F.; Arnholt, A. *Probability and Statistics with R*; Chemical Rubber Company (CRC) Press: Boca Raton, FL, USA, 2008.

30. De Groot, M.H.; Schervish, M.J. *Probability and Statistics*; Pearson: Boston, MA, USA, 2012.

31. Yinghui, L.; Michaels, J.E. Feature extraction and sensor fusion for ultrasonic structural health monitoring under changing environmental conditions. *IEEE Sens. J.* **2009**, *9*, 1462–1471.

energies

MDPI

Article

Representational Learning for Fault Diagnosis of Wind Turbine Equipment: A Multi-Layered Extreme Learning Machines Approach †

Zhi-Xin Yang *,‡, Xian-Bo Wang ‡ and Jian-Hua Zhong

Department of Electromechanical Engineering, Faculty of Science and Technology, University of Macau, Macau SAR 999078, China; xb_wang@live.com (X.-B.W.); zjheme@gmail.com (J.-H.Z.)

* Correspondence: zxyang@umac.mo; Tel.: +853-8822-4456

† This paper is an extended version of our paper published in ELM Based Representational Learning for Fault Diagnosis of Wind Turbine Equipment. In Proceedings of the Extreme Learning Machines 2015, Hangzhou, China, 15–17 December 2015.

‡ These authors contributed equally to this work.

Academic Editor: Lance Manuel
Received: 18 March 2016; Accepted: 13 May 2016; Published: 24 May 2016

Abstract: Reliable and quick response fault diagnosis is crucial for the wind turbine generator system (WTGS) to avoid unplanned interruption and to reduce the maintenance cost. However, the conditional data generated from WTGS operating in a tough environment is always dynamical and high-dimensional. To address these challenges, we propose a new fault diagnosis scheme which is composed of multiple extreme learning machines (ELM) in a hierarchical structure, where a forwarding list of ELM layers is concatenated and each of them is processed independently for its corresponding role. The framework enables both representational feature learning and fault classification. The multi-layered ELM based representational learning covers functions including data preprocessing, feature extraction and dimension reduction. An ELM based autoencoder is trained to generate a hidden layer output weight matrix, which is then used to transform the input dataset into a new feature representation. Compared with the traditional feature extraction methods which may empirically wipe off some "insignificant' feature information that in fact conveys certain undiscovered important knowledge, the introduced representational learning method could overcome the loss of information content. The computed output weight matrix projects the high dimensional input vector into a compressed and orthogonally weighted distribution. The last single layer of ELM is applied for fault classification. Unlike the greedy layer wise learning method adopted in back propagation based deep learning (DL), the proposed framework does not need iterative fine-tuning of parameters. To evaluate its experimental performance, comparison tests are carried out on a wind turbine generator simulator. The results show that the proposed diagnostic framework achieves the best performance among the compared approaches in terms of accuracy and efficiency in multiple faults detection of wind turbines.

Keywords: fault diagnosis; wind turbine; classification; extreme learning machines (ELM); autoencoder (AE)

---

1. Introduction

Wind turbine generator systems (WTGS) are the fastest-growing applications in renewable power industry. The structure of WTGS is complex, its reliability becomes an important issue. As wind power generators are widely mounted on high mountains or offshore islands, it is costly for routine maintenance [1]. Continuously condition monitoring and fault diagnosis technologies are therefore necessary so as to reduce unnecessary maintenance cost and keep system working reliably without unexpected shutdown. A typical WTGS includes gearbox, power generator, control cabinet and rotary motor, *etc.*, (as shown in Figure 1a), where the gearbox is statistically more vulnerable compared with other components. It shed light upon the importance of the condition monitoring of gearbox. More specifically, the faults of gearbox mainly results from two major components: gears and bearings, which include broken tooth, chipped tooth, wear-off of outer race or rolling elements of bearing, *etc.* [2]. Real-time monitoring and fault diagnosis aim to detect and identify any potential abnormalities and faults, so as to take corresponding actions to avoid serious component damage or system disaster.

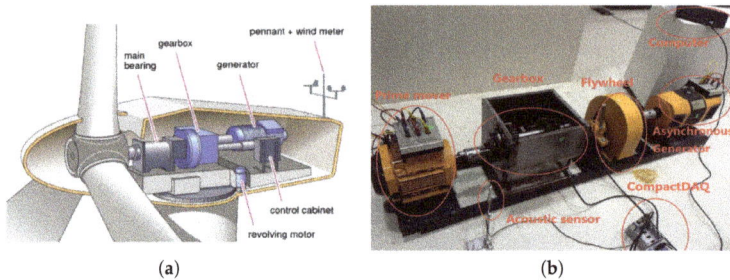

(a) (b)

Figure 1. Diagram and fault simulator for wind turbine generator system (WTGS). (a) The diagram of WTGS; (b) the simulation platform of WTGS.

Nowadays, a large body of research shows that fault detection based on a machine learning-based approach is feasible. Machine learning methods, such as neural networks (NNs), support vector machine (SVM) and deep learning (DL) may be promising solutions to classify the normal and abnormal patterns. A brief workflow of machine learning methods for fault diagnosis includes analog signal acquisition, data pre-processing and pattern recognition. Regarding the feature information sources (e.g., vibration signals, acoustic and temperature signals), the vibration signals are often adopted for their ease of acquisition and sensitivity to a wide range of faults. Moreover, intelligent fault diagnosis of WTGS relies on the effectiveness of signal processing and classification methods. The raw vibration signals contain high-dimensional information and abundant noise (includes irrelevant and redundant signals), which cannot be feasibly fed into the fault diagnostic system directly [3]. Many studies focus on the improvement of data pre-processing and feature extraction from the raw vibration signals [4,5]. Generally, a good intermediate representation method is required to retain the information of its original input, while at the same time being consistent to a given form (*i.e.*, a real-valued vector of a given size in case of an autoencoder) [6]. Therefore, it is essential to extract the compact feature information from raw vibration signals. The data processing method simplifies the computational expense and benefits the improvement of the generation performance. Some typical feature extraction methods, such as wavelet packet transform (WPT) [7–10], empirical mode decomposition (EMD) [11], time-domain statistical features (TDSF) [12,13] and independent component analysis (ICA) [14–17] have been proved to be equivalent to a large-scale matrix factorization problem (*i.e.*, there may be still some irrelevant or redundant noise in the extracted features) [18]. In order to resolve this problem, a feature selection method could be employed to wipe off irrelevant and redundant information so that the dimension of extracted feature is reduced. Typical feature selection approaches include compensation distance evaluation technique (CDET) [19], principal component analysis (PCA) and kernel principal component analysis (KPCA) [16,20,21] and the genetic algorithm (GA) based methods [22–24]. However, these linear methods have a common shortcoming in an attempt to extract nonlinear characteristics, which may result in a weak performance in the downstream pattern recognition process.

The raw vibration signals obtained from WTGS are characterized with high dimensional and nonlinear patterns, which is difficult for direct classification. In order to extract the features from the raw vibration signals, this paper introduces the concept of autoencoder and explores its application. Unlike PCA and its variants, autoencoder does not impose the dictionary elements be orthogonal, which makes it flexible to be adapted to the fluctuation in data representation [18]. In the structure of autoencoder, each layer in the stack architecture can be treated as an independent module [25]. The procedure shows briefly as follows. Each layer is firstly trained to produce a new (hidden) representation of the observed patterns (input data). It optimizes a local supervised criterion based on its received input representation from forehead layer. Each layer L_i produces a new representation that is more abstract than the previous level L_{i-1} [6]. After representational learning for a feature mapping that produces a high level intermediate representations (e.g., a high-dimensional intermediate matrix) of the input pattern, whereas, it is still complex and hard to compute directly. Therefore, it is necessary to decode the high-dimensional representations into a relatively low-dimensional and simple representations. Currently, there are only very limited algorithms that could work well for this purpose: restricted boltzmann machines (RBMs) [26–28] trained with contrastive divergence on one hand, and various types of autoencoders on the other. Regarding algorithms for classification, artificial neural networks (ANN) and multi-layer perception (MLP), are widely used for fault diagnosis of rotating machinery [3,29,30]. However, MLP has inevitable drawbacks which mainly reflects in local minima, time-consuming and over-fitting. Additionally, most classifiers are designed for binary classification. Regarding multiclass classification, the common methods actually employ a combination of binary classifiers with one-*versus*-all (1va) or one-*versus*-one (1v1) strategies [3]. Obviously, the combination with many binary classifiers increases computational burden and training time. Nowadays, researches show that SVM works well at recognizing the rotating equipment faults [31,32]. Compared with early machine learning methods, global optimum and relatively high generalization performance are the obvious advantages of SVM, while it has the same demerits with MLP, namely, time-consuming and local minimal. Considering that more than one type of fault may co-exist at the same time, it may be significant to propose a classifier which could offer the probabilities of all possible faults. In order to realize this assumption, the probabilistic neural network (PNN) [33,34] is employed as a probabilistic classifier. It is testified that the performance of PNN is superior to the SVM based method [29]. It trained a probabilistic classifier with a model using the Bayesian framework. However, the work [29] failed to explain clearly the principle of decision threshold. The value of decision threshold depends on some specific validation datasets and is not generally applicable for other areas.

Recent studies show that extreme learning machines (ELM) has better scalability and achieves much faster learning speed than SVM [35,36]. From the structural point of view, ELM is a multi-input and multi-output or single-output structure with single-hidden layer feedforward networks (SLFNs). Thus, ELM algorithm is more appropriate to multiclass classification. This paper extends the capability of ELM to the scope of feature learning, and proposes a multi-layered ELM network for feature learning and fault diagnosis. This paper proposes an ELM based autoencoder for feature mapping, and then the new representations are fed into the ELM based classifier for multi-label recognition. The proposed multi-layered ELM network consists of an ELM based autoencoder, dimension transformation, and supervised feature recognition. The autoencoder and dimension transform reconstruct the raw data into three types of representation (*i.e.*, compressed, equal and sparse dimension). The original ELM classifier is applied for the final decision making.

The paper is organized as follows. Section 2 presents the structure of the proposed fault diagnostic framework and the involved algorithms. Experimental rig setup and signals sample data acquisition with a simulated WTGS are implemented in Section 3. Section 4 discusses the experimental results of the framework and its comparisons with other methods including SVM and ML-ELM. Section 5 concludes the study.

2. The Proposed Fault Diagnostic Framework

As shown in Figure 2, the proposed fault diagnosis framework is divided into three submodules: (a) ELM based autoencoder for feature extraction; (b) Matrix compression for dimension reduction; and (c) ELM based classifier for fault classification. The ELM based autoencoder enables three scales of data transformations and representations. Firstly, the raw dataset, which is usually in the form of a high-dimensional matrix, is fed into autoencoder. The autocoder network could be trained using multi-layered ELM networks, each of which is set with a different number of hidden layer nodes L. The dimension of the output layer is set to equal with the input dataset. The output weight vector β is calculated in the ELM output mapping. Secondly, the dimensional transformation compresses the output of autoencoder with a simple matrix transform. The raw dataset is thus converted into a low-dimensioned feature matrix, which is described in detail in Section 2.2. In order to optimize the number of hidden-layer nodes L, a method using multiple sets of contrast tests is introduced in Section 3.2. Finally, one classic ELM classification slice is applied for the final decision making with the input of the converted feature matrix. It is notable that the number of hidden-layer nodes is the only adjustable parameter in the proposed method. The two weighting vectors β and δ are independent in two ELM networks (ELM-based autoencoder and ELM classifier). The former one is applied for regression, while the later is used for classification.

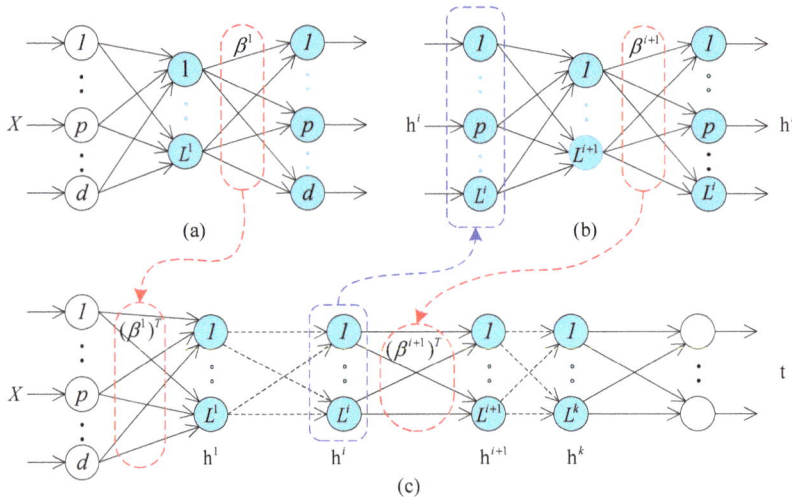

Figure 2. The proposed fault diagnostic framework using an extreme learning machine (ELM) based autoencoder and classifier. (**a**) ELM-autoencoder (AE) output weights β^1 with respect to input data x are the 1st layer weights of Multilayered (ML)-ELM; (**b**) The output weights β^{i+1} of ELM-AE, with respect to *ith* hidden layer output hi of ML-ELM are the $i+1th$ layer weights of Multilayered-ELM; (**c**) The Multilayered-ELM output layer weights are calculated using regularized least squares.

2.1. Extreme Learning Machines Based Autoencoder

ELM is a recently prevailing machine learning method that has been successfully adopted for various applications [37,38]. ELM is characterized with its single hidden layer structure, of which the parameters are initialized randomly. The parameters of the hidden layer are independent upon the target function and the training dataset [39,40]. The output weights which link the hidden layer to output layer are determined analytically through a Moore-Penrose generalized inverse [37,41]. Benefited from its simple structure and efficient learning algorithm, ELM owns very good generalization capability superior to the traditional ANN and SVM. The basics of ELM is summarized as follows:

$$f_L(\mathbf{x}) = \sum_{i=1}^{L} \beta_i h_i(\mathbf{x}) = \mathbf{h}(\mathbf{x})\beta$$
$$\begin{cases} \beta = [\beta_1, ..., \beta_L]^T \\ \mathbf{h}(\mathbf{x}) = [g_1(\mathbf{x}), ..., g_L(\mathbf{x})] \end{cases} \tag{1}$$

where β_i is the output weight vector between the hidden nodes and the output nodes. $\mathbf{h}(\mathbf{x})$ is the hidden nodes output for the input $\mathbf{x}$ and $g_i(\mathbf{x})$ is the i-th hidden node activation function.

Given the N training samples $\{(x_i, t_i)\}_{i=1}^{N}$, the ELM is to resolve the follow problems:

$$\mathbf{H}\beta = \mathbf{T} \tag{2}$$

where $\mathbf{T} = [t_1, ..., t_N]^T$ is the target labels, the matrix $\mathbf{H} = [\mathbf{h}^T(x_1), ..., \mathbf{h}^T(x_N)]^T$ is the hidden nodes output.

The output weight vector β can be calculated by Equation (3),

$$\beta = \mathbf{H}^\dagger \mathbf{T} \tag{3}$$

where $\mathbf{H}^\dagger$ is the Moore-Penrose generalized inverse of matrix $\mathbf{H}$.

In order to obtain better generalization performance and to make the solution more robust, a positive constraint parameter $\frac{1}{C}$ is added to the diagonal of $\mathbf{H}^T\mathbf{H}$ in the calculation of the output weight, as shown in Equation (4).

$$\beta = (\frac{1}{C} + \mathbf{H}^T\mathbf{H})^{-1}\mathbf{H}^T\mathbf{T} \tag{4}$$

To perform autoencoding and feature representation, the ELM algorithm is modified as follows: the target matrix is set equally to the input data, namely, $\mathbf{t} = \mathbf{x}$. The random assigned input weights and biases of the hidden nodes are chosen to be orthogonal. Widrow [42] introduced a Least Mean Square (LMS) method implementation for ELM and a corresponding ELM based autoencoder in which non-orthogonal random hidden parameters (*i.e.*, weights and biases) are used. However, orthogonalization of randomly generated hidden parameters tends to improve the generalization performance of ELM autoencoder. Generally, the objective of ELM based autoencoder is to represent the input features meaningfully in the following three optional representations:

(1) **Compressed representation**: represent features from a high dimensional input data space to a low dimensional feature space;
(2) **Sparse representation**: represent features from a low dimensional input data space to a high dimensional feature space;
(3) **Equal representation**: represent features from an input data space dimension equal to feature space dimension.

Figure 3 shows the structure of random feature mapping. In an ELM based autoencoder, the orthogonal random weight matrix and biases of the hidden nodes project the input data to a different or equal dimensional space as shown by Johnson-Lindenstrauss Lemma and calculated by Equation (5):

$$\mathbf{h} = g(\mathbf{ax} + \mathbf{b}), \quad \mathbf{a}^T\mathbf{a} = \mathbf{I}, \mathbf{b}^T\mathbf{b} = 1 \tag{5}$$

where $\mathbf{a} = [\mathbf{a}_1, ..., \mathbf{a}_L]^T$ are the orthogonal weight vector and $\mathbf{b} = [\mathbf{b}_1, ..., \mathbf{b}_L]^T$ is the orthogonal random bias vector between the input layer and hidden layer.

The output weight β of ELM based autoencoder is applied for learning the transformation from input dataset to the feature space. For sparse and compressed representation, output weight β is calculated by Equation (6),

$$\beta = (\frac{1}{C} + \mathbf{H}^T\mathbf{H})^{-1}\mathbf{H}^T\mathbf{X} \tag{6}$$

where the vector $\mathbf{X} = [\mathbf{x}_1, ..., \mathbf{x}_N]^T$ is the input data and output data, the input data equals to the output in proposed autoencoder.

For equal dimension representations, the output weight β is calculated by Equation (7),

$$\beta = \mathbf{H}^{-1}\mathbf{X} \ \ and \ \ \beta^T\beta = \mathbf{I} \tag{7}$$

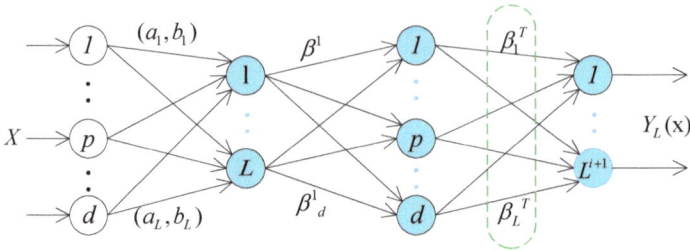

Figure 3. ELM orthogonal random feature mapping.

2.2. Dimension Compression

This paper adopts the regression method to train the parameters of the autoencoder. However, the above transform is not enough for the data preprocessing, because the dimension of input data does not decrease (see Equation (2) and let $\mathbf{t} = \mathbf{x}$). The output data with equal dimension of input data cannot reduce the complexity of the post classifier. After all the parameters of autoencoder are identified, this paper applies a transform to represent the input data. The eventual representation vector shows in Equation (8),

$$Y_L(\mathbf{x}) = (\beta f_L(\mathbf{x}))^T = (\beta \mathbf{X})^T \tag{8}$$

where $Y_L(\mathbf{x})$ is the final output of autoencoder. The dimension of $Y_L(\mathbf{x})$ is shown as Equation (9). The subscripts N and L represent the number of input samples and hidden layer nodes, respectively.

$$Y_L(\mathbf{x}) = \begin{bmatrix} Y_1(\mathbf{x}) & ... & Y_L(\mathbf{x}) \end{bmatrix} = \begin{bmatrix} Y_1(x_1) & ... & Y_L(x_1) \\ \vdots & \vdots & \vdots \\ Y_1(x_N) & ... & Y_L(x_N) \end{bmatrix} \tag{9}$$

From the Equations (8) and (9), the procedure from the high-dimensional vector to the low-dimensional vector can be explained that each element in sample data $x_{i(i \in N)}$ has relationship with β, in other words, β can be seen as a weight distribution of $x_{i(i \in N)}$. The procedure from $x_{i(i \in N)}$ to $Y_L(\mathbf{x})$ is an unsupervised learning as the parameters have been identified in the first part as shown in Figure 2.

Unlike the concept of DL-based autoencoder, the proposed ELM-based autoencoder shows differences at the following four aspects:

(1) The proposed autoencoder is a ELM based network composing of a set of single-hidden-layer slice, whereas the DL-based autoencoder is a multiple hidden layers network.
(2) DL tends to adopt BP algorithm to train all parameters of autoencoder, differently, this paper employs the ELM to configure the network with supervised learning (*i.e.*, Let the output data equal to input data, $\mathbf{t} = \mathbf{x}$). We can get the final output weight β so as to transform input data into a new representation through Equation (8). The dimension of converted data is much smaller than the raw input data.
(3) The DL-based autoencoder tends to map the input dataset into high-dimensional sparse features. While this research applies a compressed representation of the input data.

(4)   The DL-based autoencoder trained with BP algorithm is a really time-consuming process as it requires intensive parameters setting and iterative tuning. On the contrary, each ELM slice in the multi-layered ELM based autocoder can be seen as an independent feature extractor, which relies only on the feature output of its previous hidden layer. The weights or parameters of the current hidden layer could be assigned randomly.

2.3. ELM for Classification

Regarding the binary classification problems, the decision function of ELM is:

$$f_L(\mathbf{x}) = \text{sign}[\mathbf{h}(\mathbf{x})\delta] \tag{10}$$

Unlike other learning algorithms, ELM tends to reach not only the smallest training error but also the smallest norm of output weights. Bartlett's theory [7,43] mentioned that if a neural network is used for a pattern recognition problem, the smaller size of weights brings a smaller square error during the training process, and then realizes a better generalization performance, which doesn't relate directly to the number of nodes. In order to reach smaller training error, the smaller the norms of weights tend to have a better generalization performance. For a m-label classification case, ELM aims to minimize the training error as well as the norm of the output weights. The problem can be summarized as:

$$\textbf{Minimize}: \ \ ||\mathbf{H}\delta - T||^2 \ \ and \ \ ||\delta|| \tag{11}$$

where $\delta = [\delta_1, ..., \delta_l]^T$ is the vector of the output weight between the hidden layer of l-nodes and the output nodes.

$$\mathbf{H} = \begin{bmatrix} \mathbf{h}(\mathbf{x}_1) \\ \vdots \\ \mathbf{h}(\mathbf{x}_N) \end{bmatrix} = \begin{bmatrix} h_1(\mathbf{x}_1) & \cdots & h_l(\mathbf{x}_1) \\ \vdots & \vdots & \vdots \\ h_1(\mathbf{x}_N) & \cdots & h_l(\mathbf{x}_N) \end{bmatrix} \tag{12}$$

where $\mathbf{h}(\mathbf{x}) = [h_1(\mathbf{x}), ..., h_l(\mathbf{x})]^T$ is the output vector of the hidden layer which maps the data from the d-dimensional input space to the l-dimensional hidden-layer space $\mathbf{H}$, $\mathbf{T}$ is the training data target matrix.

$$\mathbf{T} = \begin{bmatrix} \mathbf{t}_1^T \\ \vdots \\ \mathbf{t}_N^T \end{bmatrix} = \begin{bmatrix} t_{11} & \cdots & t_{1m} \\ \vdots & \vdots & \vdots \\ t_{N1} & \cdots & t_{Nm} \end{bmatrix} \tag{13}$$

In the binary classification case, ELM has just a single output node. The optimal output value is chosen as the predicted output label. However, for a multiclass identification problem, this binary classification method could not be applied directly. There are two conditions for multilabel classification:

(1)   If the ELM only has a single-output node, among the multiclass labels, ELM selects the most closed value as the target label. In this case, the ELM solution to the binary classification case becomes a specific case of multiclass solution.
(2)   If the ELM has multi-output nodes, the index of the output node with the highest output value is considered as the label of the input data.

According to the conclusion of study [35], the single-output node classification can be considered a specific case of multi-output nodes classification when the number of output nodes is set to 1. This paper discuss only the multi-output case.

If the original number of class labels is P, the expected output vector of the M-output nodes is $\mathbf{t}_i = [0, ..., 0, \overset{P}{1}, ..., \overset{M}{0}]^T$. In this case, only the P-th elements of $\mathbf{t}_i = [t_{i,1}, ..., t_{i,M}]^T$ is set to 1, while the rest is set to 0. The classification problem (see Equation (11)) for ELM with multi-output nodes can be formulated as Equation (14),

$$\text{Minimize}: \ L_{P_{ELM}} = \tfrac{1}{2}||\delta||^2 + C\tfrac{1}{2}\sum_{i=1}^{N}||\xi_i||^2 \tag{14}$$
$$\text{s.t.} \quad \mathbf{h}(\mathbf{x}_i)\delta = \mathbf{t}_i^T - \xi_i^T, (i=1,...,N)$$

where $\xi_i = [\xi_{i,1},...,\xi_{i,M}]^T$ is the training error vector of the M-output nodes with respect to the training sample $\mathbf{x}_i$.

Based on the **Karush-Kuhn-Tucker** (KKT) theorem, to train ELM is equivalent to solve the following dual optimization problem:

$$L_{D_{ELM}} = \frac{1}{2}||\delta||^2 + C\frac{1}{2}\sum_{i=1}^{N}||\xi_i||^2 - \sum_{i=1}^{N}\sum_{j=1}^{M}\alpha_{i,j}(\mathbf{h}(\mathbf{x}_i)\delta_j - t_{i,j} + \xi_{i,j}) \tag{15}$$

We can have the KKT corresponding optimality conditions as follows:

$$\frac{\partial L_{D_{ELM}}}{\partial \beta_j} = 0 \rightarrow \delta_j = \sum_{i=1}^{N}\alpha_{i,j}(\mathbf{h}(\mathbf{x}_i)^T \rightarrow \delta = \mathbf{H}^T a \tag{16}$$

$$\frac{\partial L_{D_{ELM}}}{\partial \xi_i} = 0 \rightarrow \alpha_i = C\xi_i, \ i=1,...,N \tag{17}$$

$$\frac{\partial L_{D_{ELM}}}{\partial \alpha_i} = 0 \rightarrow \mathbf{h}(\mathbf{x}_i)\delta - t_i^T + \xi_i^T, \ i=1,...,N \tag{18}$$

where $a_i = [\alpha_{i,1},...,\alpha_{i,M}]^T$. In this case, by substituting Equations (16) and (17) into Equation (18), the aforementioned equations can be equivalently written as:

$$(\frac{1}{C} + \mathbf{HH}^T)\alpha = \mathbf{T} \tag{19}$$

From Equations (16)–(19), we have:

$$\delta = \mathbf{H}^T(\frac{1}{C} + \mathbf{HH}^T)^{-1}\mathbf{X} \tag{20}$$

The output function of ELM classifier shows:

$$\mathbf{f}(\mathbf{x}) = \mathbf{h}(\mathbf{x})\delta = \mathbf{h}(\mathbf{x})\mathbf{H}^T\left(\frac{1}{C} + \mathbf{HH}^T\right)^{-1}\mathbf{T} \tag{21}$$

For multiclass cases, the predicted class label of a given testing sample is the index number of the output node which has the highest output value for the given testing sample. Let $f_j(\mathbf{x})$ denoted the output function of the j-th output node (i.e., $f_j(\mathbf{x}) = [f_1(\mathbf{x}),...,f_M(\mathbf{x})]^T$), and then the predicted class label of sample $\mathbf{x}$ is:

$$\text{Label}(\mathbf{x}) = arg\max_{i\in\{1,...,M\}}f_i(\mathbf{x}) \tag{22}$$

In short, there are very few parameters required to set in ELM algorithm. If the feature mapping $\mathbf{h}(\mathbf{x})$ is already known, only one parameter C needs to be specified. The generalization performance of ELM is not sensitive to the dimensionality l of the feature space (i.e., the number of hidden nodes) as long as l is not set to be too small. Different from SVM which usually requests to specify two parameters (C,γ), single-parameter setting makes ELM easy and efficient in the computation for feature representation.

2.4. General Workflow

Table 1 summarizes the ELM training algorithm. The flowchart of the proposed fault diagnostic system for WTGS shows in Figure 4. It consists of three components, namely, (a) signal acquisition module; (b) feature extraction module; (c) fault identification module. For the signal acquisition module, the real-time dataset x_{new} acquisition model uses accelerometers to record the vibration signals of the WTGS. Two tri-axial accelerometers are mounted on the outboard of the gearbox along with the shaft transmission, in order to acquire the vibration signals along the horizontal and vertical directions respectively. The training dataset D_{train} and testing dataset D_{test} are recorded from experiment by accelerometers. In this paper, the real-time signal is processed by the data pre-processing approaches (*i.e.*, x_{new} is converted into x_{proc}), which is identified by the simultaneous fault diagnostic model. In feature extraction module, ELM based autoencoder is employed to generate the most important information (D_{AE_train} and D_{AE_test}) of the input dataset (D_{train} and D_{test}). In order to avoid domination of largest feature values, D_{AE_train} and D_{AE_test} are normalized into D_{nor_train} and D_{nor_test} which are within [0, 1]. After feature extraction and normalization, the datasets D_{nor_train} and D_{nor_test} are sent into classifier for fault recognition. Regarding the real application of this method, the proposed scheme can be seen as a fault pattern indicator in the whole wind forms protection system. First, the real-time vibration signals are collected by accelerators installed on transmission case, and then the vibrations signals are converted into voltage signals and sent into sampling unit. The sampling unit modulates these signals and sends the processed signals into recognition unit. Compared with the proposed scheme, the functions of vibration signals acquisition unit and sampling unit equal to the module (a) in Figure 4. Second, the pattern recognition unit extracts the input signals and classifies them into different labels, and then outputs single or multilabels to the decision making unit. The function of pattern recognition unit equals to the modules (b) and (c) in Figure 4.

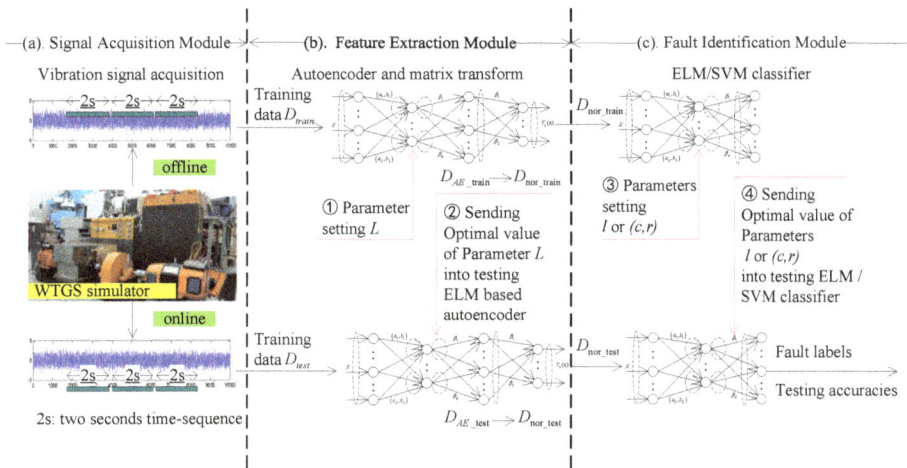

Figure 4. The proposed real-time fault diagnostic scheme for WTGS.

Table 1. The ELM training algorithm.

The ELM Training Algorithm
Step1, Initializing the hidden nodes L;
Step2, Randomly assign input weight ω and bias $\mathbf{b}$;
Step3, Calculate the hidden layer output matrix $\mathbf{H}$;
Step4, Calculate the output weight vector β;
Step5, Calculate the matrix $Y_L(x_N)$ (as shown in eqs (8) and (9));
Step6, Initializing the hidden nodes l;
Step7, Randomly assign input weight $\hat{\omega}$ and bias $\hat{\mathbf{b}}$;
Step8, Calculate the hidden layer output matrix $\hat{\mathbf{H}}$;
Step9, Calculate the output weight matrix δ.

3. Case Study and Experimental Setup

To verify the effectiveness of the proposed fault diagnostic framework for WTGS, experimental test rig is constructed to acquire representative sample data for model construction and analysis. The details of the experiments are discussed in the following subsections, followed by the corresponding results and comparisons. All the proposed methods mentioned are implemented with MATLAB R2015b and executed on a computer with an Intel Core i7-930CPU@ 2.8GHz/12GB RAM.

3.1. Test Rig and Signals Acquisition

The experiments are implemented on a test rig as shown in Figure 1b, which is constructed as the simulation platform of WTGS. The simulator is consisted of a prime mover, a gearbox, a flywheel and an asynchronous generator. Because of the high complexity in real WTGS, it is infeasible for the fault diagnostic system to detect all of the real-time states for all components in the simulation platform. As described in the first section, the gearbox is the core component of the whole WTGS. The gearbox of the test rig is selected in this case study as the valuable component for fault detection. Two tri-axial accelerometers are mounted on the outboard of the gearbox along with the shaft transmission, in order to acquire the vibration signals along the horizontal and vertical directions respectively. A computer is connected with data acquisition board for data analysis. The test rig can simulate both systematic malfunctions, such as unbalance, mechanical misalignment, and looseness, and component faults in terms of periodic patterns and irregular models, including gear crack, broken tooth, chipped tooth, wearing of bearing (as shown in Figure 5). Table 2 presents a total of 13 cases (including normal case, eight single-fault cases and four simultaneous-fault cases) that can be simulated in the test rig for acquisition of dataset for training and test. It should be noted that some cases can be realized by specific tools, For example, the mechanical misalignment of the gearbox is simulated by adjusting height of the gearbox with shims, and the mechanical unbalance case is simulated by adding one eccentric mass on the output shaft. For data acquisition, as the vibration signal along the axial direction is not obvious for the fault detection compared with the other direction, the vibration signal along the axial direction is ignored in the test rig. In the diagnostic model, each simulated single fault is repeated two hundred times and one hundred times for each simultaneous-fault under various random electric loads. Each time, vibration signals in a two second window are recorded with a sampling frequency at 10240 Hz. From a feasible data requisition point of view, the sample frequency must be much higher than the gear meshing frequency, which can ensure no missing signals during the process of sampling. In other words, each sampling dataset records 40,960 points (2 accelerometers × 2 s × 10,240) in each 2 s time window. Table 3 presents that there are 1800 sample dataset (*i.e.*, (1 normal care + 8 kinds of single-fault cases) × 200 samples) and 280 simultaneous-fault sample data (*i.e.*, four kinds of simultaneous-fault data × 100 samples). Table 3 gives the description of the volume of different kinds of data. Some samples for single-fault and simultaneous-fault patterns are shown in Figures 5 and 6, respectively. Figure 6 shows that the signal waveforms of simultaneous-faults are very similar.

3.2. Feature Extraction and Dimension Reduction

The procedure of autoencoder for features extraction and dimension reduction shows in Figure 2. According to the parameters involved in Table 3, the structure of autoencoder in this paper is set as $40,960 \times L \times 40,960$ and $40,960 \times L$. The output of the first part equals to dimension representation of the input matrix, it is a supervised learning. In the second part, this study applies an unsupervised learning for dimension-reduced transform. Furthermore, statistic feature indicators are extracted from the original signal as they are important in analyzing the vibration signals. This paper employs 10 types of statistic features as shown in Table 4.

Table 2. Sample single-faults and simultaneous-fault.

Case No.	Condition	Fault Description
C0	Normal	Normal
C1		Unbalance
C2		Looseness
C3		Mechanical misalignment
C4	Single	Wear of cage and rolling elements of bearing
C5	fault	Wear of outer race of bearing
C6		Gear tooth broken
C7		Gear crack
C8		Chipped tooth
C9		Gear tooth broken and chipped tooth
C10	Simultaneous	Chipped tooth and wear of outer race of bearing
C11	fault	Gear tooth broken and wear of cage and rolling elements of bearing
C12		Gear tooth broken and wear of cage and rolling elements of bearing and wear of outer race of bearing

(a) C6   (b) C7   (c) C8

(d) C3   (e) C4   (f) C1

Figure 5. Singular Component Failure in WTGS. (**a**) Gear tooth broken fault; (**b**) gear crack fault; (**c**) chipped tooth fault; (**d**) mechanical misalignment fault; (**e**) wear of cage and rolling elements of bearing; (**f**) Unbalance.

Table 3. Division of the sample dataset into different subsets.

Dataset	Type of Dataset	Single Fault	Simultaneous Fault
Raw sample data	Training dataset	$D_{\text{train}_t}(1600)$	$D_{\text{train}_s}(200)$
	Test dataset	$D_{\text{test}_t}(200)$	$D_{\text{test}_s}(80)$
Feature extraction (SAE)	Training dataset	$D_{\text{proctrain}_t}(1600)$	$D_{\text{proctrain}_s}(200)$
	Test dataset	$D_{\text{proctest}_t}(200)$	$D_{\text{proctest}_s}(80)$

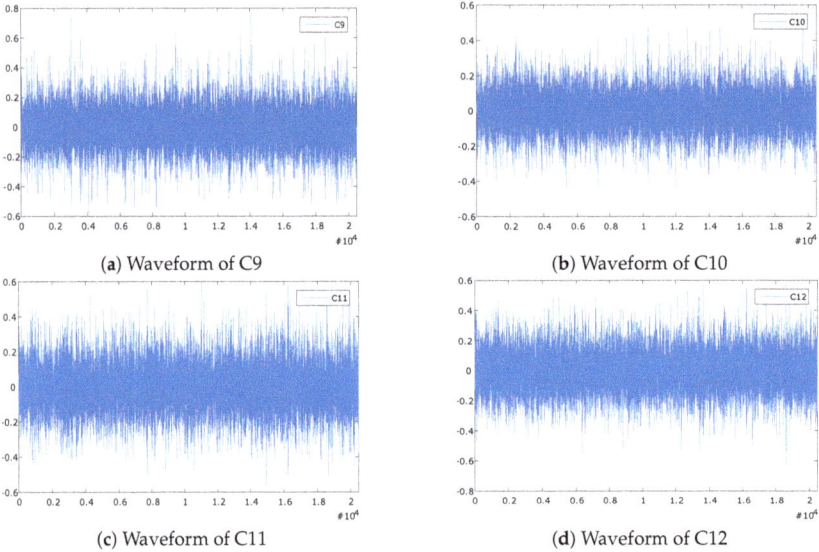

(a) Waveform of C9

(b) Waveform of C10

(c) Waveform of C11

(d) Waveform of C12

Figure 6. Sample normalized simultaneous-fault patterns of WTGS. (**a**) Gear tooth broken and chipped tooth; (**b**) chipped tooth and wear of outer race of bearing; (**c**) gear tooth broken and wear of cage and rolling elements of bearing; (**d**) gear tooth broken and wear of cage and rolling elements of bearing and wear of outer race of bearing.

Table 4. Definition of the selected statistical features for acoustic signal. (Note: x_i represents a signal series for $i = 1, ...N$. where N is the number of data points of a raw signal.)

Features	Equation	Features	Equation
Mean	$x_m = \frac{1}{N}\sum\limits_{i=1}^{N} x_i$	Kurtosis	$x_{\text{kur}} = \dfrac{\sum\limits_{i=1}^{N}(x_i - x_m)^4}{(N-1)x_{\text{std}}^4}$
Standard deviation	$x_{\text{std}} = \sqrt{\dfrac{\sum\limits_{i=1}^{N}(x_i - x_m)^2}{N-1}}$	Crest factor	$CF = \dfrac{x_{\text{pk}}}{x_{\text{rms}}}$
Root mean square	$x_{\text{rms}} = \sqrt{\dfrac{1}{N}\sum\limits_{i=1}^{N} x_i^2}$	Clearance factor	$CLF = \dfrac{x_{\text{pk}}}{(\frac{1}{N}\sum\limits_{i=1}^{N}\sqrt{|x_i|})^2}$
Peak	$x_{\text{pk}} = \max|x_i|$	Shape factor	$SF = \dfrac{x_{\text{rms}}}{\frac{1}{N}\sum\limits_{i=1}^{N}|x_i|}$
Skewness	$x_{\text{ske}} = \dfrac{\sum\limits_{i=1}^{N}(x_i - x_m)^3}{(N-1)x_{\text{std}}^3}$	Impulse factor	$IF = \dfrac{x_{\text{pk}}}{\frac{1}{N}\sum\limits_{i=1}^{N}|x_i|}$

To ensure that all the features have even contribution, all reduced features should go through normalization. The interval of normalization is restrained in [0, 1]. Each extracted feature is normalized by Equation (23):

$$y = \frac{(x - x_{min})}{(x_{max} - x_{min})} \tag{23}$$

where x is the output feature, y is the result after normalization.

After normalization, a processed dataset D_{porc} is obtained. The classifier can be trained by using D_{porc_train}.

4. Experimental Results and Discussion

In order to verify the effectiveness of the proposed scheme, this paper applies various combinations of methods to realize the contrast experiments. Testing accuracy and testing time are introduced to evaluate the prediction performance of the classifier. As suggested in Section 3, the ELM based autoencoder can convert the input data space into three types of output data space. In this paper, we choose the compressed dimensional representation and use the ELM learning method to train the parameters. The function of autoencoder is to get an optimal matrix β, and the function of matrix transform is to reduce the dimension of input X. Before the experiments, it is not clear how many dimensions it is appropriate to cut down. In other words, the model needs proper values of L and β to improve the testing accuracies. In order to get a set of optimal parameters (*i.e.*, hidden layer nodes L in autoencoder, hidden layer nodes l in classifier), D_{train} (D_{train} includes dataset D_{train_l} and D_{train_s}) is applied to train the networks. As shown in Figure 7a,b,we set the number of hidden layer nodes $L = 800$, when the number of hidden layer nodes l increase from 1 to 2000 at 10 interval, the largest accuracy is 95.62% at single fault condition and 93.22% simultaneous-fault condition, respectively. The optimal hidden layer nodes in the classifier is set as $l = 600$.

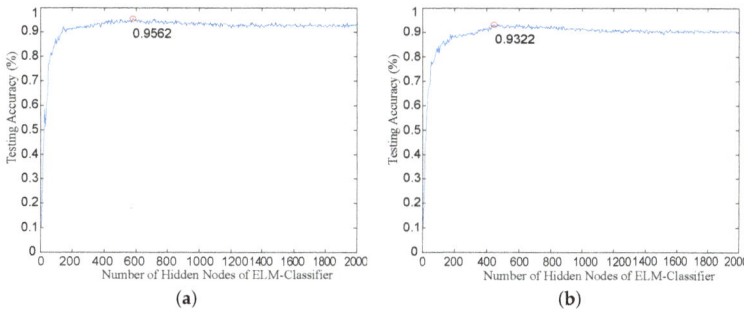

Figure 7. Testing accuracy in l subspace for the multi-layered ELM when L = 800. (**a**) Testing accuracy of single-fault; (**b**) Testing accuracy of simultaneous-fault.

As suggested in Table 5, a total of 16 kinds of combinations of method are implemented to compare the generation performance. According to the feature extraction, this paper takes three kinds of methods as references. They are WPT+TDSF+KPCA combination, EMD+SVD combination and LMD+TDSF combination, respectively. This paper takes the Db4 (Daubechies) wavelet as the mother wavelet and sets the level of decomposition at the range from 3 to 5. The radial basis function (RBF) acts as kernel function for KPCA. In order to reduce the number of trials, the hyperparameter R of RBF based on 2^v is tried for v ranged from -3 to 3. In the KPCA processing, this paper selects the polynomial kernel with $d = 4$ and the RBF kernel with $R = 2$. After dimension reduction, a total of 80 principal components are obtained. After feature extraction, the next step is to optimize parameters of classifiers. This paper takes four kinds of methods, namely PNN, RVM, SVM and ELM. As mentioned previously, probabilistic based classifiers have their own hyperparameters for tuning.

PNN uses spread s and RVM employs width ω. In this case study, the value of s is set from 1 to 3 at an interval of 0.5, and the values of ω is selected from 1 to 8 at an interval of 0.5. In order to find the optimal decision threshold, this paper sets the search region at the range from 0 to 1 at an interval of 0.01. For the configuration of ELM, this paper takes the sigmoid function as the activation function and sets the number of hidden modes l as 600 for a trial. According to the experimental results in Table 4, a total of 80 components are obtained from the feature extractor. It is clear that the accuracies with autoencoder are higher than those with WPT+TDSF+KPCA. The results can be explained because the ELM based autoencoder holds all information of the input data during the representational learning. However, KPCA tends to hold the important information and inevitably lose some unimportant information.

Table 5. Evaluation of different combinations of methods using the optimal model parameters. Wavelet packet transform: WPT; time-domain statistical features: TDSF; kernel principal component analysis: KPCA; empirical mode decomposition: EMD.

Feature Extraction	Classifier	Accuracies for Test Case (%)		
		Single-Fault	Simultaneous-Fault	Overall Fault
WPT+TDSF+KPCA	PNN	83.64	83.64	83.76
	RVM	82.99	74.64	81.21
	SVM	**92.88**	**89.73**	90.78
	ELM	91.29	89.72	**90.89**
EMD+TDSF	PNN	85.64	84.64	84.52
	RVM	83.99	77.64	83.21
	SVM	**95.83**	**92.87**	**94.35**
	ELM	96.20	92.44	94.32
LMD+TDSF	PNN	85.64	84.64	84.52
	RVM	83.99	77.64	83.21
	SVM	95.25	92.87	93.27
	ELM	**95.83**	**93.04**	**94.44**
ELM-AE	PNN	85.64	84.64	84.52
	RVM	83.99	77.64	83.21
	SVM	95.83	92.87	93.27
	ELM	**95.62**	**93.22**	**94.42**

In order to compare the performances of classifiers, this paper sets the contrast experiments with the same ELM based autoencoder and different classifiers. As shown in Table 5, the number of hidden nodes L in autoencoder is 800, the last dimensions of training data D_{train} and testing data D_{test} are 1800×800 and 280×800, respectively. As suggested in Figure 7, this optimal value of l is 600. According to the experimental results not listed here, SVM employed polynomial kernel with $C = 10$ and $d^* = 4$ show the best accuracy. Table 6 shows that the fault detection accuracy of ELM is similar to that of SVM, while the fault identification time of ELM and SVM take 20 ms and 157 ms respectively. The performance of ELM is much faster than SVM. Quick recognition is necessary for real-time fault diagnosis system. In actual WTGS application, the real-time fault diagnostic system is required to analyze signals for 24 hours per day. In terms of fault identification time, ELM is faster than SVM by 88.46%. The test results show that ELM and SVM have relatively high testing accuracies, but the advantage of ELM is embodied in testing time, which is very significant in real situation because a practical real-time WTGS diagnostic system will analyze more sensor signals than the two sensor signals used in this case study.

Table 6. Evaluation of methods using ELM or SVM with ELM based autoencoder (Note: Dimension reduction from 20,480 to 80).

Feature Extraction	Fault Type	Accuracies for Test Case (%)		Time for Test Case (ms)	
		SVM	ELM	SVM	ELM
ELM-AE	Single-fault	95.72 ± 2.25	**95.62 ± 2.25**	156 ± 0.9	**18 ± 0.8**
	Simultaneous-fault	92.98 ± 1.25	**93.22 ± 3.25**	158 ± 0.8	**20 ± 0.5**
	Overall fault	93.55 ± 3.15	**94.42 ± 2.75**	157 ± 0.4	**20 ± 0.75**

5. Conclusions

This paper proposes a new application of ELM to the real-time fault diagnostic system for rotating machinery. The framework is successfully applied on recognizing fault patterns coming from the WTGS. At the stage of data preprocessing, this study applies an ELM based autoencoder for data representational learning, which train a network of ELM slices to acquire the feature reconstruction, and then the ELM network generates a new low-dimensional representation. Unlike the well adopted data preprocessing methods using a combination of WPT, TDSF and KPCA, the proposed ELM based autoencoder could leverage the down-streamed classification accuracy in around 5%–10% for different corresponding classifiers. Compared with the widely-applied classifiers (e.g., SVM and RVM), ELM algorithm searches optimal solution from the feature space without any other constraints. Therefore, ELM network is superior to SVM at producing lightly higher diagnostic accuracy. Besides, ELM aims to generate a smaller weights and norms, and then gets a faster generalization performance than SVM. This study makes contributions at the following four aspects: (1) It is the first research to analyze the ELM based autoencoder as a tool for compressed representation; (2) It is the first application of ELM based autoencoder to the fault diagnosis for rotating machinery; (3) It is the original application of the proposed scheme to fault diagnosis of WTGS; (4) It is the first study to solely use ELM method as a combination of two different training processes in terms of regression and classification, to realize autoencoding and classification respectively. Since the proposed framework for fault diagnosis of wind turbine equipment is general, it is suitable to apply to other industrial problems.

Acknowledgments: The authors would like to thank the University of Macau for its funding support under Grants MYRG2015-00077-FST.

Author Contributions: Zhi-Xin Yang and Xian-Bo Wang conceived and designed the experiments; Jian-Hua Zhong performed the experiments and contributed analysis tools; Xian-Bo Wang analyzed the data and wrote the paper.

Conflicts of Interest: The authors declare no conflict of interest.

References

1. Amirat, Y.; Benbouzid, M.E.H.; Al-Ahmar, E.; Bensaker, B.; Turri, S. A brief status on condition monitoring and fault diagnosis in wind energy conversion systems. *Renew. Sustain. Energy Rev.* **2009**, *13*, 2629–2636.
2. Yin, S.; Luo, H.; Ding, S.X. Real-time implementation of fault-tolerant control systems with performance optimization. *IEEE Trans. Ind. Electron.* **2014**, *61*, 2402–2411.
3. Wong, P.K.; Yang, Z.; Vong, C.M.; Zhong, J. Real-time fault diagnosis for gas turbine generator systems using extreme learning machine. *Neurocomputing* **2014**, *128*, 249–257.
4. Yan, R.; Gao, R.X.; Chen, X. Wavelets for fault diagnosis of rotary machines: A review with applications. *Signal Process.* **2014**, *96*, 1–15.
5. Fan, W.; Cai, G.; Zhu, Z.; Shen, C.; Huang, W.; Shang, L. Sparse representation of transients in wavelet basis and its application in gearbox fault feature extraction. *Mech. Syst. Signal Process.* **2015**, *56*, 230–245.
6. Vincent, P.; Larochelle, H.; Bengio, Y.; Manzagol, P.A. Extracting and composing robust features with denoising autoencoders. In Proceedings of the 25th International Conference on Machine Learning, Helsinki, Finland, 5–9 July 2008; pp. 1096–1103.

7. Bianchi, D.; Mayrhofer, E.; Gröschl, M.; Betz, G.; Vernes, A. Wavelet packet transform for detection of single events in acoustic emission signals. *Mech. Syst. Signal Process.* **2015**, doi:10.1016/j.ymssp.2015.04.014 .

8. Keskes, H.; Braham, A.; Lachiri, Z. Broken rotor bar diagnosis in induction machines through stationary wavelet packet transform and multiclass wavelet SVM. *Electric Power Syst. Res.* **2013**, *97*, 151–157.

9. Li, N.; Zhou, R.; Hu, Q.; Liu, X. Mechanical fault diagnosis based on redundant second generation wavelet packet transform, neighborhood rough set and support vector machine. *Mech. Syst. Signal Process.* **2012**, *28*, 608–621.

10. Wang, Y.; Xu, G.; Liang, L.; Jiang, K. Detection of weak transient signals based on wavelet packet transform and manifold learning for rolling element bearing fault diagnosis. *Mech. Syst. Signal Process.* **2015**, *54*, 259–276.

11. Ebrahimi, F.; Setarehdan, S.K.; Ayala-Moyeda, J.; Nazeran, H. Automatic sleep staging using empirical mode decomposition, discrete wavelet transform, time-domain, and nonlinear dynamics features of heart rate variability signals. *Comput. Methods Prog. Biomed.* **2013**, *112*, 47–57.

12. Khorshidtalab, A.; Salami, M.J.E.; Hamedi, M. Robust classification of motor imagery EEG signals using statistical time–domain features. *Physiol. Meas.* **2013**, *34*, doi:10.1088/0967-3334/34/11/1563.

13. Li, W.; Zhu, Z.; Jiang, F.; Zhou, G.; Chen, G. Fault diagnosis of rotating machinery with a novel statistical feature extraction and evaluation method. *Mech. Syst. Signal Process.* **2015**, *50*, 414–426.

14. Allen, E.A.; Erhardt, E.B.; Wei, Y.; Eichele, T.; Calhoun, V.D. Capturing inter-subject variability with group independent component analysis of fMRI data: a simulation study. *Neuroimage* **2012**, *59*, 4141–4159.

15. Du, K.L.; Swamy, M. Independent component analysis. In *Neural Networks and Statistical Learning*; Springer: New York, NY, USA, 2014; pp. 419–450.

16. Shlens, J. A tutorial on principal component analysis. Available online: http://arxiv.org/pdf/1404.1100v1.pdf (accessed on 7 April 2014).

17. Waldmann, I.P.; Tinetti, G.; Deroo, P.; Hollis, M.D.; Yurchenko, S.N.; Tennyson, J. Blind extraction of an exoplanetary spectrum through independent component analysis. *Astrophys. J.* **2013**, *766*, doi:10.1088/0004-637X/766/1/7.

18. Mairal, J.; Bach, F.; Ponce, J. Task-driven dictionary learning. *IEEE Trans. Pattern Anal. Mach. Intell.* **2012**, *34*, 791–804.

19. Lei, Y.; He, Z.; Zi, Y.; Chen, X. New clustering algorithm-based fault diagnosis using compensation distance evaluation technique. *Mech. Syst. Signal Process.* **2008**, *22*, 419–435.

20. Bro, R.; Smilde, A.K. Principal component analysis. *Anal. Methods* **2014**, *6*, 2812–2831.

21. Xanthopoulos, P.; Pardalos, P.M.; Trafalis, T.B. Principal component analysis. In *Robust Data Mining*; Springer: New York, NY, USA, 2013; pp. 21–26.

22. Hoque, M.S.; Mukit, M.; Bikas, M.; Naser, A. An implementation of intrusion detection system using genetic algorithm. *Int. J. Netw. Secur.* **2012**, *4*, 109–120.

23. Johnson, P.; Vandewater, L.; Wilson, W.; Maruff, P.; Savage, G.; Graham, P.; Macaulay, L.S.; Ellis, K.A.; Szoeke, C.; Martins, R.N. Genetic algorithm with logistic regression for prediction of progression to Alzheimer's disease. *BMC Bioinform.* **2014**, *15*, doi:10.1186/1471-2105-15-S16-S11.

24. Whitley, D. An executable model of a simple genetic algorithm. *Found. Genet. Algorithms* **2014**, *2*, 45–62.

25. Tang, J.; Deng, C.; Huang, G.B. Extreme Learning Machine for Multilayer Perceptron. *IEEE Trans. Neural Netw. Learn. Syst.* **2015**, *27*, 809–821.

26. Hinton, G.E. A practical guide to training restricted boltzmann machines. In *Neural Networks: Tricks of the Trade*; Springer: Lake Tahoe, NV, USA, 2012; pp. 599–619.

27. Srivastava, N.; Salakhutdinov, R.R. Multimodal learning with deep boltzmann machines. In *Advances in Neural Information Processing Systems*; Springer: Lake Tahoe, NV, USA, 2012; pp. 2222–2230.

28. Fischer, A.; Igel, C. An introduction to restricted Boltzmann machines. In *Progress in Pattern Recognition, Image Analysis, Computer Vision, and Applications*; Springer: Buenos Aires, Argentina, 2012; pp. 14–36.

29. Yang, Z.; Wong, P.K.; Vong, C.M.; Zhong, J.; Liang, J. Simultaneous-fault diagnosis of gas turbine generator systems using a pairwise-coupled probabilistic classifier. *Math. Probl. Eng.* **2013**, *2013*, doi:10.1155/2013/827128.

30. Vong, C.M.; Wong, P.K. Engine ignition signal diagnosis with wavelet packet transform and multi-class least squares support vector machines. *Expert Syst. Appl.* **2011**, *38*, 8563–8570.

31. Abbasion, S.; Rafsanjani, A.; Farshidianfar, A.; Irani, N. Rolling element bearings multi-fault classification based on the wavelet denoising and support vector machine. *Mech. Syst. Signal Process.* **2007**, *21*, 2933–2945.

32. Widodo, A.; Yang, B.S. Application of nonlinear feature extraction and support vector machines for fault diagnosis of induction motors. *Expert Syst. Appl.* **2007**, *33*, 241–250.

33. Sankari, Z.; Adeli, H. Probabilistic neural networks for diagnosis of Alzheimer's disease using conventional and wavelet coherence. *J. Neurosci. Methods* **2011**, *197*, 165–170.

34. Othman, M.F.; Basri, M.A.M. Probabilistic neural network for brain tumor classification. In Proceedings of the 2011 Second International Conference on Intelligent Systems, Modelling and Simulation (ISMS), Kuala Lumpur, Malaysia, 25–27 January 2011; pp. 136–138.

35. Huang, G.B.; Zhou, H.; Ding, X.; Zhang, R. Extreme learning machine for regression and multiclass classification. *IEEE Trans. Syst. Man Cybern. Part B Cybern.* **2012**, *42*, 513–529.

36. Huang, G.B.; Ding, X.; Zhou, H. Optimization method based extreme learning machine for classification. *Neurocomputing* **2010**, *74*, 155–163.

37. Huang, G.B.; Zhu, Q.Y.; Siew, C.K. Extreme learning machine: theory and applications. *Neurocomputing* **2006**, *70*, 489–501.

38. Huang, G.B. What are Extreme Learning Machines? Filling the Gap Between Frank Rosenblatt's Dream and John von Neumann's Puzzle. *Cognit. Comput.* **2015**, *7*, 263–278.

39. Wong, P.K.; Wong, K.I.; Vong, C.M.; Cheung, C.S. Modeling and optimization of biodiesel engine performance using kernel-based extreme learning machine and cuckoo search. *Renew. Energy* **2015**, *74*, 640–647.

40. Luo, J.; Vong, C.M.; Wong, P.K. Sparse bayesian extreme learning machine for multi-classification. *IEEE Trans. Neural Netw. Learn. Syst.* **2014**, *25*, 836–843.

41. Cambria, E.; Huang, G.B.; Kasun, L.L.C.; Zhou, H.; Vong, C.M.; Lin, J.; Yin, J.; Cai, Z.; Liu, Q.; Li, K.; *et al.* Extreme learning machines [trends and controversies]. *IEEE Intell. Syst.* **2013**, *28*, 30–59.

42. Widrow, B.; Greenblatt, A.; Kim, Y.; Park, D. The No-Prop algorithm: A new learning algorithm for multilayer neural networks. *Neural Netw.* **2013**, *37*, 182–188.

43. Bartlett, P.L. The Sample Complexity of Pattern Classification with Neural Networks: The Size of the Weights is More Important than the Size of the Network. *IEEE Trans. Inf. Theory* **1998**, *44*, 525–536.

Article

Optimal Maintenance Management of Offshore Wind Farms

Alberto Pliego Marugán [1], Fausto Pedro García Márquez [1],* and Jesús María Pinar Pérez [2]

[1] Ingenium Research Group, Universidad Castilla-La Mancha, 13071 Ciudad Real, Spain;
    alberto.pliego@uclm.es
[2] CUNEF-Ingenium, Colegio Universitario de Estudios Financieros, 28040 Madrid, Spain;
    jesusmaria.pinar@cunef.edu
* Correspondence: faustopedro.garcia@uclm.es; Tel.: +34-926-295300 (ext. 6230)

Academic Editor: Frede Blaabjerg
Received: 30 October 2015; Accepted: 24 December 2015; Published: 15 January 2016

Abstract: Nowadays offshore wind energy is the renewable energy source with the highest growth. Offshore wind farms are composed of large and complex wind turbines, requiring a high level of reliability, availability, maintainability and safety (RAMS). Firms are employing robust remote condition monitoring systems in order to improve RAMS, considering the difficulty to access the wind farm. The main objective of this research work is to optimise the maintenance management of wind farms through the fault probability of each wind turbine. The probability has been calculated by Fault Tree Analysis (FTA) employing the Binary Decision Diagram (BDD) in order to reduce the computational cost. The fault tree presented in this paper has been designed and validated based on qualitative data from the literature and expert from important European collaborative research projects. The basic events of the fault tree have been prioritized employing the criticality method in order to use resources efficiently. Exogenous variables, e.g., weather conditions, have been also considered in this research work. The results provided by the dynamic probability of failure and the importance measures have been employed to develop a scheduled maintenance that contributes to improve the decision making and, consequently, to reduce the maintenance costs.

Keywords: offshore; wind turbines; maintenance management; fault tree analysis; binary decision diagrams

---

1. Introduction

The renewable energy industry is in continuous development to achieve the energy framework targets established by governments [1]. Nowadays, the most developed countries are focused on improving the technology for offshore wind energy. The main advantages of the offshore wind farms are [2]:

-   The wind power captured by wind turbines (WTs) is more than onshore.
-   The size of offshore wind farms can be larger than onshore.
-   The environmental impact for offshore is less than in onshore.

    The main disadvantages are:

-   It is more complex to evaluate the wind characteristics.
-   Larger investment costs. The offshore installation cost is 1.44 million €/MW, where the onshore is 0.78 million €/MW [3].
-   Operation and maintenance (O & M) tasks are more complex and expensive than onshore. The offshore O & M costs tasks are from 18% to 23% of the total system costs, being 12% for onshore wind farms [4].

The objective of this paper is to develop a novel maintenance management approach in order to establish a proper strategy for the maintenance task by using a predictive maintenance method based on statistical studies. This approach provides information about the WTs with high fault probability, a ranking of components of WTs to repair or replace according to the state of the system over the time, and when those maintenance tasks must be carried out. An adequate maintenance planning to ensure the right operation of an offshore wind farm is required. For this purpose, different techniques and methods of condition monitoring (CM) are employed for detection and diagnosis of faults of WTs [5]. Most of the research papers consider the CM in WTs referred to blades [6], gearboxes [7], electrical or electronic components [8] and tower [9]. CM leads to improve RAMS and to increase the productivity of wind farms.

2. CM Applied to WT

The main components of WTs are illustrated in Figure 1. WTs are usually three-blade units [10]. Once the wind drives the blades, the energy is transmitted via the main shaft through the gearbox to the generator. At the top of the tower, assembled on a base or foundation, the housing or nacelle is mounted and the alignment with the direction of the wind is controlled by a yaw system. There is a pitch system in each blade. This mechanism controls the wind power and sometimes is employed as an aerodynamic brake. Finally, there is a meteorological unit that provides information about the wind (speed and direction) to the control system.

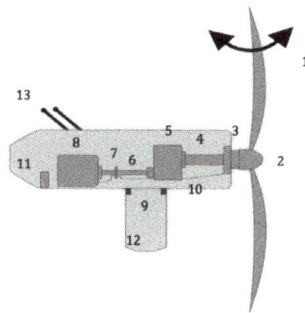

Figure 1. Components of the wind turbine (WT) where: 1—pitch system; 2—hub; 3—main bearing; 4—low speed shaft; 5—gearbox; 6—high speed shaft; 7—brake system; 8—generator; 9—yaw system; 10—bedplate; 11—converter; 12—tower; 13—meteorological unit.

CM systems are composed of different types of sensors and signal processing equipment applied on the main components of WTs such as blades, gearboxes, generators, bearings and towers. The choice and location of the right type and number of sensors are a key factor. The acquisition of accurate data is critical to determine the occurrence of a fault and to address the solution to apply. Nowadays, different techniques are available for CM: vibration analysis [3,11], acoustic emission [3,12], ultrasonic testing techniques [13,14], oil analysis [15], thermography [3,13] and other methods [16].

The first step of the CM process is the choice of an adequate technique for data acquisition, including electronic signals with the measurement of the required conditions, e.g., sound, vibration, voltage, temperature, speed. Then, a correct signal processing method is applied, e.g., fast Fourier transform, wavelet transforms, hidden Markov models, statistical methods and trend analysis. Fault detection and diagnosis (FDD) involves both CM techniques and the signal processing methods.

The frequency of occurrence, *i.e.*, the probability of failure, of an event is necessary to study in order to improve the application of CMS for WTs [17]. This paper employed the Fault Tree Analysis (FTA) technique to calculate the probability of failure of the WT. FTA is a graphical representation of logical relationships between events. A Binary Decision Diagram (BDD) has been used to provide an alternative to the traditional technique in order to reduce the computational cost. BDD is an approach that determines the probability of failure of a system by examining the probability of failure of the components. The BDD method does not analyses the FTA directly. The Boolean equation represents the main event to analyse, e.g., the wind turbine failure, and it is obtained by BDDs that come from the fault tree. The ordering algorithm for the construction of the BDD has a crucial effect on its size, and therefore the computational cost. The algorithms are heuristics, and this is the reason that in this paper has been considered several in order to compare the results, being: Top-down-left-right, Depth First Search, AND, Breath First Search, and Level.

Finally, in order to optimize the resources, e.g., human, material, economic resources, *etc.*, proper and accurate prioritization of the basic events, based on importance measurement, has been done according to the criticality method [18]. The information provided by the aforementioned method leads to establish an optimal maintenance management for offshore wind farms, considering both endogenous and exogenous variables.

3. FTA and BDD

A Fault Tree (FT) is a graphical structure formed by the causes of a certain type of failure mode (Top Event) and the failure mode of the components (basic events) connected by logical operators such as AND/OR gates [19]. The probability vector **p** represents the failure probabilities of the basic events q_i, $i \in \{1, \dots, n\}$, being n the total number of events [20,21].

Then, the system failure probability Q_{sys} can be obtained via FTA according to **q**:

$$\mathbf{q} = \begin{pmatrix} q_1 \\ \vdots \\ q_n \end{pmatrix}$$

Complex systems analysis produce thousands of combinations of events (minimal cut sets) that would cause the failure of the system and are used in the calculation of Q_{sys} [21]. The determination of these minimal cut sets can be a large and time-consuming process, even on modern high speed computers. When the FT has many minimal cut sets, the determination of the exact failure probability of the top event also requires a high calculation costs. For many complex FTs, this requirement may be beyond the capability of the available computers. Therefore, some approximation techniques have been introduced with a loss of accuracy.

The BDD method does not analyze the FT directly. The conversion of the FT to a BDD make possible to calculate the probability of the top event by determining the Boolean equation of the top event. The conversion process from FT to BDD presents several problems, e.g., the ordering scheme chosen for the construction of the BDD has a crucial effect on its resulting size. A wrong ordering scheme may result in large BDD that presents high computational costs [19]. In order to improve the resource deployment in an existing system, proper and accurate ranking of the basic events is necessary [23,24]. Some prioritizations of the basic events of the FT have been considered in this paper. For further details of FTA and ranking methods, consultation of references [18,25] is recommended. BDDs have been successfully used in the literature as an efficient way to simulate FTs. BDDs were introduced by Lee [26], and further popularized by Akers [27], Moret [25] and Bryant [22]. These decision diagrams are composed by a data structure that can represent a Boolean function [28].

A BDD is a directed acyclic graph representation (**V**, **N**), with vertex set **V** and index set **N**, of a Boolean function where equivalent Boolean sub-expressions are uniquely represented [29]. A directed acyclic graph is a directed graph, *i.e.*, to each vertex v there is no possible directed path that starts and finishes in v. It is composed of some interconnected nodes with two vertices. Each vertex is possible to be a terminal or non-terminal vertex. Each single variable has two branches: 0-branch corresponds to the cases where the variable has not occur and it is graphically represented by a dashed line (Figure 2); on the other hand, 1-branch cases are those where the event is being carried out and corresponds to the occurrence of the variable, and it is represented by a solid line (Figure 2). It allows to obtain an analytical expression depending on the probability of failure of the basic events and the topology of the FT. Paths starting from the top event to a terminal 1 provide a state in which the top event will occur. These paths are named cut sets.

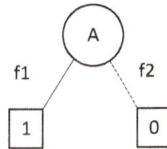

Figure 2. Binary Decision Diagram (BDD).

ITE (If-Then-Else) conditional expression is employed in this research work as an approach for the BDD's cornerstones, based on the approach presented in reference [30]. Figure 2 shows an example of an ite done in a BDD that is described as: "If event A occurs, Then f_1, Else f_2" [31]. The solid line always belongs to the ones and the dashed lines to the zeros, explained above.

The following expression is obtained from Figure 2, considering Shannon's theorem:

$$f = b_i \cdot f_1 + \bar{b}_i \cdot f_2 = ite\,(b_i, f_1, f_2)$$

The size of the BDD, equivalent to CPU runtime, has a strong dependence on the ordering of the events. Different ranking methods can be used in order to reduce the number of cut sets, and consequently, to reduce the computational runtime. Note that there is no method that provides the minimum size of BDD in all cases. The following methods have been considered in this paper: Top-down-left-right, Depth First Search, AND, Breath First Search, and Level. The AND method has chosen for ranking the events because it provides the best results in this case. For further information about BDDs readers are recommended to see references [20,22,26,27].

The quantitative analysis also takes into account the importance of each basic event within the global system. With this purpose, different importance measures (IMs) are considered in this paper. IMs are used in reliability and risk analysis to quantifying the impact of single component on a system failure [32]. In order to determine the importance of a component, it is necessary to consider all the related basic events as a group [33]. A complete importance analysis of all groups is therefore impractical for large systems, and it is necessary to focus on the most important groups of components [34]. In this work Birnbaum and Criticality IMs are presented.

Birnbaum IM introduced, for an event k, a measure of importance based on the fault probability of the system caused by the failure of the component k [35]. The priority of the event k is given by its Birnbaum IM and is calculated as follows:

$$I_k^{Birn} = \frac{\partial Q_{sys}}{\partial q_k}$$

where q_k is the failure probability assigned to the k event, and Q_{sys} is the probability of the top event. A drawback related to the Birnbaum IM is that it does not consider the failure probability of the k event and, therefore, a high importance can be assigned to rare events.

Criticality IM [18], in contrast to Birnbaum, takes into account the failure probability of a certain component. It rectifies the drawback presented in Birnbaum IM, balancing the values obtained. It is defined as:

$$I_k^{Crit} = \frac{q_k}{Q_{sys}} \cdot \frac{\partial Q_{sys}}{\partial q_k} = \frac{q_k}{Q_{sys}} \cdot I_k^{Birn}$$

where I_k^{Crit} is the Criticality IM of the k event, q_k is the probability assigned to the k event and Q_{sys} is the top event probability. Criticality IM provides a different perspective than the Birnbaum IM, even though both are connected providing a measurement of the criticality of each components. Therefore, the Criticality IM has been employed in the following sections to carry out a simulation as realistic as possible.

4. FTA for WTs

A study of failure modes and effects analysis (FMEA) for WTs in 2010 (RELIAWIND project) collected the causes of failure and failure modes of a specific WT of 2MW with a diameter of 80 m [35]. Some causes of failures (or root causes) are summarized in [36]. These main causes of the failures can be due to environmental conditions (e.g., lightning, ice, fire, strong winds, *etc.*) or to defects, malfunctions or failures in the components of the WT (e.g., braking system failure, or be struck by blade, *etc.*) [37,38]. The causes of failures (or root causes) of the components of a wind turbine can be summarized as follows [35,39]: structural (design fault, external damage, installation defect, maintenance fault, manufacturing defect, mechanical overload, mechanical overload–collision, mechanical overload–wind, presence of debris); wear (corrosion, excessive brush wear, fatigue, pipe puncture, vibration fatigue, overheating, insufficient lubrication); electrical (calibration error, connection failure, electrical overload, electrical short, insulation failure, lightning strike, loss of power input, conducting debris, software design fault). Some of the principal component failure modes of WTs are [35,39]: mechanical (rupture, uprooting, fracture, detachment, thermal, blockage, misalignment, scuffing); electrical (electrical insulation, electrical failure, output inaccuracy, software fault, intermittent output); material (fatigue, structural, ultimate, buckling, deflection).

In this work, the construction of the illustrative FT has been focused on a three blades, pitch system and geared WT. The turbine has been divided into four major groups of elements for a better FTA: The foundation and tower; the blades system; the electrical components (including generator, electrical and electronic components), and; the power train (including speed shafts, bearings and a gearbox). The elements of the FT are connected by AND and OR gates, and their fault probability is unknown. The failures considered in this paper are set by an exhaustive review of the literature and the support of experts from the NIMO and OPTIMUS FP7 European projects [40,41].

Table 1 shows a summary of the failures from the literature taken into account for this paper. It can be seen that gearboxes, generators, blades and electric and control systems have been extensively studied in the literature, but there are not many references about other components such as brakes, hydraulic and yaw systems.

Table 1. Failures of the main elements of a WT.

Foundation and Tower Failure	Structural fault [17,38,42–45]	
	Yaw system failure [46]	
Critical Rotor Failure	Blade failure	Structural failure [17,34,47–53]
		Pitch system failure [54,55]
		Hydraulic system fault [50,56]
		Meteorological unit failure [50,57]
	Rotor system failure	Rotor hub [42,46]
		Bearings [45–47]
Power Train Failure	Low speed train failure [17,46,48]	
	Critical gearbox failure [7,46,53,58–62]	
	High speed train failure	Shaft [6,46,58]
		Critical brake failure [6,56]
Electrical Components Failure	Critical generator failure [6,46,58,60,63–65]	
	Power electronics and electric controls failure [17,56,58,60]	

The following sub-sections show the events or components considered to build the FT presented in Appendix 1. This FT is built from the different sub-trees that correspond to the four main parts of a WT aforementioned (see also the first column of Table 1). The components and faults that are involved in system failures are obtained from the NIMO and OPTIMUS European Projects. The interrelation between these faults is also done considering the literature. The FT in Appendix 1 is composed by the following four main sub-trees:

- g001 corresponds to a "Foundation and Tower Failure" described in Section 4.1.
- g002 corresponds to a "Critical Rotor Failure" depicted in Section 4.2.
- g003 corresponds to a "Power Train Failure" showed in Section 4.4.
- g004 corresponds to a "Electrical Components Failure" presented in Section 4.3.

4.1. Foundation and Tower

The tower supports the nacelle that is located at a suitable height in order to minimize the influence of turbulence and to maximize the wind energy. The tower is assembled by thin-wall cylindrical elements welded together along their perimeters in three sections that are joined by bolts. This is done in order to enable the transportation of the large structural elements to the wind farm where they need to be assembled [66]. The base section of the tower is installed on a reinforced concrete foundation comprising a square base [67].

Structural defects associated with the tower, foundation, blades and hub, in the form of fatigue cracks, delamination *etc.*, can initiate and evolve with time [44]. The main causes for structural failures are fatigue induced crack initiation and propagation, extreme wind speeds and distribution, extreme turbulences, maximum flow inclination and terrain complexity [39], and also the fire, ice accumulation or lightning bolt strikes. Material fatigue [38] (tower-based fatigue damage has been shown to decrease significantly when using active pitch for the blades [40,43]), impact of blades on the tower, faulty welding and failure of the brakes [45] are the main representative failure modes.

The literature shows that the major faults found on WT towers are: cracks in the concrete base, corrosion, gaps in the foundation section, loosen studs joining the foundation and the first section, loosen bolts joining first/second and second/third sections and welding damages [38].

On the top of the tower, the yaw system turns the nacelle in an optimum angle with respect to the wind direction. Powered by electrical or hydraulic mechanisms (this paper the electrical is considered), the yaw systems can fail due to the failure of the yaw motor or the meteorological unit [46] resulting in a wrong yaw angle. Structural failures could appear when the yaw motor is damaged or it does not have power supply, in addition to extreme wind speed or turbulences and some structural faults. These structural failures can cause the collapse of the tower [38].

4.2. Blade System

The rotor is located inside the nacelle. The blades are attached to the rotor shaft by the hub and they are mounted on bearings in the rotor hub. The blades are the components of the WT with the highest percentage of failures and downtimes [68,69]. Ciang *et al.* reviewed damage detection methods [70] in 2008, considering in particular the blades [42]. The rotor hub supports heavy loads that can lead to faults such as clearance loosening at the blade root, imbalance, cracks and surface roughness [46]. Bearings between blades and hub can be damaged by wear produced by pitting, deformation of outer face and rolling elements of the bearings [46], spalling and overheating [56]. Cracks can appear due to the fatigue [56]. Faults in lubrication and corrosion of pins are typically the main failure cause of bearings.

The blades faults are predominantly related to structural failures, e.g., strength [47] and fatigue of the fibrous composite materials. Other faults, e.g., cracks, erosions, delamination and debonding, could appear in the leading and trailing edges of the blades [48,69]. Delamination and debonding or cracks are found in the shell [49,50], and also in the root section of the blades [51]. The tip deflections (a structural failure of the blade [46]) increase drag near the end of the blades [53].

A common fault of the blades is associated with the failure of the pitch control system [54]. In pitch-controlled turbines, the pitch system is a mechanism that turns the blade, or part of the blade, in order to adjust the angle of attack of the wind. Turbulence of wind is an important cause for pitch system faults [71]. Pitching motion can be done by hydraulic actuators or electric motors. The hydraulic system leads stiffness of bearings, a little backlash and a higher reliability than the electric motors [52]. The hydraulic system can suffer from possible defects such as leakages, overpressure and corrosion [56].

The weather station or meteorological unit provides information about some characteristics of the wind (direction and speed) to the control system of the WT. The main failures found in the WT weather station are related to the vane and the anemometer faults [57]. These can be the cause of a wrong blade angle [50,55].

4.3. Generator, Electrical and Electronic Components

The generator, electrical and electronic components are installed inside the nacelle. The high speed shaft drives the rotational torque to the generator, where the mechanical energy is converted to electrical energy. This conversion needs a specific input speed, or a power electronic equipment to adapt the output energy from the generator to the characteristics of the grid.

Faults in generators can be the result of electrical or mechanical causes [65]. The main electrical faults are due to open-circuits or short-circuit of the winding in the rotor or stator [58] that could cause overheating [46]. Many research works have demonstrated that bearings, rotors and stators involve a high failure rate in WTs [63]. The bearing failures of the generator are usually caused by cracks, asymmetry and imbalance [72]. The rotor and stator failures can be produced by broken bars [64], air-gap eccentricities and dynamic eccentricities, among other failures [58]. Rotor imbalance and aerodynamic asymmetry can have their origin in the non-uniform accumulation of ice and dirt over the blades system [58]. Short-circuit faults, open-circuit faults and gate drive circuit faults are the three major electrical faults of the power electronics and electric controls in WTs [58]. Corrosion, dirt and terminal damage are the main mechanical defects [56]. The group formed by generator, electrical system and control system, has a relevant rate of failures and downtime in WTs.

4.4. Power Train

The power train, or drive train, is installed in the nacelle and is compound by the low speed train, the gearbox and the high speed train. Through the main bearing, the rotor is attached to the low speed shaft that drives the rotational energy to the gearbox. The rotational speed of the rotor is generally between 5 and 30 rpm, and the generator speed is from 750 to 1500 rpm, depending on the type and size of generator. A gearbox is mounted between the rotor and the generator in order to increase the rotational speeds. The gearbox output is driven to the generator through the high speed train. A mechanical brake powered by a hydraulic system is usually mounted in the high speed train as a secondary safe breaking system.

The low speed train failure includes main bearing [56] and low speed shaft defects. Severe vibrations can appear due to impending cracks in any component, or to the mass imbalance in the low speed shaft [58]. The gearbox failure is one of the most typical failures [53]. There are many studies about gearboxes in the literature because their failure causes significant downtimes in the system [73]. The most common faults were found in gear teeth and bearings due to lubrication faults [58], e.g., contamination due to defective sealing [54] or loss of oil [60], wear or fatigue damage which can generate pitting, cracking, gear eccentricity, gear tooth deterioration, offset or other potential faults [46,53].

Overheating can appear in shafts due to the rotational movement of the high speed train. The wear and fatigue, that can initiate cracks [46] and mass imbalance [58], are the principal source of failures in the high speed shaft. The main failure causes of brakes are overpressure or oil leakages [6], cracking of the brake disc and calipers [56].

5. Maintenance Management Approach

The maintenance management proposed in this paper aims to maximise the RAMS of the offshore wind farms optimising the resources such as human or material, conditioned to exogenous variables, e.g., weather conditions [74]. This approach is based on the probability of failure of each WT. The operation of the WT will be focused on a set of components collected by a FT (see Appendix 1). The fault probability of any component is simulated by a statistical function of failure probability over the time (see Appendix 2). Then, the failure probability of a WT is set by the Boolean expression obtained from the BDD. Therefore, according to the resources, the maintenance task will be done in the WTs that present more fault probability over a threshold set. It will lead to predict any preventive/predictive maintenance task over the time. The importance measurements will determine the components that need a maintenance task. A low probability threshold is set to determine if the fault probability of the WT is under control or not. The importance measurement is calculated with the Criticality IM method. The downtime can be defined as the period of time that is required to carry out the corresponding maintenance task. Each event of the fault tree has associated one maintenance task with a specific downtime. The downtime depends on endogenous and exogenous variables. Figure 3 shows the flowchart of the procedure maintenance management.

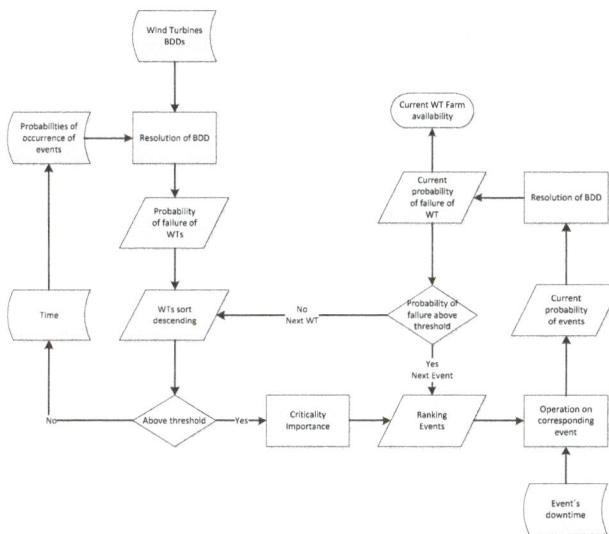

Figure 3. The maintenance management procedure.

6. Case Study

An offshore wind farm composed by 20 WTs has been taken into account. The offshore wind farm has been designed taking into account considerations from expert of the NIMO and OPTIMUS research projects. It has been designed in order to demonstrate and validate the approach proposed in this paper. The WTs are the same type, with the same FT, given in Appendix 1. Different mathematical models have been defined for each event (see Appendix 2). These models have been based on time-dependent probability functions to describe the behavior of events over the time. These probability models are not intended to match exactly the real behavior of the events because there is no dataset to validate it, therefore it they have been set by the aforementioned expert. For example, the event e006 corresponds to the corrosion of the foundation or tower, where a linear increasing probability have been assigned to this event, this is due to the salinity that is assumed to be constant over the time. The main novelty lies in the procedure to elaborate qualitatively and quantitatively a preventive maintenance planning process based on the knowledge of the WTs and on statistical data that, for example, could be collected through condition monitoring systems [75,76]. The probability functions employed are:

I   *Constant probability*

In this model the probability of the event is constant over the time:

$$q(t) = K, (K \in \mathbb{R}/0 \leqslant K \leqslant 1)$$

II   *Exponential increasing probability*

In this model, the probability function assigned is:

$$q(t) = 1 - e^{-\lambda t}, (\lambda \in \mathbb{R}/\lambda \geqslant 0)$$

where λ determines the rising velocity of the probability.

III   *Linear increasing probability*

In this model, the probability function is:

$$q(t) = \begin{cases} mt & mt < 1 \\ 1 & mt \geqslant 1 \end{cases} ; \forall m > 1$$

where m determines the rising velocity of the probability.

IV   *Periodic probability*

This model represents those components that need to be replaced, repaired, and zeroed in a periodical way. In this model, the events have a periodic behavior following the next expression:

$$q(t) = 1 - e^{-\lambda(t - n\alpha)}, n = 1, 2, 3$$

where λ is a positive parameter and determines the rising velocity of the probability, and α is a parameter that defines the size of the time period.

Figure 4 shows the probability of the events of one WT over the time taken into account the probability function assigned to each event. The simulation has been carried out for 600 samples, where each sample can be considered as a period of one day. The objective is to propose an algorithm able to collect stochastic information of the failure probability of a complex system.

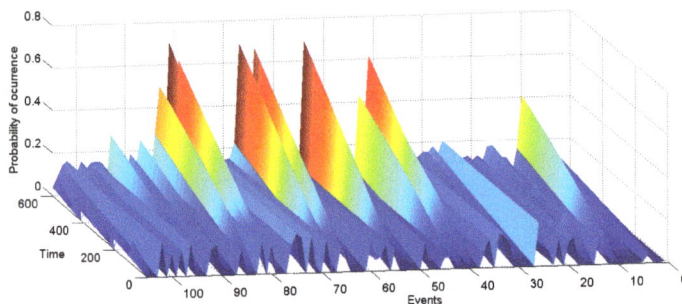

Figure 4. Occurrence probabilities of events.

Considering the last probabilities obtained for each event and the analytical expression of the system failure provided by the BDD, the probability of failure for all WTs of the offshore wind farm can be achieved. Figure 5 presents the failure probability of each WT over the time. The probability of failure for each WT is different among them and over the time, because the values of the parameters that represent the occurrence function of each event are not exactly the same.

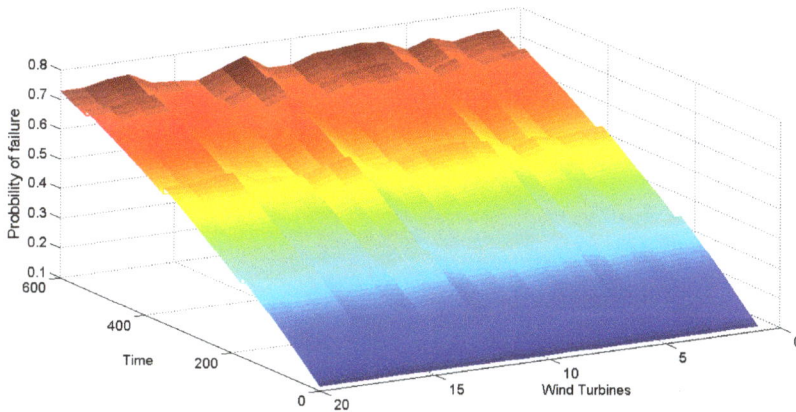

Figure 5. Probabilities of failure of each WT over the time.

The components that require any maintenance task have been set by the importance measurements, specifically by the Criticality IM method. Figure 6 shows the criticality importance of the events of all WTs considered in this case study in a period of time (in this case the study has been considered for a total of 600 days).

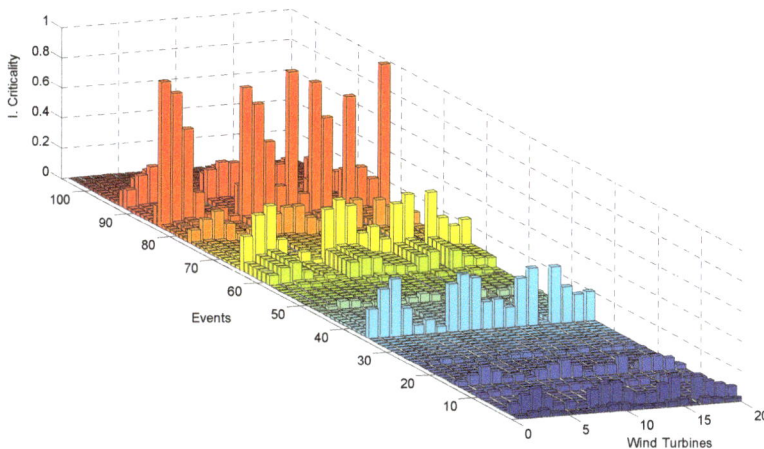

Figure 6. Criticality importance of the events in a given time.

7. Results

The exogenous conditions such as maintenance budget, human and material resources and weather conditions will determine the downtimes, together with the time required to carry out any maintenance task. Figure 7 shows the fault probability over the time of a WT considering different maintenance polices. An upper probability threshold of 0.20 has been established to suggest when the maintenance must be started. Moreover, a lower threshold of 0.15 has been set indicating when the maintenance should be finished. The availability of resources will lead to attend to one or several WTs at the same time.

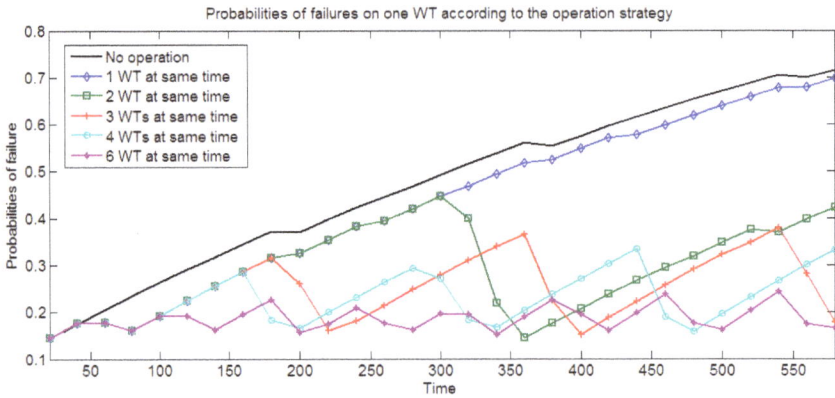

Figure 7. Probabilities of failure of a WT.

The average fault probability of the offshore wind farm according to the resource employed is illustrated in Figure 8. The probability decreases when the potential of maintenance tasks is bigger. In this case study, the average fault probability of the offshore wind farm decreases faster when it is attended at the same time two instead of one WT, than four instead of three WTs. The main conclusion is that a correct resources use could optimize the average fault probability of the offshore wind farm.

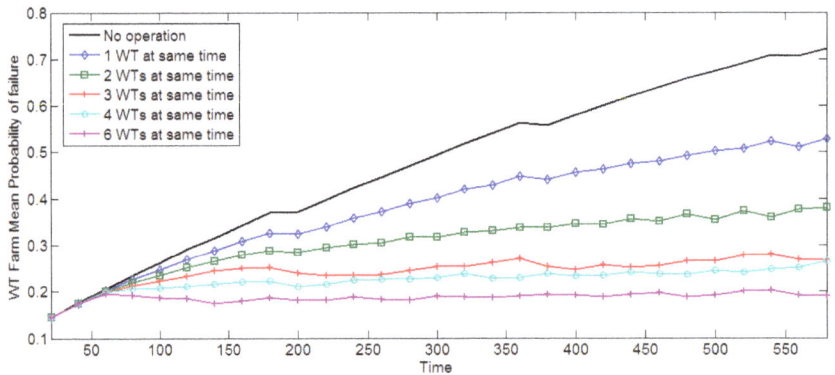

Figure 8. Average fault probability of the offshore wind farm.

The boxplots of Figure 9 show the behavior of the offshore wind farm for different maintenance management policies. The approach lead to control the average probability of failure by a correct maintenance police, and the boxes to be smaller, *i.e.*, presenting a homogeneous probability distribution in all WTs.

The maintenance management performance for offshore wind farms is subject to several uncertainties related to the randomness of exogenous conditions, e.g., weather conditions [77]. Therefore, the approach presented requires weather forecasting. Weather forecasting depends on the temperature, dew point, wind velocity, pressure, visibility, cloud height and quantity [4]. In addition, the state of the sea, the wind and the wave heights need to be considered. There are some probabilistic models based on historical wave height data that are used to determine the conditions of the sea in a certain moment, e.g., the Markovian wave height model [78], forecasting of safe sea-state using finite elements method and artificial neural networks [79], short-term predictions based on nonlinear deterministic time series analysis [80], Gaussian processes [81], resampling methods, parametric models, *etc.*

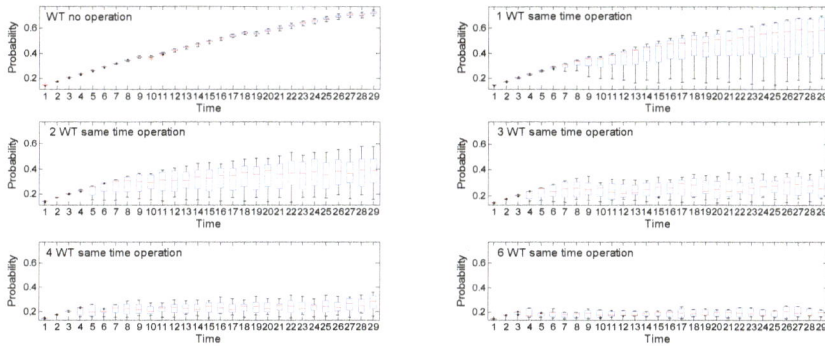

Figure 9. Boxplot of the fault probability of the offshore wind farm for WT operated at the same time.

The maintenance task will be carried out when certain permission value is reached. This dimensionless value, which varies from 0 to 1, will be given by a weighting of the weather conditions and external permissions. It has been simulated in this paper and validated by experts. Figure 10 shows the maximum allowed value assigned to each event. The maximum allowed valued is randomly generated for this case. It is due to the goal of this study is to clarify how the proposed methodology should be applied, taking into account that the method is close to the reality only from the qualitative point of view. This value is compared with a predicted value given randomly in this paper in order to consider the stochastic of the system. If the value assigned to the task is bigger than the predicted value, the maintenance task must be carried out, in other case, it must be necessary to wait for a suitable value from the forecasting.

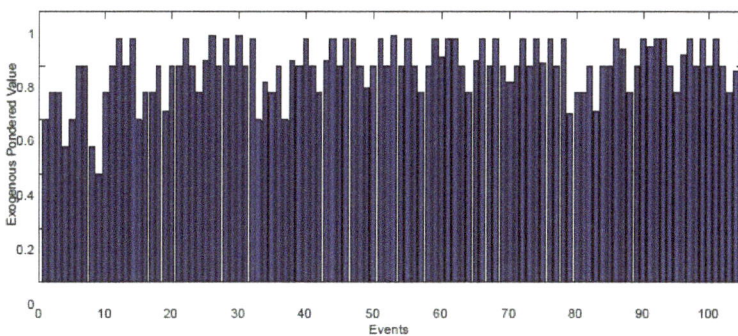

Figure 10. Maximum allowed exogenous pondered value for each maintenance tasks.

Figure 11 shows a randomized forecasting value of the weather conditions given for each day (sample) evaluated in the example. This figure can be used to determine the tasks that can be performed according to the exogenous variables. For example, in the 100th day (green circle) there is a value of 0.2 (this value is a ponderation between temperature, dew point, wind velocity, pressure, visibility,

179

etc.), *i.e.*, any maintenance task can be carried out because this value is lower than all the maximum allowed exogenous pondered values. However, in the 300th day (red circle) none of the tasks can be carried out because the value is higher than the allowable value in all the cases.

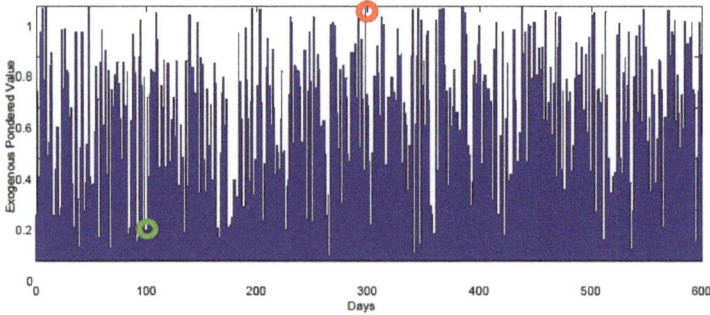

Figure 11. Representative exogenous pondered value forecasting per day.

Figure 12 represents the weather influence on the distribution of the failure probabilities of the WTs over the time. Different weather scenarios have been taken into account randomly in order to evaluate the weather conditions and the influence to the maintenance tasks.

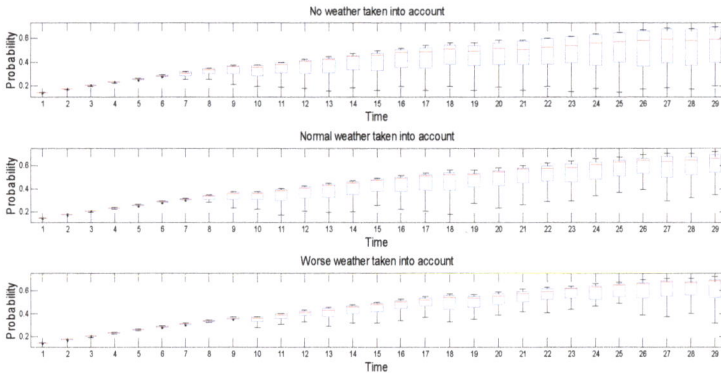

Figure 12. Influence of exogenous variables on the state of the offshore wind farm.

In the top boxplot of Figure 12, the weather conditions have not been taken into account. In the second one, the weather forecasting presented in Figure 11 has been considered. In the last one boxplot, an adverse weather conditions have been established. The presence of adverse weather conditions makes to increase the average fault probability of the offshore wind farm, and the size of the boxes of boxplot decreases because the maintenance tasks that can be done are minimum.

8. Conclusions

The offshore wind energy is being supported by the international community. Offshore wind farms employ large and complex wind turbines that generate more power electricity than onshore. The farms are located in places with difficulty to access that depends of the weather conditions. These conditions have leaded the development of robust remote condition monitoring system in order to increase the RAMS of the offshore wind farms.

This paper presents the BDD in order to evaluate qualitatively the FTA of a WT. The approach is based on the fault probabilities of each component of the WT, that depend of a statistical function of probability of occurrence over the time. The fault probability of the WT has been set by the Boolean expression obtained by the BDD. An optimal ranking of the events has been done for minimising the computational cost.

The IMs have been employed in order to facilitate the improvement of the maintenance management and the resources deployment in an offshore wind farm, where a proper and accurate prioritization of the basic events has been elaborated according to Criticality IM method.

The maintenance management approach proposed in this paper maximise the RAMS of the offshore wind farm, optimising the resources as human, materials, *etc.* The maintenance task will be carried out in the WTs that present more fault probability over a threshold. It will lead to establishment of preventive/predictive maintenance tasks over time. A low probability threshold has also been set to determine when the fault probability of the WT is under control. The time to carry out a maintenance task has been established by the downtime associated to each failure. The downtime depends on the time to repair or replace the component, human resourcesstate of the sea, *etc.*

It has been demonstrated that the average fault probability of the offshore wind farm decreases more when two instead of one WT can be attended at the same time than between four instead of three. The main conclusion is that there is a reasonable amount of resources that allow controlling the average fault probability of the offshore wind farm, and this method can be used to calculate this value.

The weather conditions have been also considered. The average fault probability of the offshore wind farm increases when there is a presence of adverse weather conditions. The adverse weather increases the gap between the failure probabilities of the different WTs that compose the wind farm because the maintenance tasks that can be done are minimum.

The dynamic analysis proposed in this paper can be used to improve the maintenance planning using the fault probability of the system over the time. The fault probability and the IMs determine when the maintenance tasks must be carried out and to set the tasks over the events.

The qualitative data used in this paper is gathered from several research projects and the results have been validated by experts involve in the research projects. The main novelty of the paper is the procedure to analyse endogenous and exogenous data using graphical tools.

Acknowledgments: The work reported herewith has been financially supported by the European Commission under the European FP7 OPTIMUS project [41], and the Spanish Ministerio de Economía y Competitividad, under Research Grant DPI2012-31579.

Author Contributions: Alberto Pliego, Fausto García and Jesús Pinar conceived and designed the experiments; Alberto Pliego, Fausto García and Jesús Pinar performed the experiments; Alberto Pliego, Fausto García and Jesús Pinar analyzed the data Alberto Pliego, Fausto García and Jesús Pinar contributed reagents/materials/ analysis tools; Alberto Pliego, Fausto García and Jesús Pinar wrote the paper.

Conflicts of Interest: The authors declare no conflict of interest.

Appendix 1. FT for a Wind Turbine

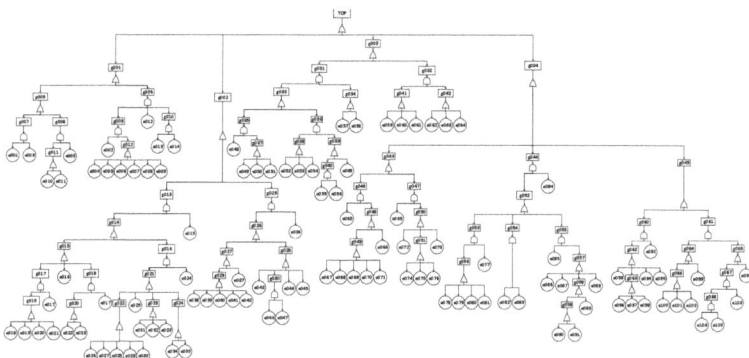

Appendix 2. Events and Probabilistic Models

Fault Tree 1 Foundation and Tower Failure				Probabilistic Model Assignment
Intermediate Event	Code	Final Event	Code	
Yaw System Failure	g005	Yaw motor fault	e001	Constant
Critical Structural Failure	g006	Abnormal Vibration I	e002	Linear Increasing
yaw motor failure	g007	Abnormal Vibration H	e003	Linear Increasing
Wrong Yaw Angle	g008	Cracks in concrete base	e004	Constant
Structural Failure (Foundation and tower)	g009	Welding damage	e005	Constant
No electric power for yaw motor	g010	Corrosion	e006	Linear Increasing
Metereologhical Unit Failure	g011	Loosen studs in joining foundation and first section	e007	Linear Increasing
Structural Fault (Foundation and tower)	g012	Loosen bolts in joining different sections	e008	Linear Increasing
		Gaps in the foundation section	e009	Exponential Increasing
		Vane damage	e010	Exponential Increasing
		Anemometer damage	e011	Exponential Increasing
		High wind speed	e012	Periodic
		No power supply from generator	e013	Constant
		No power supply from grid	e014	Constant
Fault Tree 2 Critical Rotor Failure				Probabilistic Model Assignment
Intermediate Event	Code	Final Event	Code	
Critical blade failure	g013	High wind speed	e015	Periodic
Blade Failure	g014	Blade Angle asymmetry	e016	Exponential Increasing
Pitch System Failure	g015	Abnormal Vibration A	e017	Exponential Increasing
Critical structural Failure (Blades)	g016	Motor failure	e018	Exponential Increasing
Hydraulic system Failure	g017	Leakages	e019	Constant
Wrong Blade Angle	g018	Over pressure	e020	Constant
Hydraulic system Fault	g019	Corrosion	e021	Exponential Increasing
Metereologhical Unit Failure	g020	Vane damage	e022	Constant
Structural Failure (Blades)	g021	Anemometer damage	e023	Constant
Leading and trailling edges	g022	Abnormal Vibration B	e024	Constant
Shell	g023	Root Cracks	e025	Constant
Tip	g024	Cracks	e026	Constant
Rotor System Failure	g025	Erosion	e027	Exponential Increasing
Rotor System Fault	g026	Delamination in leading edges of blades	e028	Exponential Increasing
Bearings (Rotor)	g027	Delamination in trailing edges of blades	e029	Exponential Increasing
Rotor Hub	g028	Debonding in edges of blades	e030	Exponential Increasing
Wear	g029	Delamination in shell	e031	Exponential Increasing
Imbalance	g030	Crack with structural damage	e032	Constant
		Crack on the beam-shell joint	e033	Constant
		Open tip	e034	Constant
		Lightning strike	e035	Periodic
		Abnormal Vibration C	e036	Constant
		Cracks	e037	Constant
		Corrosion of Pins	e038	Exponential Increasing
		Abrasive Wear	e039	Exponential Increasing
		Pitting	e040	Linear Increasing
		Deformation of face & rolling element	e041	Linear Increasing
		Lubrication Fault	e042	Linear Increasing
		Clearance loosening at root	e043	Exponential Increasing
		Cracks	e044	Constant
		Surface Roughness	e045	Constant
		Mass Imbalance	e046	Exponential Increasing
		Fault in Pitch adjustment	e047	Exponential Increasing

Appendix 2. *Cont.*

Fault Tree 3 Electrical Components Failure				Probabilistic Model Assignment
Intermediate Event	Code	Final Event	Code	
Critical Generator Failure	g031	Abnormal Vibration G	e048	Exponential Increasing
Power Electronics and Electric Controls Failure	g032	Cracks	e049	Constant
Mechanical Failure (Generator)	g033	Imbalance	e050	Exponential Increasing
Electrical Failure (Generator)	g034	Asymmetry	e051	Exponential Increasing
Bearing Generator Failure	g035	Air-Gap eccentricities	e052	Linear Increasing
Rotor and Stator Failure	g036	Broken bars	e053	Linear Increasing
Bearing Generator Fault	g037	Dynamic eccentricity	e054	Linear Increasing
Rotor and Stator Fault	g038	Sensor T error	e055	constant
Abnormal Signals A	g039	T above limit	e056	Periodic
Overwarming generator	g040	Short Circuit (Gen)	e057	Constant
Electrical Fault (PE)	g041	Open Circuit (Gen)	e058	Constant
Mechanical Fault (PE)	g042	Short Circuit	e059	Constant
		Open Circuit	e060	Constant
		Gate drive circuit	e061	linear increasing
		Corrosion	e062	Periodic
		Dirt	e063	Periodic
		Terminals damage	e064	linear increasing

Fault Tree 4 Power Train Failure				Probabilistic Model Assignment
Intermediate Event	Code	Final Event	Code	
Low speed train Failure	g043	Abnormal Vibration D	e065	Constant
Critical Gearbox Failure	g044	Cracks in main bearing	e066	Constant
High speed train Failure	g045	Spalling	e067	Linear Increasing
Main Bearing failure	g046	Corrosion of Pins	e068	Linear Increasing
Low speed shaft failure	g047	Abrasive Wear	e069	Constant
Main Bearing fault	g048	Deformation of face & rolling element	e070	Linear Increasing
Wear main bearing	g049	Pitting	e071	exponential increasing
Low speed shaft fault	g050	Imbalance	e072	Constant
Wear low shaft	g051	Cracks in l.s. shaft	e073	Linear Increasing
Gearbox Fault	g052	Spalling	e074	Constant
Bearings failure(Gearbox)	g053	Abrasive Wear	e075	Constant
Lubrication fault	g054	Pitting	e076	Constant
Gear Failure	g055	Abnormal Vibration F	e077	Linear Increasing
Wear bearing gearbox	g056	Corrosion of Pins	e078	Exponential Increasing
Gear Fault	g057	Abrasive Wear	e079	Linear Increasing
Tooth Wear	g058	Pitting	e080	Constant
Offset	g059	Deformation of face & rolling element	e081	Linear Increasing
High speed shaft Failure	g060	Oil Filtration	e082	Constant
Critical Brake Failure	g061	Particle Contamination	e083	Exponential Increasing
High speed structural damage	g062	Overwarming gearbox	e084	Linear Increasing
Wear high shaft	g063	Abnormal Vibration E	e085	Periodic
Brake Fault	g064	Eccentricity	e086	Constant
Abnormal Signals B	g065	Pitting	e087	Linear Increasing
Hydraulic brake system Fault	g066	Cracks in gears	e088	Exponential Increasing
Abnormal Signals C	g067	Gear tooth deterioration	e089	Exponential Increasing
Overwarming brake	g068	Poor design	e090	Periodic
		Tooth surface defects	e091	Constant
		Abnormal Vibration J	e092	Constant
		Cracks in h.s. shaft	e093	Linear Increasing
		Imbalance	e094	Periodic
		Overwarming	e095	Exponential Increasing
		Spalling	e096	Constant
		Abrasive Wear	e097	Linear Increasing
		Pitting	e098	Constant
		Cracks in brake disk	e099	Exponential Increasing
		Motor brake fault	e100	Constant
		Oil Leakage	e101	Linear Increasing
		Over pressure	e102	Constant
		Abnormal speed	e103	Linear Increasing
		T sensor error	e104	Periodic
		T above limit	e105	Periodic

References

1. Márquez, F.P.G.; Tobias, A.M.; Pérez, J.M.P.; Papaelias, M. Condition monitoring of wind turbines: Techniques and methods. *Renew. Energy* **2012**, *46*, 169–178. [CrossRef]
2. Esteban, M.D.; Diez, J.J.; López, J.S.; Negro, V. Why offshore wind energy? *Renew. Energy* **2011**, *36*, 444–450. [CrossRef]
3. *Guidelines for the Certification of Condition Monitoring Systems for Wind Turbines*; Germanisher LLoyd: Hamburg, Germany, 2007.
4. Tavner, P. *Offshore Wind Turbines Reliability, Availability and Maintenance*; The Institution of Engineering and Technology: London, UK, 2012.
5. Novaes Pires, G.; Alencar, E.; Kraj, A. Remote Conditioning Monitoring System for a Hybrid Wind Diesel System-Application at Fernando de Naronha Island. Brasil. Available online: http://www.globalislands.net/userfiles/_brazil_FdNpdf2.pdf (accessed on 10 July 2015).
6. Tsai, C.-S.; Hsieh, C.-T.; Huang, S.-J. Enhancement of damage-detection of wind turbine blades via CWT-based approaches. *IEEE Trans. Energy Convers.* **2006**, *21*, 776–781. [CrossRef]
7. Guo, P.; Bai, N. Wind turbine gearbox condition monitoring with AAKR and moving window statistic methods. *Energies* **2011**, *4*, 2077–2093. [CrossRef]
8. Chen, Z.; Guerrero, J.M.; Blaabjerg, F. A review of the state of the art of power electronics for wind turbines. *IEEE Trans. Power Electronics* **2009**, *24*, 1859–1875. [CrossRef]
9. Jiang, W.; Fan, Q.; Gong, J. Optimization of welding joint between tower and bottom flange based on residual stress considerations in a wind turbine. *Energy* **2010**, *35*, 461–467. [CrossRef]
10. Pérez, J.M.P.; Márquez, F.P.G.; Tobias, A.; Papaelias, M. Wind turbine reliability analysis. *Renew. Sustain. Energy Rev.* **2013**, *23*, 463–472. [CrossRef]
11. Soua, S.; van Lieshout, P.; Perera, A.; Gan, T.-H.; Bridge, B. Determination of the combined vibrational and acoustic emission signature of a wind turbine gearbox and generator shaft in service as a pre-requisite for effective condition monitoring. *Renew. Energy* **2013**, *51*, 175–181. [CrossRef]
12. Chacon, J.L.F.; Andicoberry, E.A.; Kappatos, V.; Asfis, G.; Gan, T.-H.; Balachandran, W. Shaft angular misalignment detection using acoustic emission. *Appl. Acoust.* **2014**, *85*, 12–22. [CrossRef]
13. Park, S.; Inman, D.J.; Yun, C.-B. An outlier analysis of MFC-based impedance sensing data for wireless structural health monitoring of railroad tracks. *Eng. Struct.* **2008**, *30*, 2792–2799. [CrossRef]
14. De la Hermosa González, R.R.; Márquez, F.P.G.; Dimlaye, V.; Ruiz-Hernández, D. Pattern recognition by wavelet transforms using macro fibre composites transducers. *Mech. Syst. Signal Proc.* **2014**, *48*, 339–350. [CrossRef]
15. Nie, M.; Wang, L. Review of condition monitoring and fault diagnosis technologies for wind turbine gearbox. *Procedia CIRP* **2013**, *11*, 287–290. [CrossRef]
16. Zeng, Z.; Tao, N.; Feng, L.; Li, Y.; Ma, Y.; Zhang, C. Breakpoint detection of heating wire in wind blade moulds using infrared thermography. *Infrared Phys. Technol.* **2014**, *64*, 73–78. [CrossRef]
17. García Márquez, F.P.; Pinar Pérez, J.M.; Pliego Marugán, A.; Papaelias, M. Identification of critical components of wind turbines using FTA over the time. *Renew. Energy* **2016**, *87*, 869–883. [CrossRef]
18. Lambert, H.E. *Measures of Importance of Events and Cut Sets. Reliability and Fault Tree Analysis*; SIAM: Philadelphia, PA, USA, 1975; pp. 77–100.
19. Pliego Marugán, A.; García, F.P. A novel approach to diagnostic and prognostic evaluations applied to railways: A real case study. *J. Rail Rapid Transit* **2015**. [CrossRef]
20. Sinnamon, R.M.; Andrews, J.D. Fault tree analysis and binary decision diagrams. In Proceedings of the Reliability and Maintainability Symposium, Las Vegas, NV, USA, 22–25 January 1996; pp. 215–222.
21. Jinglun, Z.; Quan, S. Reliability analysis based on binary decision diagrams. *J. Qual. Maint. Eng.* **1998**, *4*, 150–161. [CrossRef]
22. Bryant, R.E. Graph-based algorithms for Boolean function manipulation. *IEEE Trans. Comput.* **1986**, *100*, 677–691. [CrossRef]
23. Remenyte, R.; Andrews, J.D. Qualitative analysis of complex modularized fault trees using binary decision diagrams. *Proc. Inst. Mech. Eng. O* **2006**, *220*, 45–53. [CrossRef]

24. Prescott, D.R.; Remenyte-Prescott, R.; Reed, S.; Andrews, J.; Downes, C. A reliability analysis method using binary decision diagrams in phased mission planning. *Proc. Inst. Mech. Eng. Part O* **2009**, *223*, 133–143. [CrossRef]
25. Moret, B.M. Decision trees and diagrams. *ACM Comput. Surv.* **1982**, *14*, 593–623. [CrossRef]
26. Lee, C.-Y. Representation of switching circuits by binary-decision programs. *Bell Syst. Technol. J.* **1959**, *38*, 985–999. [CrossRef]
27. Akers, S.B. Binary decision diagrams. *IEEE Trans. Comput.* **1978**, *100*, 509–516. [CrossRef]
28. Pliego Marugán, A.; García Márquez, F.P.; Lorente, J. Decision making process via binary decision diagram. *Int. J. Manag. Sci. Eng. Manag.* **2015**, *10*, 3–8. [CrossRef]
29. Fujita, M.; Fujisawa, H.; Kawato, N. Evaluation and improvements of Boolean comparison method based on binary decision diagrams. In Proceedings of the Computer-Aided Design IEEE International Conference (ICCAD-88), Santa Clara, CA, USA, 7–10 November 1988; pp. 2–5.
30. Márquez, F.P.G.; Mangurán, A.P.; Zaman, N. For information systems design. *Softw. Dev. Technol. Constr. Inf. Syst. Des.* **2013**, *1*, 308–318.
31. Brace, K.S.; Rudell, R.L.; Bryant, R.E. Efficient implementation of a BDD package. In Proceedings of the 27th ACM/IEEE Design Automation Conference, Orlando, FL, USA, 24–28 June 1991; pp. 40–45.
32. Liu, Q.; Homma, T. A new computational method of a moment-independent uncertainty importance measure. *Reliab. Eng. Syst. Saf.* **2009**, *94*, 1205–1211. [CrossRef]
33. Cheok, M.C.; Parry, G.W.; Sherry, R.R. Use of importance measures in risk-informed regulatory applications. *Reliab. Eng. Syst. Saf.* **1998**, *60*, 213–226. [CrossRef]
34. Birnbaum, Z.W. *On the Importance of Different Components in a Multicomponent System*; Washington University Seattle Lab of Statistical Research: Washington, DC, USA, 1968.
35. Arabian-Hoseynabadi, H.; Oraee, H.; Tavner, P. Failure modes and effects analysis (FMEA) for wind turbines. *Int. J. Electr. Power Energy Syst.* **2010**, *32*, 817–824. [CrossRef]
36. RELIAWIND Project. European Union's Seventh Framework Programme for RTD (FP7). Available online: http://www.reliawind.eu/ (accessed on 22 January 2014).
37. Lotsberg, I. Structural mechanics for design of grouted connections in monopile wind turbine structures. *Mar. Struct.* **2013**, *32*, 113–135. [CrossRef]
38. Chou, J.-S.; Tu, W.-T. Failure analysis and risk management of a collapsed large wind turbine tower. *Eng. Fail. Anal.* **2011**, *18*, 295–313. [CrossRef]
39. International Electrotechnical Commission. *Wind Turbine—Part 1: Design Requirements, IEC 61400-1*; International Electrotechnical Commission: Geneva, Switzerland, 2005.
40. Development and Demonstration of a Novel Integrated Condition Monitoring System for Wind Turbines, NIMO Project. (NIMO, Ref.:FP7-ENERGY-2008-TREN-1: 239462). Available online: http://www.nimoproject.eu (accessed on 30 January 2012).
41. Demonstration of Methods and Tools for the Optimisation of Operational Reliability of Large-Scale Industrial Wind Turbines, OPTIMUS Project. (OPTIMUS, Ref.: FP-7-Energy-2012-TREN-1: 322430). Available online: http://www.optimusproject.eu (accessed on 25 February 2014).
42. Ciang, C.C.; Lee, J.-R.; Bang, H.-J. Structural health monitoring for a wind turbine system: A review of damage detection methods. *Meas. Sci. Technol.* **2008**, *19*. [CrossRef]
43. Stol, K.A. Disturbance tracking control and blade load mitigation for variable-speed wind turbines. *J. Sol. Energy Eng.* **2003**, *125*, 396–401. [CrossRef]
44. Caithness Windfarm Information Forum. Available online: http://www.caithnesswindfarms.co.uk/ (accessed on 30 January 2012).
45. Cotton, I.; Jenkins, N.; Pandiaraj, K. Lightning protection for wind turbine blades and bearings. *Wind Energy* **2001**, *4*, 23–37. [CrossRef]
46. Hameed, Z.; Hong, Y.; Cho, Y.; Ahn, S.; Song, C. Condition monitoring and fault detection of wind turbines and related algorithms: A review. *Renew. Sustain. Energy Rev.* **2009**, *13*, 1–39. [CrossRef]
47. Padgett, W. A multiplicative damage model for strength of fibrous composite materials. *IEEE Trans. Reliab.* **1998**, *47*, 46–52. [CrossRef]
48. Jørgensen, E.R.; Borum, K.K.; McGugan, M.; Thomsen, C.; Jensen, F.M.; Debel, C.; Sørensen, B.F. *Full Scale Testing of Wind Turbine Blade to Failure-Flapwise Loading*; RISØ National Laboratory: Copenhagen, Denmark, 2004.

49. Jensen, F.M.; Falzon, B.; Ankersen, J.; Stang, H. Structural testing and numerical simulation of a 34m composite wind turbine blade. *Compos. Struct.* **2006**, *76*, 52–61. [CrossRef]

50. Borum, K.K.; McGugan, M.; Brondsted, P. Condition monitoring of wind turbine blades. In Proceedings of the 27th Riso International Symposium on Materials Science: Polymer Composite Materials for Wind Power Turbines, Denmark, 4–7 September 2006; pp. 139–145.

51. Van Leeuwen, H.; van Delft, D.; Heijdra, J.; Braam, H.; Jorgensen, E.; Lekou, D.; Vionis, P. *Comparing Fatigue Strength from Full Scale Blade Tests with Coupon-Based Predictions*; American Society of Mechanical Engineers: New York, NY, USA, 2002; pp. 1–9.

52. Griffin, D.A.; Zuteck, M.D. Scaling of composite wind turbine blades for rotors of 80 to 120 meter diameter. *J. Sol. Energy Eng.* **2001**, *123*, 310–318. [CrossRef]

53. Herbert, G.J.; Iniyan, S.; Sreevalsan, E.; Rajapandian, S. A review of wind energy technologies. *Renew. Sustain. Energy Rev.* **2007**, *11*, 1117–1145. [CrossRef]

54. Gray, C.S.; Watson, S.J. Physics of failure approach to wind turbine condition based maintenance. *Wind Energy* **2010**, *13*, 395. [CrossRef]

55. Maughan, J.R. Technology and reliability improvements in GE's 1.5 MW WT fleet. In Proceedings of the 2nd WT Reliability Workshop, Albuquerque, NM, USA, 17–18 September 2007.

56. Liu, W.; Tang, B.; Jiang, Y. Status and problems of wind turbine structural health monitoring techniques in china. *Renew. Energy* **2010**, *35*, 1414–1418. [CrossRef]

57. Parent, O.; Ilinca, A. Anti-icing and de-icing techniques for wind turbines: Critical review. *Cold Reg. Sci. Technol.* **2011**, *65*, 88–96. [CrossRef]

58. Lu, B.; Li, Y.; Wu, X.; Yang, Z. A review of recent advances in wind turbine condition monitoring and fault diagnosis. In Proceedings of the Power Electronics and Machines in Wind Applications (PEMWA), Lincoln, NM, USA, 24–26 June 2009; pp. 1–7.

59. Ribrant, J. Reliability Performance and Maintenance—A Survey of Failures in Wind Power Systems. Ph.D. Thesis, KTH School of Electrical Engineering, Stockholm, Sweden, 2006.

60. Fischer, K.; Besnard, F.; Bertling, L. A limited-scope reliability-centred maintenance analysis of wind turbines. In Proceedings of the European Wind Energy Conference and Exhibition EWEA 2011, Brussels, Belgium, 14–17 March 2011; pp. 89–93.

61. Feng, Y.; Qiu, Y.; Crabtree, C.J.; Long, H.; Tavner, P.J. Use of SCADA and CMS signals for failure detection and diagnosis of a wind turbine gearbox. In Proceedings of the European Wind Energy Conference and Exhibition 2011, Sheffield, UK, 2011; pp. 17–19.

62. Entezami, M.; Hillmansen, S.; Weston, P.; Papaelias, M. Fault detection and diagnosis within a WT mechanical braking system. In Proceedings of the International Conference on Condition Monitoring and Machinery Failure Prevention Technologies (CM 2012 and MFPT 2011), Cardiff, UK, 20–22 June 2011.

63. Popa, L.M.; Jensen, B.-B.; Ritchie, E.; Boldea, I. Condition monitoring of wind generators. In Proceedings of the Industry Applications Conference (38th IAS Annual Meeting), Salt Lake City, UT, USA, 12–16 October 2003; pp. 1839–1846.

64. Douglas, H.; Pillay, P.; Ziarani, A. Broken rotor bar detection in induction machines with transient operating speeds. *IEEE Trans. Energy Convers.* **2005**, *20*, 135–141. [CrossRef]

65. Hansen, A.D.; Michalke, G. Fault ride-through capability of DFIG wind turbines. *Renew. Energy* **2007**, *32*, 1594–1610. [CrossRef]

66. Bazeos, N.; Hatzigeorgiou, G.; Hondros, I.; Karamaneas, H.; Karabalis, D.; Beskos, D. Static, seismic and stability analyses of a prototype wind turbine steel tower. *Eng. Struct.* **2002**, *24*, 1015–1025. [CrossRef]

67. Scottishpower SP Transmission Ltd. Black Law Wind Farm Extension Grid Connection Environmental Statement. Available online: http://www.spenergynetworks.co.uk/userfiles/file/Black_Law_Environmental_Statement_Windfarm_Extension_Grid_Connection.pdf (accessed on 20 July 2015).

68. Van Bussel, G.; Zaaijer, M. *Estimation of Turbine Reliability Figures within the DOWEC Project*; DOWEC Report Nr. 10048; The Netherlands; Issue 4, October; 2003.

69. García, F.P.; Pedregal, D.J.; Roberts, C. Time series methods applied to failure prediction and detection. *Reliab. Eng. Syst. Saf.* **2010**, *95*, 698–703. [CrossRef]

70. Márquez, F.P.; Chacón Muñoz, J.M.; Tobias, A.M. B-spline approach for failure detection and diagnosis on railway point mechanisms case study. *Qual. Eng.* **2015**, *27*, 177–185. [CrossRef]

71. Tavner, P.; Qiu, Y.; Korogiannos, A.; Feng, Y. The Correlation between Wind Turbine Turbulence and Pitch Failure. In Proceedings of European Wind Energy Conference & Exhibition, Brussels, Belgium, 14–17 March 2011.

72. Wu, A.P.; Chapman, P.L. Simple expressions for optimal current waveforms for permanent-magnet synchronous machine drives. *IEEE Trans. Energy Conver.* **2005**, *20*, 151–157. [CrossRef]

73. Spinato, F.; Tavner, P.J.; van Bussel, G.J.W.; Koutoulakos, E. IET Reliability of WT subassemblies. *Renew. Power Gener.* **2009**, *3*, 387–401. [CrossRef]

74. De la Hermosa González, R.R.; Márquez, F.P.G.; Dimlaye, V. Maintenance management of wind turbines structures via mfcs and wavelet transforms. *Renew. Sustain. Energy Rev.* **2015**, *48*, 472–482. [CrossRef]

75. Marquez, F.P.G. An approach to remote condition monitoring systems management. In Proceedings of the Institution of Engineering and Technology International Conference on Railway Condition Monitoring, Birmingham, UK, 29–30 November 2006; pp. 156–160.

76. Márquez, F.P.G.; Pedregal, D.J.; Roberts, C. New methods for the condition monitoring of level crossings. *Int. J. Syst. Sci.* **2015**, *46*, 878–884. [CrossRef]

77. Vasquez, T. *Weather Forecasting Handbook*; Weather Graphics Technologies: Garland, TX, USA, 2002; ISBN: 0970684029.

78. Sørensen, J.D. Framework for risk-based planning of operation and maintenance for offshore wind turbines. *Wind Energy* **2009**, *12*, 493–506. [CrossRef]

79. Rothkopf, M.H.; McCarron, J.K.; Fromovitz, S. A weather model for simulating offshore construction alternatives. *Manag. Sci.* **1974**, *20*, 1345–1349. [CrossRef]

80. Yasseri, S.; Bahai, H.; Bazargan, H.; Aminzadeh, A. Prediction of safe sea-state using finite element method and artificial neural networks. *Ocean Eng.* **2010**, *37*, 200–207. [CrossRef]

81. Härdle, W.; Horowitz, J.; Kreiss, J.P. Bootstrap methods for time series. *Int. Stat. Rev.* **2003**, *71*, 435–459. [CrossRef]

energies

MDPI

Article

Optimal Coordinated Control of Power Extraction in LES of a Wind Farm with Entrance Effects

Jay P. Goit, Wim Munters and Johan Meyers *

Department of Mechanical Engineering, University of Leuven, Celestijnenlaan 300A, Leuven B3001, Belgium;
goitjay@gmail.com (J.P.G.); wim.munters@kuleuven.be (W.M.)
*   Correspondence: johan.meyers@kuleuven.be; Tel.: +32-0-1632-2502; Fax: +32-0-1632-2985

Academic Editor: Frede Blaabjerg
Received: 4 November 2015; Accepted: 29 December 2015; Published: 6 January 2016

Abstract: We investigate the use of optimal coordinated control techniques in large eddy simulations of wind farm boundary layer interaction with the aim of increasing the total energy extraction in wind farms. The individual wind turbines are considered as flow actuators, and their energy extraction is dynamically regulated in time, so as to optimally influence the flow field. We extend earlier work on wind farm optimal control in the fully-developed regime (Goit and Meyers 2015, J. Fluid Mech. 768, 5–50) to a 'finite' wind farm case, in which entrance effects play an important role. For the optimal control, a receding horizon framework is employed in which turbine thrust coefficients are optimized in time and per turbine. Optimization is performed with a conjugate gradient method, where gradients of the cost functional are obtained using adjoint large eddy simulations. Overall, the energy extraction is increased 7% by the optimal control. This increase in energy extraction is related to faster wake recovery throughout the farm. For the first row of turbines, the optimal control increases turbulence levels and Reynolds stresses in the wake, leading to better wake mixing and an inflow velocity for the second row that is significantly higher than in the uncontrolled case. For downstream rows, the optimal control mainly enhances the sideways mean transport of momentum. This is different from earlier observations by Goit and Meyers (2015) in the fully-developed regime, where mainly vertical transport was enhanced.

Keywords: large eddy simulations; wind farm; turbulent boundary layers; wind farm control; optimization; adjoints

1. Introduction

The size of wind farms has increased rapidly in recent years, and the power production of some of the largest farms is comparable to that of conventional power plants. The largest offshore wind farm to date is the 630-MW London Array with 175 turbines spread over an area of 100 km^2. At these sizes, the efficiency of individual turbines in the wind farms differs considerably from that of a lone-standing turbine. It is well known that wake accumulation and the interaction of the wind farm with the atmospheric boundary layer lead to a decrease in energy extraction downstream in the farm that can amount up to 40% and more [1,2]. Moreover, increased turbulence intensities and wake meandering also lead to higher turbine loading. The current work investigates coordinated optimal control of wind turbines in a wind farm, focusing on improving energy extraction. To this end, large eddy simulations (LESs) of a wind farm boundary layer are performed, where the LES model itself is used as a control model in a receding horizon optimal control framework. Individual turbines are considered as flow actuators, whose energy extraction can be regulated dynamically in time and per turbine. Such an approach is infeasible as a real wind farm controller, as computational costs are prohibitive. Instead, the methodology is used as a means to explore the potential of coordinated control in wind farms, without excluding *a priori* any of the turbulent flow physics that could potentially improve performance. Recently, this approach was used by Goit and Meyers [3] for fully-developed 'infinite' wind farm boundary layers. In the current work, we extend this approach to a 'finite' farm in which entrance effects and boundary layer development play an important role.

In the past, studies on increasing energy extraction in wind farms have mainly focused on optimization of static power set-points of individual turbines throughout the farm. One of the first studies of this kind was performed by Steinbuch *et al.* [4] They proposed a concept of downrating the power output from upwind turbines in a farm, so that the wind speed in their wake would be higher, leading to higher energy extraction by downwind turbines and possibly an overall increase of power extraction. Later, many other studies followed this approach [5–9]. Unfortunately, in most of these studies, validation of the controllers in experiments or high fidelity turbulence-resolving flow simulations (such as LES) were impossible. Instead, simple wake engineering models were mostly used. Only recently, Gebraad [9] performed a detailed analysis using large eddy simulations, finding that static downrating of upstream turbines is not effective, as the reduction of wake deficit diffuses too much to be fully captured by the downstream turbines. Another static approach to increase energy extraction in wind farms is the use of non-zero yaw angles for turbines, redirecting their wake away from downstream turbines [10–12]. This approach has been shown to be successful both in experiments and in large eddy simulations. In the current work, however, our focus is not on yaw control.

Studies on dynamic control of turbine set points, in which turbine operational conditions are changed at a much faster rate, are more scarce. Most work in this area has focused on mitigation of turbine loads (see, e.g., [13,14]). A major challenge for increasing energy extraction is the formulation of fast enough control models that predict the complex high-dimensional interaction between control actions and the three-dimensional turbulent flow structures in the turbine wakes and in the atmospheric boundary layer. Due to the high dimensionality and complex physics of turbulent flows, such models currently do not exist. A number of model-free approaches to wind farm control have been considered recently [15–18], but for reasons of the convergence speed of the algorithm, *a priori* choices were required with respect to the structure and dimensionality of the controller. In fact, only relatively slow changes in turbine set points are practicable in such approaches. Moreover, to the authors knowledge, none of these methods were tested in experiments or in a high-fidelity simulation environment, such as LES, instead relying on simple wake accumulation models, such as the Jensen model [19].

The challenge in developing and evaluating dynamic wind farm control approaches is related to the very high dimensionality and complexity of the turbulent flow state with which the controls should interact. In addition, simulating the evolution of this turbulent flow state with large eddy simulations is very expensive. As a result, designing controllers based on intuition or simple first principle-based physical insights is nontrivial. The mixed successes of approaches as simple as the static set-point optimization discussed above illustrate this. In this context, Goit and Meyers [3] developed a method that allows one to explore optimal control actions in large parameter spaces taking into account the coupled interaction with the instantaneous turbulent motions. They found that energy extraction could be potentially increased by up to 16%. The main mechanism was related to improved wake recovery by increasing wake mixing and the vertical transport of energy. This was established for the limit of a fully-developed 'infinite' wind farm boundary layer [3]. In the current work, we extend the approach to a regular 'finite' wind farm, in which entrance effects in the first rows are expected to play an important role. Including entrance effects is of particular importance, since in the case of static set-point optimization (*cf.* the discussion above), it was shown that the potential of increased energy extraction in downstream rows does not compensate for the related losses due to downregulating the first row of turbines [9]. Finally, we remark that in the current work, we consider a wind farm optimal control study under neutral atmospheric conditions. On the simulation side, this is similar to many neutral wind farm simulation studies, such as, e.g., [20,21]. Note that wind farm performance is strongly influenced by stratification in the boundary layer and the atmosphere aloft (see, e.g., [22–24]), and these conditions can be potentially very interesting for an optimal control study. However, this is not in the scope of the current work.

The paper is further organized as follows. In Section 2, the governing flow equations and the optimal control framework are introduced. The optimization approach is also briefly discussed. Results are presented in Section 3, and optimized power output and time-averaged flow profiles are discussed. Finally, in Section 4, the main conclusions are presented.

2. Numerical Method

In Section 2.1, the governing equations for large eddy simulations are introduced, including implementation details on boundary conditions and wind turbines. Next, the optimal control approach and optimization method are presented in Section 2.2. Finally, the case set-up is discussed in Section 2.3.

2.1. Governing Flow Equations and Discretization

Large eddy simulations are performed in SP-Wind, an in-house research code that was developed in a series of earlier studies, wind farm simulations and flow optimization (see, e.g., [25–27]). The governing equations are the filtered incompressible Navier–Stokes equations for neutral flows and the continuity equation, *i.e.*,

$$\nabla \cdot \tilde{u} = 0 \tag{1}$$

$$\frac{\partial \tilde{u}}{\partial t} + \tilde{u} \cdot \nabla \tilde{u} = -\frac{1}{\rho} \nabla \tilde{p} + \nabla \cdot \tau_M + f + \lambda(x)(\tilde{u}_{\text{in}} - \tilde{u}) \tag{2}$$

where $\tilde{u} = [\tilde{u}_1, \tilde{u}_2, \tilde{u}_3]$ is the resolved velocity field, $\tilde{p}$ is the pressure field, the term $\lambda(x)(\tilde{u}_{\text{in}} - \tilde{u})$ relates to our fringe-region implementation (*cf.* the further discussion below) and τ_M is the subgrid-scale (SGS) model. We use a standard Smagorinsky model [28] with Mason and Thomson's wall damping [29] to model the SGS stress. Furthermore, f represents the forces (per unit mass) introduced by the turbines on the flow. This turbine-induced force is modeled using an actuator-disk model (ADM) and is written for turbine i as:

$$f^{(i)} = -\frac{1}{2} C'_{T,i} \hat{V}_i^2 \mathcal{R}_i(x) e_\perp \quad i = 1 \cdots N_t \tag{3}$$

where $C'_{T,i}$ is the disk-based thrust coefficient. It should be noted that, unlike the conventional thrust coefficient C_T, which is based on the undisturbed velocity far upstream of a turbine, $C'_{T,i}$ is defined using the velocity at the turbine disk. It results from integrating lift and drag coefficients over the turbine blades, taking design geometry and flow angles into account (*cf.* Appendix A in [3] for a detailed formulation). Moreover, in an undisturbed uniform flow field, momentum theory can be employed to find $C'_T = C_T/(1-a)^2$ (*cf.*, e.g., [30]). Further, $\hat{V}_i$ is the average axial flow velocity at the turbine rotor disk, obtained by disk averaging and time filtering the axial velocity at the turbine disk level, *i.e.*, [3,27]:

$$V_i(t) = \frac{1}{A} \int_\Omega \tilde{u}(x,t) \cdot e_\perp \mathcal{R}_i(x) \, dx, \text{ and } \frac{d\hat{V}_i}{dt} = \frac{1}{\tau}(V_i - \hat{V}_i) \tag{4}$$

where we use $\tau = 5$ s. In the above equations, $e_\perp$ represents the unit vector perpendicular to the turbine disk, and $\mathcal{R}_i(x)$ is a geometrical smoothing function that distributes the uniform surface force of the turbine over surrounding LES grid cells, with $\int_\Omega \mathcal{R}_i(x) dx' = A$, where A is the turbine disk area. For more details regarding the implementation of the ADM in the SP-Wind code, the reader is referred to Meyers and Meneveau [27] and Goit and Meyers [3]. The total farm power, extracted by all turbines, can be expressed as:

$$P = -\int_\Omega f \cdot \tilde{u} dx = \int_\Omega \sum_{i=1}^{N_t} \frac{1}{2} C'_{T,i} \hat{V}_i^2 \tilde{u} \cdot e_\perp \mathcal{R}_i(x) \, dx = \sum_{i=1}^{N_t} \frac{1}{2} C'_{T,i} \hat{V}_i^2 V_i A \tag{5}$$

Finally, we remark that nowadays, other turbine models exist, such as the actuator line models (ALM). In particular, on sufficiently fine meshes, these allow a much finer description of the vortex dynamics in the near wake of the turbine (e.g., the effect of tip and root vortices, *etc.*). However, as shown in numerous studies, ADMs do provide a good description of the far wake dynamics and wake mixing, and the overall wind farm–boundary layer interaction (see, e.g., [31,32]). In view of the complexity of ALMs in an optimal control framework and the required high resolution (*cf.* also the discussion at the end of Section 4), we do not further consider them in the current study and focus on the use of the simpler ADM.

An overview of the computational domain is schematically shown in Figure 1. Inflow boundary conditions are used for the plane Γ_1^-; a classical high Reynolds number wall stress boundary condition is used on Γ_3^- [33,34]; periodic boundary conditions are used for Γ_2^+ and Γ_2^-; and a symmetry condition is used on Γ_3^+. SP-Wind uses a pseudo-spectral discretization in the horizontal directions. Therefore, the inflow boundary condition cannot be straightforwardly implemented as a Dirichlet condition. Instead, a fringe region technique [35] is used that smoothly forces the outflow region in the fringe region towards a desired inlet profile. To this end, we select:

$$\lambda(x) = \lambda_{\max} \left[S \left(\frac{x - x_s}{\Delta_s} \right) - S \left(\frac{x - x_e}{\Delta_e} + 1 \right) \right]$$ (6)

where:

$$S(x) = \begin{cases} 0 & x \le 0, \\ 1/\left[1 + e^{\frac{1}{x-1} + \frac{1}{x}} \right] & 0 < x < 1, \\ 1 & x \ge 1, \end{cases}$$ (7)

with x_s, x_e the start and end of the fringe region and where Δ_s, Δ_e control the widths of the increase and decrease regions of the function $\lambda(x)$.

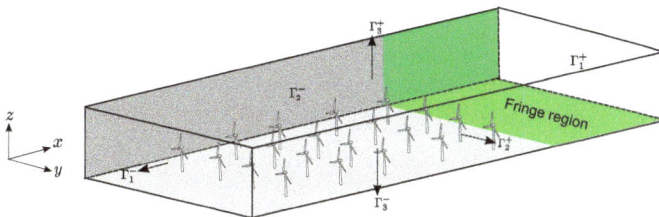

Figure 1. Computational domain with the fringe region.

The desired turbulent inflow field $\tilde{u}_{\text{in}}(x, t)$ is generated using a precursor method in which a separate simulation is performed, and stored to disk. This precursor simulation comprises a classical fully-developed boundary layer simulation (without a wind farm) that can be straightforwardly run on periodic domains. The method is visualized in Figure 2, showing velocity snapshots from the precursor boundary layer simulation and the main domain wind farm simulation, including the coupling from precursor simulation to the main domain fringe region.

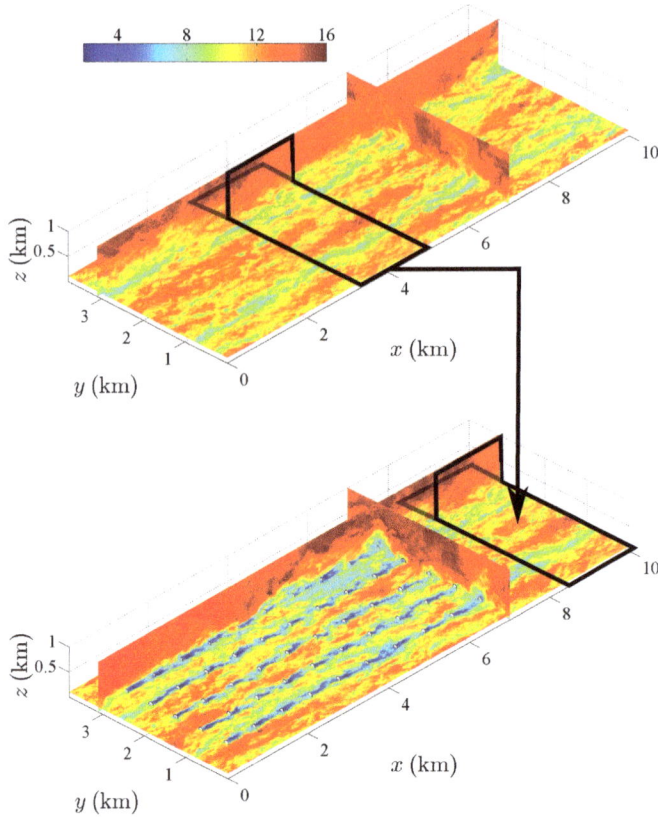

Figure 2. Snapshots representing instantaneous streamwise velocity fields from the precursor boundary layer simulation (**top**) and from the finite farm simulation (**bottom**). The horizontal planes in the figures are taken at the hub height.

As mentioned above, SP-Wind uses a pseudo-spectral discretization in the horizontal directions, applying the $3/2$ rule for dealiasing [36]. In the vertical direction, a fourth-order energy-conservative finite difference discretization scheme is used [37]. Mass is conserved by using a Poisson equation for the pressure that is solved using a direct solver. Finally, time integration is performed using a classical four-stage fourth-order Runge–Kutta scheme. For the simulations discussed in this paper, a fixed time step corresponding to a Courant–Friedrichs–Lewy (CFL) number of approximately 0.4 is used.

2.2. Optimal Control and Optimization Approach

A classical receding horizon optimal control approach is employed for the control of wind farm boundary layer interaction. In this approach, a control time horizon T is selected, and the control parameters are optimized as a function of time over this control horizon given the full interaction with the turbulent flow field as described by the LES equations (*cf.* Figure 3). Here, the control parameters $\varphi(t)$ correspond to all disk-based turbine thrust coefficients $\varphi \equiv [C'_{T,1}(t), C'_{T,2}(t), \cdots, C'_{T,N_t}(t)]$ in the control horizon. The optimization problem per time horizon is solved iteratively using a gradient-based approach (*cf.* further details below). Once an optimal set of controls is obtained, they are applied during a control time $T_A < T$. Subsequently, a new optimization problem is formulated for the next time horizon, *etc.* In the current work, we use $T_A = T/2$, similar to Goit and Meyers [3].

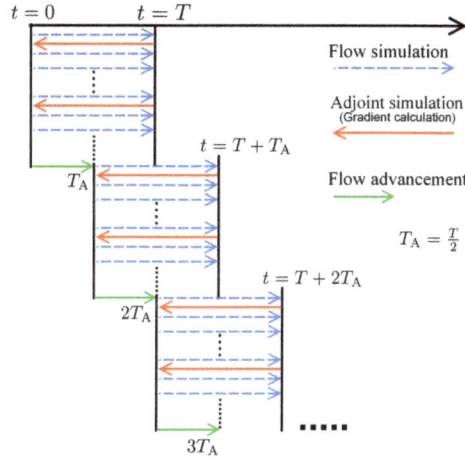

Figure 3. Schematic of the receding horizon optimal control approach.

For the optimization problem per time horizon, we aim at maximizing the total wind farm energy extraction. To this end, a cost functional is defined as:

$$\mathcal{J}(\varphi, q) = -\int_0^T P(t)\,\mathrm{d}t = \int_0^T \int_\Omega f \cdot \tilde{u}\,\mathrm{d}x\mathrm{d}t$$

$$= -\int_0^T \int_\Omega \sum_{i=1}^{N_t} \frac{1}{2} C'_{\mathrm{T},i} \hat{V}_i^2\, \mathcal{R}_i(x) \tilde{u}(x,t) \cdot e_\perp\,\mathrm{d}x\,\mathrm{d}t = -\int_0^T \sum_{i=1}^{N_t} \frac{1}{2} C'_{\mathrm{T},i} \hat{V}_i^2 V_i A\,\mathrm{d}t \qquad (8)$$

and where $q \equiv [\tilde{u}(x,t), \tilde{p}(x,t), \hat{V}(t)]$ are the state variables corresponding to the LES velocity field, pressure field and the time-filtered turbine disk velocity fields. The optimization problem is then formulated in its reduced form [3], *i.e.*,

$$\min_{\varphi} \tilde{\mathcal{J}}(\varphi) \equiv \mathcal{J}(\varphi, q(\varphi)) \qquad (9)$$

where $q(\varphi)$ is the solution to the state equations given the control inputs φ. This is obtained by solving the LES equations.

In order to solve Equation (9), a Polak–Ribière conjugate gradient method is used in combination with the Brent line search algorithm [38–40]. Implementation details are given in Delport, Baelmans and Meyers [26]. An important aspect of the algorithm is the determination of the gradient of the reduced cost functional $\nabla \tilde{\mathcal{J}}$ in the high-dimensional control space φ. To this end, the solution of the adjoint LES equations is required (*cf.* below), such that the gradient can be expressed as [3]:

$$\nabla \tilde{\mathcal{J}} = \frac{1}{2} \int_\Omega \hat{V}^{\circ 2} \circ \mathcal{R}(x)\,[(-\tilde{u} + \xi) \cdot e_\perp]\,\mathrm{d}x \qquad (10)$$

with $\mathcal{R} \equiv [\mathcal{R}_1, \cdots, \mathcal{R}_{N_t}]$ and where $\circ$ is used to denote the entry-wise product (or Hadamard product), and $\hat{V}^{\circ 2}$ is the entry-wise square of $\hat{V}$. Furthermore, $\xi(x,t)$ is the adjoint velocity field that is obtained by solving the adjoint equations. The derivation of the above relation is quite lengthy, but we refer the reader to [3] (Appendix C) for details.

The adjoint solution, required for the evaluation of Equation (10), is obtained by solving the following adjoint wind farm LES equations (*cf.* [3] for details of the derivation):

$$-\frac{\partial \boldsymbol{\xi}}{\partial t} - \tilde{\boldsymbol{u}} \cdot \nabla \boldsymbol{\xi} - (\nabla \boldsymbol{\xi})^T \cdot \tilde{\boldsymbol{u}} = -\frac{1}{\rho} \nabla \pi + \nabla \cdot \boldsymbol{\tau}_M^* + \boldsymbol{f}^* - \lambda(x)\boldsymbol{\xi}$$

$$\nabla \cdot \boldsymbol{\xi} = 0 \tag{11}$$

$$-\frac{d\chi_i}{dt} = \frac{1}{\tau}\left[-\chi_i + C_{T,i}'\widehat{V}_i \int_\Omega \mathcal{R}_i(\mathbf{x})\,(\tilde{\boldsymbol{u}} - \boldsymbol{\xi}) \cdot \boldsymbol{e}_\perp \, d\mathbf{x}\right], \quad \text{for } i = 1 \cdots N_t.$$

Here, $(\boldsymbol{\xi}, \pi, \chi)$ are the adjoint variables associated with each state variable $q = (\tilde{\boldsymbol{u}}, \tilde{p}, \widehat{V})$. Further, $\boldsymbol{f}^*$ and $\boldsymbol{\tau}_M^*$ are the adjoint forcing term and the adjoint of the SGS model, respectively. They are given by:

$$\boldsymbol{f}^* = \sum_{i=1}^{N_t} \left(\frac{1}{2}C_{T,i}'\widehat{V}_i^2 + \frac{\chi_i}{A}\right)\mathcal{R}_i(\mathbf{x})\boldsymbol{e}_\perp, \tag{12}$$

$$\boldsymbol{\tau}_M^* = 2\ell_s^2 \left(\frac{2S : S^*}{(2S : S)^{1/2}}S + (2S : S)^{1/2}S^*\right), \tag{13}$$

where $S^* = (\nabla \boldsymbol{\xi} + (\nabla \boldsymbol{\xi})^T)/2$ and $S = (\nabla \boldsymbol{u} + (\nabla \boldsymbol{u})^T)/2$.

Adjoint boundary conditions are similar to those of the forward problem. In the x and y directions, periodic boundary conditions are used. At Γ_3^+, a symmetry condition is imposed, while for Γ_3^-, the adjoint of the high Reynolds number wall stress boundary condition is imposed (*cf.* [3] for details). Finally, Equation (11) slightly differs from the equations given in [3], *i.e.*, an additional term $-\lambda(x)\boldsymbol{\xi}$ appears, corresponding to the adjoint of the fringe forcing. The derivation of this term is trivial. The term dampens the outflow of the adjoint field at Γ_1^- to an inflow $\boldsymbol{\xi} = 0$. This is fully equivalent to a standard system having non-periodicity with prescribed inflow at the upstream boundary and an outflow condition at the downstream boundary. A derivation of the adjoint boundary conditions for such a system yields a Dirichlet boundary condition for the adjoint velocity at the downstream boundary (with $\boldsymbol{\xi} = 0$) and an outflow condition at the upstream boundary (see, e.g., [41]).

The adjoint equations show some similarity to the flow equations of the forward problem, e.g., time derivatives and convective terms can be recognized (albeit with different signs), continuity looks the same and there is also an adjoint pressure variable. Therefore, much of the discretization of the forward problem can be reused, with the same pseudo-spectral discretization in the horizontal directions, in combination with a fourth-order energy-conservative discretization in the vertical direction. For the time integration, a fourth-order Runge–Kutta method is also used. Due to the different signs of time and convective terms, the equations are solved backward in time, and the direction of the 'adjoint' flow is reversed. The adjoint equations themselves follow from a linearization of the governing equations around a state $(\tilde{\boldsymbol{u}}, \tilde{p}, \widehat{V})$ [3]. In the adjoint equations, this state is also required (*cf.* Equation (11)). To this end, the nonlinear forward problem is solved first, and the full space-time state is stored on disk. Subsequently, the state is used during the solution of the adjoint equations.

Details of the case set-up are summarized in Table 1. The domain size corresponds to $L_x \times L_y \times H = 10 \times 3.8 \times 1$ km^3. The computational grid corresponds to $N_x \times N_y \times N_z = 384 \times 256 \times 200$, using $576 \times 384 \times 200$ when applying the 3/2 dealiasing rule. The fringe region accounts for 15% of the streamwise length and is located at the downstream end of the domain, starting from $x = 8.5$ km. Fifty turbines with a diameter $D = 100$ m are arranged in a 10×5 matrix, with streamwise spacing $S_x = 7D$ and spanwise spacing $S_y = 6D$. The selected resolution corresponds to typical cell sizes used in other wind farm simulations; in particular, it closely resembles the resolution of Case A3 in [30] and Case 1 in [42]. We refer the reader to these studies for a detailed grid sensitivity analysis.

Table 1. Summary of the simulation set-up for the optimal control of a finite farm.

Domain size	$L_x \times L_y \times H = 10 \times 3.8 \times 1 \text{ km}^3$
Fringe size	$L_f \equiv 15\% \text{ of } L_x = 1.5 \text{ km}$
Fringe region	Start: 8.5 km; End: 10 km
Driving pressure gradient	$f_\infty = 4 \times 10^{-4} \text{ m/s}^2$
Turbine dimensions	$D = 0.1H = 100 \text{ m}, \quad z_h = 0.1H = 100 \text{ m}$
Turbine arrangement	10×5
Turbine spacing	$S_x = 7D, \text{ and } S_y = 6D$
Surface roughness	$z_0 = 10^{-4}H = 0.1 \text{ m}$
Grid size	$N_x \times N_y \times N_z = 384 \times 256 \times 200$
Cell size	$\Delta_x \times \Delta_y \times \Delta_z = 26.0 \times 14.8 \times 5.0 \text{ m}^3$
Time step	0.6 s

2.3. Case Set-up

The distance between the last row of turbines and the start of the fringe region is set to 1.5 km (equivalent to $15D$). The spanwise distance between the outer turbine columns and the (spanwise) domain boundaries corresponds to $7D$. This leads to a blockage of the wind farm frontal area to the total frontal area of the domain of around 9%. This could be further decreased by increasing the spanwise gap between the wind farm and the side boundaries, but in view of the huge computational resources required for the optimal-control study, this is not further explored in the current work. Note that the spanwise periodic boundary conditions do make our wind farm formally semi-infinite, *i.e.*, only the streamwise direction is truly non-periodic, thus representing a farm that is finite in that direction. However, apart from the 9% blockage, we did further verify that the sideways wake expansion of the turbines in the side columns does not interfere with the spanwise boundary conditions along the length of the wind farm, so that our set-up is a reasonable approximation of a finite wind farm.

The inflow velocity is generated in a separate precursor boundary layer simulation, using the same domain size and grid resolution as those of the actual wind farm simulation (*cf.* Figure 2 and the discussion in Section 2.1). The precursor simulation is driven by a constant pressure gradient and has periodic boundary conditions in the horizontal directions. After an initial statistical convergence during which the flow evolves into a fully-developed turbulent boundary layer, the instantaneous velocity data in a region of a size equal to that of the fringe region of the main domain are written to files in every time step. These precursor data are read from the database and fed into the fringe region during the forward simulations in the optimization. The wind farm simulation is initialized with the converged precursor field and uses a spin-up period of approximately two through-flow time periods. Finally, note that during the iterations of the optimization, the same precursor data are needed multiple times, so that running the precursor simulation concurrently to the main simulation without the need of storing data on the disk (as, e.g., in the approach of Stevens *et al.* [43]) causes unnecessary computational overhead.

For the optimal control, we take a time horizon $T = 240$ s. This roughly corresponds to the time for the flow to pass four turbine rows, similar to the value used in [3]. The optimization algorithm is started with $C'_{T,i}(t) = 2.0$ (*i.e.*, $\varphi^{(0)} = 2.0$) for all turbines in the farm. This corresponds to the optimal operating condition of a lone-standing turbine following the Betz theory. To limit the computational cost, the optimization is not formally converged, but terminated after four conjugate gradient iterations. One conjugate gradient iteration requires roughly eight standard LES simulations (for the line search) and one adjoint LES for the determination of the gradient, leading to a total of 36 simulations per optimal control window. In total, 17 optimal control windows are performed, leading to a total control time of $17T_A = 2040$ s and a total number of 612 simulations.

Similar to Goit and Meyers [3], we impose box constraints on the controls, *i.e.*, $0 \leq C'_{T,i}(t) \leq 4$ (these are trivially applied in the conjugate gradient algorithm). The lower constraint prevents the turbine from starting to operate as a fan, even if the optimization algorithm would ask for this. For the upper boundary, we do not *a priori* want to limit C'_T to the Betz limit ($C'_T = 2$), but at the same time, we cannot leave it free, as $C'_T \to \infty$ is impractical from a turbine construction point of view. Therefore, we select an upper limit of $C'_T = 4$. In a uniform steady flow, this corresponds to a turbine that maximizes the thrust force, leading to an axial induction factor $a = 1/2$. In practice, this requires a turbine that is designed with a larger blade chord or a larger tip-speed ratio than is usually done for the Betz limit. For instance, using the "NREL offshore 5 MW wind turbine" [44] as a baseline, $C'_T = 4$ can be attained by a design with a double chord length, keeping all other aerodynamic parameters unchanged (*cf.* [3], Appendix A.2). We remark here that such a design may be economically not feasible, depending on the size of the potential gains of the optimal control in a wind farm. However, such an analysis is not in the scope of the current study.

An appreciation of a typical LES velocity field was already shown in Figure 2. Typical adjoint velocity fields are shown in Figure 4. Since the initial condition for the adjoint equation corresponds to $\xi(x, T) = 0$, at $T - t = 22$ s (Figure 4a), the field is largely zero (see the discussion in [3]). At $T - t = 22$ s and $T - t = 70$ s, it is observed that changes to the cost functional originate from tubes upstream of the rotors. In later snapshots (e.g., at time $T - t = 175$ s, shown in Figure 4c) the tubes hit the upstream turbines, and the adjoint field becomes fully turbulent inside the farm. This indicates that the power generation in a wind farm is influenced by the interactions between the turbines, as well as by their interaction with the boundary layer. Finally, the role of the fringe region in the adjoint simulation is to suppress the upstream propagation of the field over the periodic boundary condition. This is clearly visible in Figure 4c. The adjoint field in the fringe region is almost zero, except for the very end of the domain, where the fringe function $\lambda(x)$ is small. In this way, the interaction between the adjoint field developing upstream of the farm and the downstream turbines can be avoided, thereby imposing the non-periodicity in the streamwise direction.

3. Results and Discussion

Results of the optimal control in a 'finite' wind farm are now presented, and differences or parallels to the fully-developed case of Goit and Meyers [3] are discussed.

3.1. Controls and Optimized Power Output

First of all, in Figure 5, a time series of the total wind farm power extraction per unit farm area is shown, choosing the time origin $t = 0$ s as the point where the optimal control is started. For -1000 s $\leq t < 0$ s, all turbines operate in greedy mode with $C'_{T,i} = 2.0$ (denoted as the 'uncontrolled' case). It is appreciated from the figure that the total farm power fluctuates significantly more after the start of the optimal control. This was also observed in the optimization of the infinite farm. As discussed further below, the increased variability results from the fluctuations of $C'_{T,i}$. When averaging the farm power output over the 2000 s of accumulated optimal control and comparing to the average power output of the uncontrolled case, a gain in energy extraction of 7% is achieved. This value is significantly lower than the gain of 16% in the infinite farm case [3], but a gain of this magnitude is still important. The difference between the finite and infinite case is further discussed below, but one obvious contribution is that the first row of turbines in the developing case is already operating optimally at $C'_{T,i} = 2.0$. In the current farm, the power extraction from the first row accounts for about 17% of the total farm power (in the uncontrolled case).

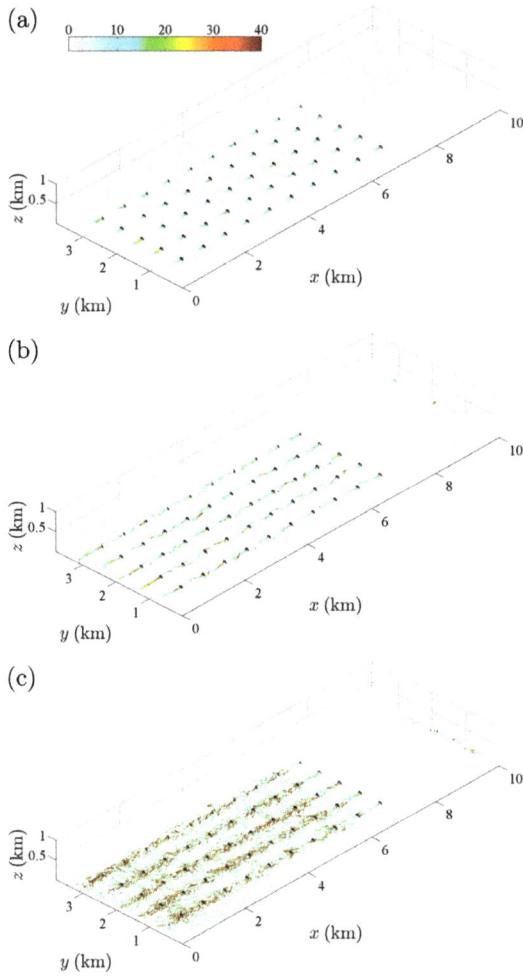

Figure 4. Contours of instantaneous streamwise adjoint fields, obtained from the first gradient calculation in Control Window 1. Horizontal planes in the figures are taken at the hub height. (**a**) $T - t = 22$ s; (**b**) $T - t = 70$ s; (**c**) $T - t = 175$ s.

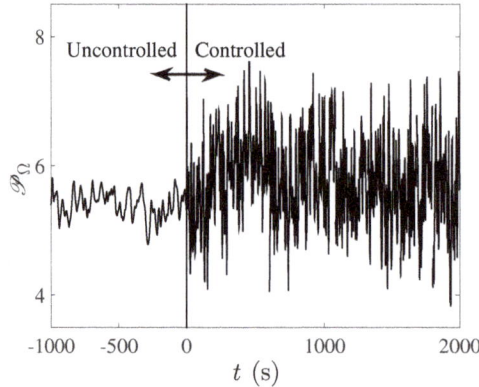

Figure 5. Time evolution of total farm power output. At time $t = 0$ s, optimal control is activated.

In Figure 6, the time evolution of the optimal thrust coefficient for one of the turbines is shown. The control changes strongly in response to the turbulent flow field. It can be seen that C'_T frequently touches the lower and upper limits imposed by the box constraints, *i.e.*, $0 \leq C'_{T,i}(t) \leq 4$. When zooming in on C'_T in Figure 6, it is however appreciated that the control is smoothly represented on the time grid (this is not *a priori* guaranteed in optimal control problems, often requiring additional regularization). Moreover, the fastest time scale with which C'_T changes from its lowest to highest value remains above 10 s. Thus, these control actions on C'_T are not too fast to realize using blade pitching. We should notice here that the current optimal controls may lead to increased structural loading and fatigue. These issues are currently not taken into account in the optimization cost functional. Taking such a detailed structural analysis into account is not in the scope of the current study. This will need further investigation in the future.

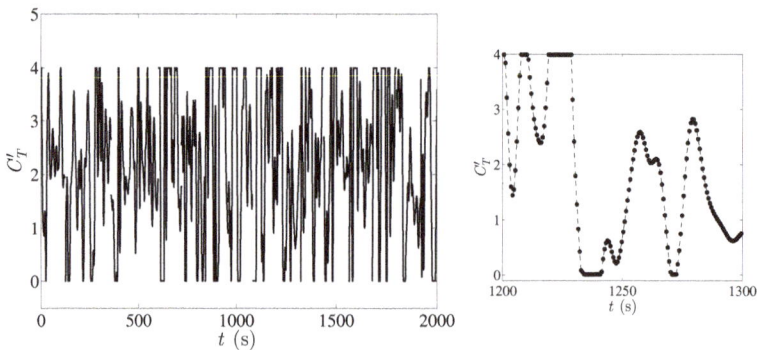

Figure 6. Time evolution of the thrust coefficient of one of the turbines in the farm.

In Figure 7, the time-averaged and row-averaged power output of different turbine rows are shown for both the optimal control and uncontrolled cases. All results are normalized by the average power output of the first row in the uncontrolled case. As a further reference, we have also included results from another (uncontrolled) LES of a similar wind farm by Porté-Agel *et al.* [20]. It is appreciated that the uncontrolled power outputs from the current study largely follow the trend observed by Porté-Agel *et al.* The uncontrolled power outputs also show a good agreement with typical field data [1,2], as well as with other LES investigations (see, e.g., [43]) of wind farms with similar configurations. A sharp drop in power production is observed between the first row and the second row. For turbine rows further downstream, the power deficit remains more or less constant.

(a)   (b)

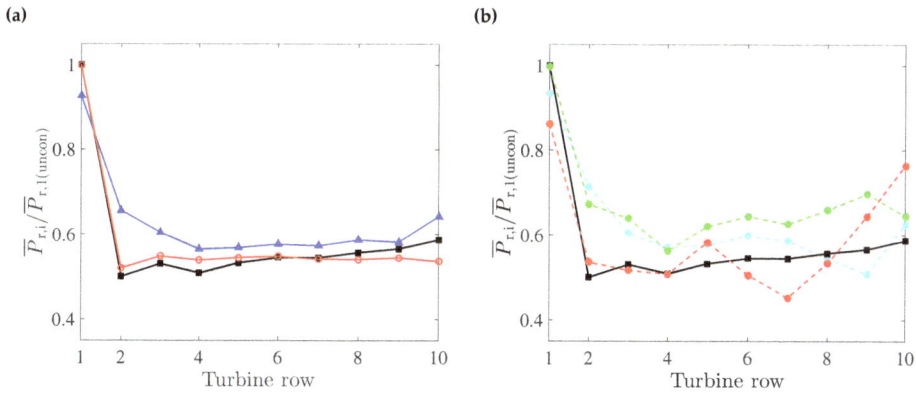

Figure 7. Comparison of the time-averaged power output for the controlled and uncontrolled farm as a function of turbine row. (■): power output for the uncontrolled case averaged over the time interval $[0, 17T_A]$. (a) (○) Power output from a prior large eddy simulation (LES) investigation with similar turbine spacing [20]; (▲) comparison to optimal control results averaged over $[0, 17T_A]$; (b) Power output for different control windows. Window 1 (●); Window 5 (●); Window 12 (●).

When looking at the optimal control results in Figure 7a, it is apparent that this typical trend changes. First of all, the power output from the first row is a bit lower compared to the uncontrolled case. However, for the later rows, the power output is significantly higher, and the transition towards a roughly constant power deficit in later rows is much smoother. Figure 7b presents power output averaged over three different control windows (instead of over the full set of 17 windows). A strong variation in the extracted power is observed from window to window, indicating that the control also depends on instantaneous turbulent flow features and that sufficient averaging over a range of control windows is required to accumulate converged statistics.

From Figure 7, it is clear that the power output in Rows 2 to 10 increases significantly. Overall, the power extraction in these rows is increased by 9.6%. The power output of the first row is decreased by about 7.7%. However, in contrast to the optimization of static turbine set-points [9], in the current dynamic optimal control study, this reduction in the first row is more than compensated by the gains in the later rows.

3.2. Averaged Flow Statistics

In order to further understand the relation between the flow field and the power production of the farm, the spatial distributions of the time-averaged mean velocity profiles and Reynolds stresses are analyzed in this subsection. The flow fields of both the uncontrolled case and the optimal control case are averaged over time window $[0, 17T_A]$. First of all, in Figure 8a,b, the mean streamwise velocity in a horizontal plane at hub height is shown. Since results are roughly invariant in the spanwise direction (except for small deviations in the two side columns), we also average the flow field of the different turbine columns, including the side rows, showing only one (averaged) column in the plots. Overall, the velocity distribution in the controlled and uncontrolled cases is the same, but distinct differences are observed in the turbine wakes. In the controlled case, most turbine rows have shorter wakes, and this is particularly true for the first row. This is also observed in Figure 8c,d, where the streamwise velocity is shown in an xz-plane through the turbines. Overall, this behavior is quite similar to the observations for the fully-developed wind farm in [3], where the increased energy extraction was also related to improved wake recovery.

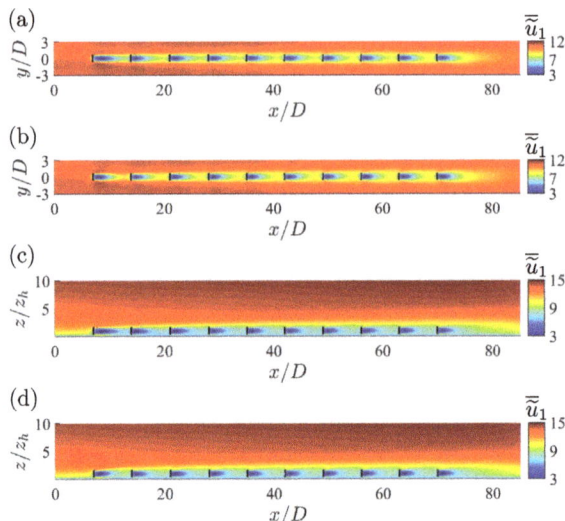

Figure 8. Contours of the mean streamwise velocity field averaged over the time window and five columns. (**a,b**) In a horizontal plane at hub height; (**c,d**) in a vertical plane through the turbine turbine. (**a,c**) Uncontrolled case; (**b,d**) optimal control case.

In Figure 9a, the turbine centerline velocity through the wind farm is shown, also illustrating the better wake recovery and higher inflow velocity of the turbines for the controlled case. Figure 9b shows the accompanying vertical Reynolds shear stress $\overline{\widetilde{u}_1'\widetilde{u}_3'}$ at the turbine tip level. For the infinite case in [3], the improved wake recovery of the turbine wakes was related to increased vertical Reynolds stresses in the wake regions. Here, we also observe a significant increase in vertical Reynolds stress after the first turbine row, explaining the faster wake recovery of the first wake. However, at downstream turbine rows, this increased Reynolds stress is not consistently observed in all wakes.

In Figures 10 and 11, the vertical $\overline{\widetilde{u}_1'\widetilde{u}_3'}$ and horizontal $\overline{\widetilde{u}_1'\widetilde{u}_2'}$ Reynolds shear stresses are shown in more detail in the vertical and horizontal planes. Apart from the wake of the first row, the vertical Reynolds shear stresses are not significantly higher in the controlled case and often even slightly lower than in the uncontrolled case. Moreover, the horizontal Reynolds shear stresses, responsible for sideways exchange of momentum, are not changed at all by the optimal control (also not in the first row). This clearly indicates that some of the mechanisms that increase energy extraction in the farm are not fully equivalent to those observed for the infinite case.

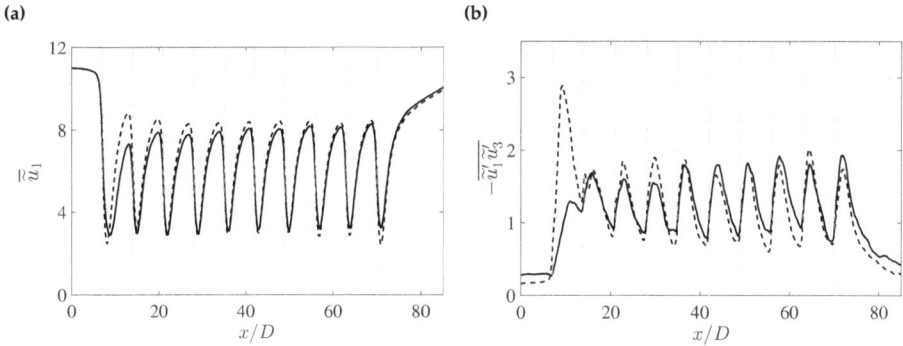

Figure 9. Time and column averaged profiles of (**a**) streamwise velocity through the rotor center and (**b**) Reynolds shear stress through the turbine tip. (——, black) uncontrolled case; (- - -, dashed) optimal control case averaged over the time interval $[0, 17T_A]$. Vertical dashed lines (light grey) represent the location of the turbines.

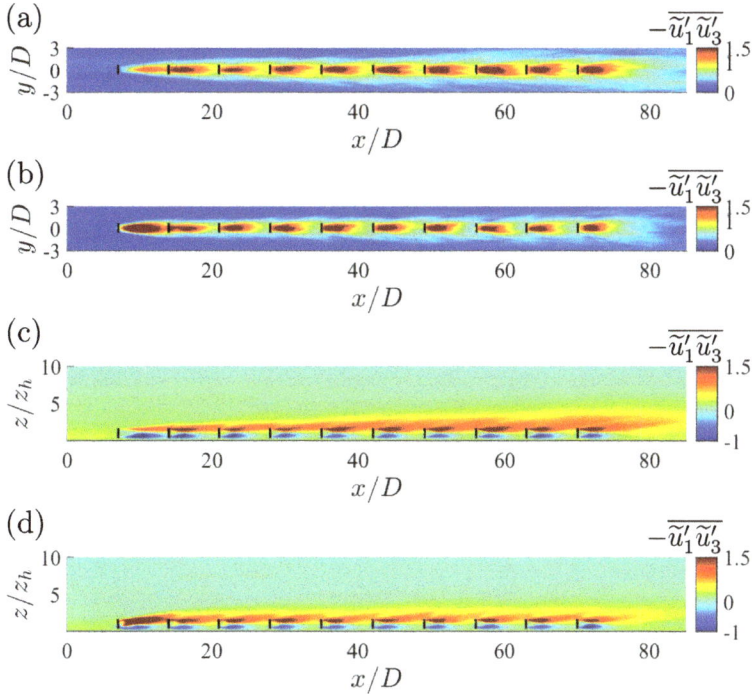

Figure 10. Contours of Reynolds shear stress $(-\widetilde{u'_1 u'_3})$ averaged over the time window and five columns. (**a**,**b**) In a horizontal plane at the turbine-tip level; (**c**,**d**) in a vertical plane through the turbine. (**a**,**c**) Uncontrolled case; (**b**,**d**) optimal control case.

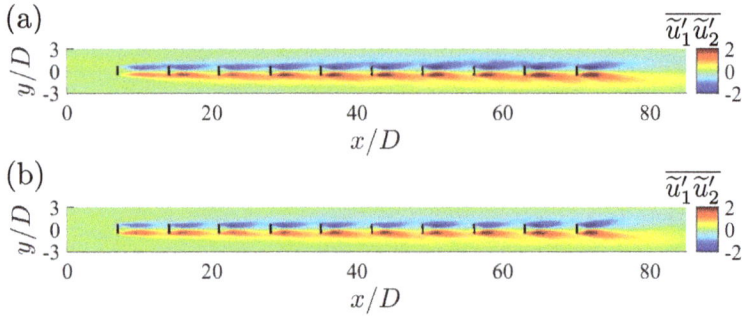

Figure 11. Contours of Reynolds shear stress $(\overline{\tilde{u}_1'\tilde{u}_2'})$ averaged over the time window and five columns in a horizontal plane at the hub level. (**a**) Uncontrolled case; (**b**) optimal control case.

Looking at the streamwise normal Reynolds stresses $\overline{\tilde{u}_1'\tilde{u}_1'}$ in Figure 12, it is observed that streamwise velocity fluctuations are even reduced in the controlled case, e.g., leading to less turbulent inflow at the next turbine row. In fact, this is similar to the trends observed for the streamwise Reynolds stresses in the infinite wind farm in [3]. This reduction in inflow turbulence at the next rows is directly related to the faster wake recovery observed above.

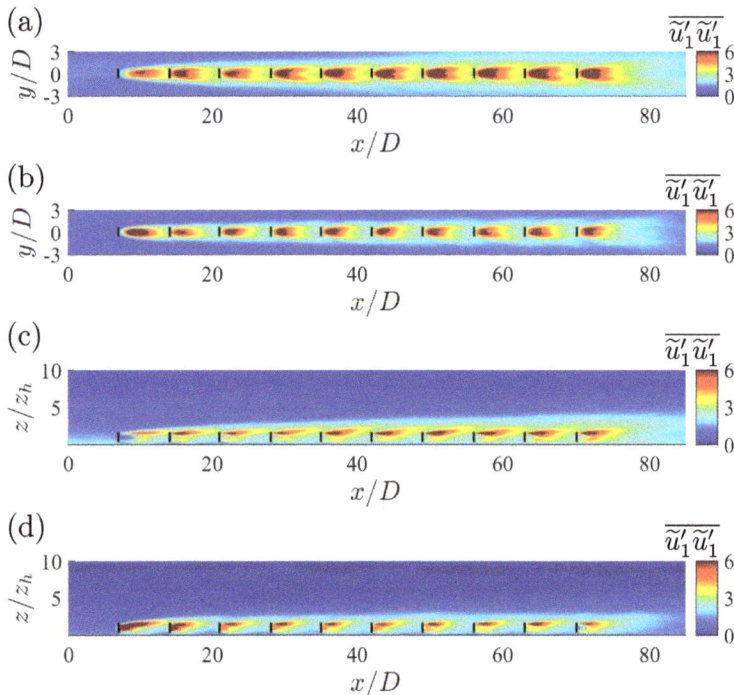

Figure 12. Contours of the mean streamwise component of the normal Reynolds stresses averaged over the time window and five columns. (**a,b**) In a horizontal plane at the turbine-tip level; (**c,d**) in a vertical plane through the turbine. (**a,c**) Uncontrolled case; (**b,d**) optimal control case.

In Figure 13, we investigate the mean transport of momentum around the wind turbines in the controlled and uncontrolled cases. The sideways mean transport $\overline{\tilde{u}_1}\overline{\tilde{u}_2}$ is shown in a horizontal plane through the turbine hub in Figure 13a,b, and the vertical transport $\widetilde{\bar{u}}_1\widetilde{\bar{u}}_3$ is shown in a vertical plane in Figure 13c,d. It is observed that mainly the horizontal sideways mean transport is significantly influenced by the optimal control, *i.e.*, the sideways transport is much bigger in the controlled case. The vertical transport remains largely unchanged from the uncontrolled to the controlled case. This is in contrast to the results for the infinite case [3], where a significant difference was found in the vertical transport, leading to much larger vertical dispersive stresses in the controlled case. For the infinite wind farm boundary layer, this is not unexpected, as all energy extracted by the turbines is balanced by vertical transport. A spatially-developing wind farm boundary layer is characterized by unattenuated high energy winds upstream of the wind farm. Given the current arrangement of turbines, a considerable fraction of this energy remains uncaptured by the front row of turbines and, hence, passes between the turbine columns (see, e.g., Figure 2). Thus, in this case, optimal control can gain a lot more from sideways transport, as long as the inflow energy is not yet depleted.

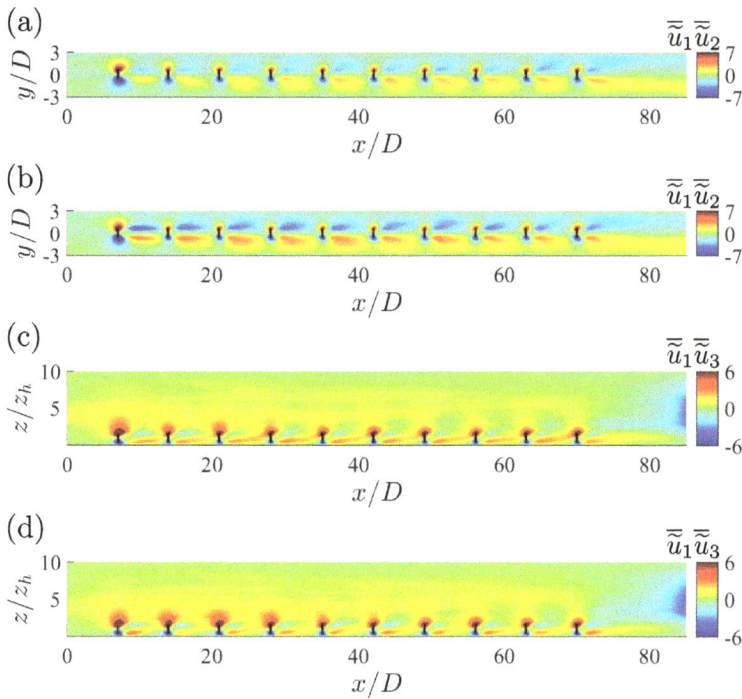

Figure 13. Contours of mean transport terms averaged over the time window and five columns. (a,b) $\widetilde{\bar{u}}_1\widetilde{\bar{u}}_2$ in a horizontal plane at the hub level; (c,d) $\widetilde{\bar{u}}_1\widetilde{\bar{u}}_3$ in a vertical plane through the turbine. (a,c) Uncontrolled case; (b,d) optimal control case.

For the fully-developed case, the increase in turbulence levels in the turbine wakes was attributed to a slight anticorrelation between the control signal C'_T and the turbine wind velocity [3]. Here, we perform the same analysis, and we expect similar findings for at least the first row of turbines. To that end, a Reynolds decomposition of the power output (*cf.* Equation (8)) is performed. We decompose the control in its time mean and fluctuating part, *i.e.*, $C'_{T,i} \equiv \overline{C'_{T,i}} + \Delta[C'_{T,i}]$. Similarly, the cubed velocity $\widehat{V}_i^2 V_i$ is decomposed as $\widehat{V}_i^2 V_i \equiv \overline{\widehat{V}_i^2 V_i} + \Delta[\widehat{V}_i^2 V_i]$. Thus,

$$\overline{P_r} = \sum_{i=1}^{N_r} \frac{1}{2}\overline{C'_{T,i}}\ \overline{\widehat{V}_i^2 V_i}A + \sum_{i=1}^{N_r} \frac{1}{2}\overline{\Delta[C'_{T,i}]\ \Delta[\widehat{V}_i^2 V_i]}A, \tag{14}$$

where $\overline{P_r}$ is the time average of the total power output from a turbine row and N_r is the number of turbines in the row. It is obvious that the second term on the right-hand side is zero in the uncontrolled case, since C'_T is constant in that case, so that $\Delta[C'_{T,i}] = 0$.

Figure 14 shows the fraction of the mean and the fluctuating terms on the right-hand side of Equation (14) to the time-averaged total extracted power $\overline{P_r}$. It is appreciated that for all but the last row, C'_T is anticorrelated with $\hat{V}_i^2 V_i$, leading to a negative value for the second term. For the last row, this is no longer the case, as the downstream development of the flow after the last row of turbines does not affect overall power output. For the first row of turbines, the negative correlation is approximately 14% of $\overline{P_r}$. This is much higher than the 6% observed in the fully-developed case [3]. Similar to the infinite case, this leads to higher Reynolds stresses and better wake mixing in the wake of the first row. For the latter rows, the negative correlation decreases to about 7%. However, as observed above, this does not directly lead to increased turbulence levels in the wake, but rather to increased sideways inflow of mean momentum. This points to more intricate correlations between C'_T and the velocity field, e.g., related to the passing of high-speed streaks in the channels between the turbines. Unfortunately, these correlations are difficult to identify. We expect them to be low, as most of the time, turbines still have to extract energy. In addition, it would also be interesting to investigate the correlations between controls from different turbines, as well as the correlations of power output and thrust coefficients with flow events. However, given the current optimal control time span (approximately 2000 seconds), the noise levels of the statistical averages remain too high to find significant correlations. Ongoing work is focusing on improving the parallelization of the adjoint equations and the speed-up of our optimization algorithms [45,46], so that in the future, longer time averaging becomes possible.

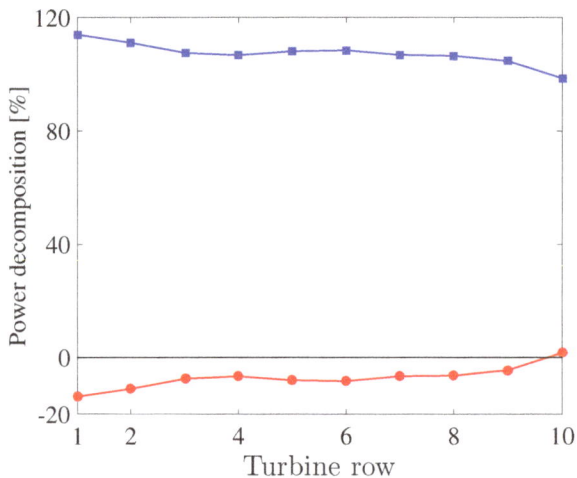

Figure 14. Reynolds decomposition of power output from different turbine rows. (■) Ratio of the mean component to the time-averaged total extracted power; (•) ratio of the fluctuating component due to C'_T to the total extracted power.

4. Conclusions

In the current paper, the application of optimal coordinated control was investigated for a finite-sized wind farm in large eddy simulations, extending the work of Goit and Meyers [3] to a regime where entrance effects are important. A receding horizon optimal control framework was considered, and optimization was performed using a gradient-based approach with adjoint simulations for the determination of the gradients. Based on this approach, the energy extraction of turbines was dynamically controlled in time so as to optimally influence the flow field in the boundary layer. Overall, an increase in energy extraction of 7% was achieved by the optimal control. The power output in the first row was a bit lower compared to the uncontrolled case, but this was compensated by increased power output in later rows. Thus, in contrast to the optimization of static turbine set-points where the loss of power in the first row is not recaptured at later rows [9], dynamic optimal control of turbine set-points allows for an overall increase of energy extraction in the farm. Note that for the infinite case, a gain of 16% was reported [3], which is significantly higher than the 7% obtained for the finite case. The difference is attributed to the fact that in the developing case, the first row of turbines, which covers a significant part of the total power output, is already operating optimally, and hence, its performance cannot be further improved.

The improved power extraction in the optimal control regime was shown to be related to improved wake recovery in all turbine rows, leading to higher inflow velocities for individual turbines. However, the mechanisms influencing this wake recovery were different for the first and subsequent rows. In the first row, the vertical Reynolds shear stresses were significantly increased, leading to better mixing from high-speed air above the farm into the wake. For the latter rows, neither Reynolds shear stresses nor vertical transport of momentum were significantly affected. Instead, sideways mean transport of momentum from high-speed channels in between turbines towards the wake regions was significantly increased.

The current study considered an aligned wind farm under neutral atmospheric conditions and further included a number of specific choices on turbine spacing, the admissible range for the controls $C_T'(t)$, the control time window, the number of iterations in the conjugate gradient method, *etc.* In further research, it will be interesting to look at other arrangement patterns, effects of stratification, larger wind farms and different constraints on the controls. Moreover, by improving the parallelism of the adjoint code and the efficiency of optimization algorithms, averaging over longer time periods will become feasible, so that a more detailed analysis of the correlations between controls and flow events becomes possible. It will also be interesting to compare current results with the optimal control using more advanced turbine representations, such as an actuator line model (ALM). In particular, when evolving computational resources allow wind farm optimal control with finer grids, an ALM inserts much more detailed near-wake physics into the flow, such as, e.g., tip vortices. Moreover, a more realistic representation of the turbine control through generator torque and blade pitching is then also possible. Nowadays, such an approach would already be feasible for smaller domains, e.g., focusing on one or two turbines, though it would still require the formulation of the adjoint actuator line model. This is the subject of ongoing research.

Acknowledgments: The authors acknowledge support from the European Research Council (FP7-Ideas, Grant No. 306471), the Flemish Science Foundation (FWO, Grant No. G.0376.12), and BOF KU Leuven (Grant No. IDO/11/012). Simulations were performed on the computing infrastructure of the VSC (Flemish Supercomputer Center), funded by the Hercules Foundation and the Flemish Government.

Author Contributions: Jay P. Goit and Johan Meyers jointly set up the study and wrote the manuscript. Jay P. Goit performed all simulations and implementations. Wim Munters implemented the fringe-region technique allowing non-periodic simulations in SP-Wind and assisted in final copy editing of the manuscript.

Conflicts of Interest: The authors declare no conflict of interest.

References

1. Barthelmie, R.; Rathmann, O.; Frandsen, S.T.; Hansen, K.; Politis, E.; Prospathopoulos, J.; Rados, K.; Cabezón, D.; Schlez, W.; Phillips, J.; *et al.* Modelling and measurements of wakes in large wind farms. *J. Phys.: Conf. Ser.* **2007**, *75*, 012049.
2. Barthelmie, R.; Pryor, S.; Frandsen, S.; Hansen, K.; Schepers, J.G.; Rados, K.; Schlez, W.; Neubert, A.; Jensen, L.E.; Neckelmann, S.; *et al.* Quantifying the impact of wind turbine wakes on power output at offshore wind farms. *J. Atmos. Ocean. Technol.* **2010**, *27*, 1302–1317.
3. Goit, J.P.; Meyers, J. Optimal control of energy extraction in wind farm boundary layers. *J. Fluid Mech.* **2015**, *768*, 5–50.

4.   Steinbuch, M.; de Boer, W.W.; Bosgra, O.H.; Peters, S.; Ploeg, J. Optimal control of wind power plants. *J. Wind Eng. Ind. Aerodyn.* **1988**, *27*, 237–246.

5.   Johnson, K.E.; Thomas, N. Wind farm control: Addressing the aerodynamic interaction among wind turbines. In Proceedings of the American Control Conference (ACC '09), St. Louis, MO, USA, 10–12 June 2009; pp. 2104–2109.

6.   Horvat, T.; Spudic, V.; Baotic, M. Quasi-stationary optimal control for wind farm with closely spaced turbines. In Proceedings of the 35th International Convention (MIPRO Croatian Society), Opatija, Croatia, 21–25 May 2012; pp. 829–834.

7.   Soleimanzadeh, M.; Wisniewski, R.; Kanev, S. An optimization framework for load and power distribution in wind farms. *J. Wind Eng. Ind. Aerodyn.* **2012**, *107–108*, 256–262.

8.   Knudsen, T.; Bak, T.; Svenstrup, M. Survey of wind farm control–power and fatigue optimization. *Wind Energy* **2014**, *8*, 1333–1351.

9.   Gebraad, P.M.O. Data-driven Wind Plant Control. Ph.D. Thesis, Delft University of Technology, Delft, The Netherlands, 2014.

10. Park, P.; Holm, R.; Medici, D. The application of PIV to the wake of a wind turbine in yaw. In Proceedings of the 4th International Symposium on Particle Image Velocimetry, Gottingen, Germany, 17–19 September 2001.

11. Fleming, P.A.; Gebraad, P.M.; Lee, S.; van Wingerden, J.W.; Johnson, K.; Churchfield, M.; Michalakes, J.; Spalart, P.; Moriarty, P. Evaluating techniques for redirecting turbine wakes using SOWFA. *Renew. Energy* **2014**, *70*, 211–218.

12. Gebraad, P.M.O.; Teeuwisse, F.W.; van Wingerden, J.W.; Fleming, P.A.; Ruben, S.D.; Marden, J.R.; Pao, L.Y. Wind plant power optimization through yaw control using a parametric model for wake effects-a CFD simulation study. *Wind Energy* **2014**, *1822*, 1–20.

13. Hansen, A.D.; Sørensen, P.; Iov, F.; Blaabjerg, F. Centralised power control of wind farm with doubly fed induction generators. *Renew. Energy* **2006**, *31*, 935–951.

14. Soleimanzadeh, M.; Wisniewski, R.; Johnson, K. A distributed optimization framework for windfarms. *J. Wind Eng. Ind. Aerodyn.* **2013**, *123*, 88–98.

15. Yang, Z.; Li, Y.; Seem, Y. Maximizing wind farm energy capture via nested-loop extremum seeking control. In Proceedings of the ASME Dynamic Systems and Control Conference, Palo Alto, CA, USA, 21–23 October 2013.

16. Marden, J.R.; Ruben, S.D.; Pao, L.Y. A model-free approach to wind farm control using game theoretic methods. *IEEE Trans. Control Syst. Technol.* **2013**, *21*, 1207–1214.

17. Gebraad, P.M.O.; van Wingerden, J.W. Maximum power-point tracking control for wind farms. *Wind Energy* **2014**, *18*, 429–447.

18. Ahmad, M.A.; Azuma, S.; Sugie, T. A model-free approach for maximizing power production of wind farm using multi-resolution simultaneous perturbation stochastic approximation. *Energies* **2014**, *7*, 5624–5646.

19. Katić, I.; Hojstrup, J.; Jensen, N.O. *A Simple Model for Cluster Efficiency*; European Wind Energy Association Conference and Exhibition: Rome, Italy, 1986; pp. 407–410.

20. Porté-Agel, F.; Wu, Y.; Chen, C. A numerical study of the effects of wind direction on turbine wakes and power losses in a large wind farm. *Energies* **2013**, *6*, 5297–5313.

21. Stevens, R.J.A.M.; Gayme, D.F.; Meneveau, C. Effects of turbine spacing on the power output of extended wind farms. *Wind Energy* **2016**, in press.

22. Abkar, M.; Porté-Agel, F. Influence of atmospheric stability on wind-turbine wakes: A large-eddy simulation study. *Phys. Fluids* **2015**, *27*, 035104.

23. Abkar, M.; Porté-Agel, F. The effect of free-atmosphere stratification on boundary-layer flow and power output from very large wind farms. *Energies* **2013**, *6*, 2338–2361.

24. Allaerts, D.; Meyers, J. Large eddy simulation of a large wind-turbine array in a conventionally neutral atmospheric boundary layer. *Phys. Fluids* **2015**, *27*, 065108.

25. Meyers, J.; Sagaut, P. Evaluation of smagorinsky variants in large-eddy simulations of wall-resolved plane channel flows. *Phys. Fluids* **2007**, *19*, 095105.

26. Delport, S.; Baelmans, M.; Meyers, J. Constrained optimization of turbulent mixing-layer evolution. *J. Turbul.* **2009**, *10*, 1–26.

27. Meyers, J.; Meneveau, C. Large eddy simulations of large wind-turbine arrays in the atmospheric boundary layer. In Proceedings of the 48th AIAA Aerospace Sciences Meeting Including the New Horizons Forum and Aerospace Exposition, Orlando, FL, USA, 4–7 January 2010; pp. 1–10.

28. Smagorinsky, J. General circulation experiments with the primitive equations: I. The basic experiment. *Mon. Weather Rev.* **1963**, *91*, 99–165.

29. Mason, P.J.; Thomson, T.J. Stochastic backscatter in large-eddy simulations of boundary layers. *J. Fluid Mech.* **1992**, *242*, 51–78.

30. Calaf, M.; Meneveau, C.; Meyers, J. Large eddy simulation study of fully developed wind-turbine array boundary layers. *Phys. Fluids* **2010**, *22*, 015110.

31. Martinez Tossas, L.A.; Leonardi, S.; Moriarty, P. *Wind Turbine Modeling for Computational Fluid Dynamics*; NREL Technical Report SR-5000-55054; National Renewable Energy Laboratory (NREL): Golden, CO, USA, 2013.

32. Wu, Y.-T.; Porté-Agel, F. Large-eddy simulation of wind-turbine wakes: Evaluation of turbine parametrisations. *Bound.-Layer Meteorol.* **2011**, *138*, 345–366.

33. Moeng, C.H. A large-eddy simulation model for the study of planetary boundary-layer turbulence. *J. Atmos. Sci.* **1984**, *41*, 2052–2062.

34. Bou-Zeid, E.; Meneveau, C.; Parlange, M.B. A scale-dependent Lagrangian dynamic model for large eddy simulation of complex turbulent flows. *Phys. Fluids* **2005**, *17*, 025105.

35. Spalart, P.R. Direct numerical study of leading edge contamination. In *AGARD Conference Proceedings, Fluid Dynamics of Three-Dimensional Turbulent Shear Flows and Transition*; Advisory Group for Aerospace Research and Development (AGARD): Neuilly-sur-Seine, France, 1988; Volume 438, pp. 5.1–5.13.

36. Canuto, C.; Hussaini, M.Y.; Quarteroni, A.; Zang, T.A. *Spectral Methods: Evolution to Complex Geometries and Application to Fluid Dynamics*; Springer: Berlin, Germany, 2007.

37. Verstappen, R.W.C.P.; Veldman, A.E.P. Symmetry-preserving discretization of turbulent flow. *J. Comput. Phys.* **2003**, *187*, 343–368.

38. Press, W.H.; Teukolsky, S.A.; Vetterling, W.T.; Flannery, B.P. *Numerical Recipes in FORTRAN77: The Art of Scientific Computing*, 2nd ed.; Cambridge University Press: Cambridge, UK, 1996.

39. Luenberger, D.G. *Linear and Nonlinear Programming*, 2nd ed.; Kluwer Academic Publishers: Boston, MA, USA, 2005.

40. Nocedal, J.; Wright, S.J. *Numerical Optimization*, 2nd ed.; Springer: Berlin, Germany, 2006.

41. Choi, H.; Hinze, M.; Kunisch, K. Instantaneous control of backward-facing step flows. *Appl. Numer. Math.* **1999**, *31*, 133–158.

42. Meyers, J.; Meneveau, C. Flow visualization using momentum and energy transport tubes and applications to turbulent flow in wind farms. *J. Fluid Mech.* **2013**, *715*, 335–358.

43. Stevens, R.J.A.M.; Graham, J.; Meneveau, C. A concurrent precursor inflow method for Large Eddy Simulations and applications to finite length wind farms. *Renew. Energy* **2014**, *68*, 46–50.

44. Jonkman, J.; Butterfield, S.; Musial, W.; Scott, G. *Definition of a 5-MW Reference Wind Turbine for Offshore System Development*; NREL Technical Report TP-500-38060; National Renewable Energy Laboratory (NREL): Golden, CO, USA, 2009.

45. Badreddine, H.; Vandewalle, S.; Meyers, J. Sequential quadratic programming (SQP) for optimal control in direct numerical simulation of turbulent flow. *J. Comput. Phys.* **2014**, *256*, 1–16.

46. Nita, C.; Vandewalle, S.; Meyers, J. On the efficiency of gradient based optimization algorithms for DNS-based optimal control in a turbulent channel flow. *Comput. Fluids* **2015**, *125*, 11–24.

Article

How Expensive Is Expensive Enough? Opportunities for Cost Reductions in Offshore Wind Energy Logistics

Thomas Poulsen [1],* and Charlotte Bay Hasager [2]

[1] Department of Mechanical and Manufacturing Engineering, Aalborg University, A. C. Meyers Vænge 15, DK-2450 Copenhagen SV, Denmark

[2] Department of Wind Energy, Technical University of Denmark, Frederiksborgvej 399, DK-4000 Roskilde, Denmark; cbha@dtu.dk

* Correspondence: tp@m-tech.aau.dk or thomas@poulsenlink.com; Tel.: +45-2383-1621 or +45-2126-6188

Academic Editor: Erik Gawel

Received: 23 March 2016; Accepted: 1 June 2016; Published: 7 June 2016

Abstract: This paper reveals that logistics may conservatively amount to 18% of the levelized cost of energy for offshore wind farms. This is the key finding from an extensive case study carried out within the organization of the world's leading offshore wind farm developer and operator. The case study aimed to, and produced, a number of possible opportunities for offshore wind cost reductions through logistics innovation; however, within the case study company, no company-wide logistics organization existed to focus horizontally on reducing logistics costs in general. Logistics was not well defined within the case study company, and a logistics strategy did not exist. With full life-cycle costs of offshore wind farms still high enough to present a political challenge within the European Union in terms of legislation to ensure offshore wind diffusion beyond 2020, our research presents logistics as a next frontier for offshore wind constituencies. This important area of the supply chain is ripe to academically and professionally cultivate and harvest in terms of offshore wind energy cost reductions. Our paper suggests that a focused organizational approach for logistics both horizontally and vertically within the company organizations could be the way forward, coupled with a long-term legislative environment to enable the necessary investments in logistics assets and transport equipment.

Keywords: offshore wind; logistics; logistics innovation; organization; levelized cost of energy; LCoE (levelized cost of energy)

---

1. Introduction

According to the Global Wind Energy Council [1], wind energy can potentially cover as much as 25%–30% of the world's electricity demand by 2050. With more than 400 giga-Watts (GW) of cumulative nominal wind energy capacity installed as of the end of 2015 [2,3], offshore wind made up a small share of the total at 11.5 GW mainly installed in Europe according to the European Wind Energy Association [3,4]. Offshore wind will, however, be very important for the global wind energy diffusion targets up to 2050. In this paper, we present new research indicating that *logistics makes up 18% of the levelized cost of energy (LCoE) for offshore wind* energy power plants. Our case study findings, conservatively, point to this number of 18% of LCoE based on a definition of logistics throughout the offshore wind farm (OWF) life-cycle, from idea conceptualization and planning through construction, operations/service and, ultimately, de-commissioning/abandonment of the OWF site.

This is the major contribution of the authors' 14-month long case study conducted at the world-leading offshore wind developer and operator [4,5], DONG Energy Wind Power (WP). Whereas our findings are derived based on a single-company case study and we recognize that different findings could possibly be found for other companies, our results are useful and significant based on the leading market position of our case study company coupled with the size and depth of their offshore wind power organization. The WP case study was conducted from July 2014–September 2015 by a group of six key researchers, supported by company representatives. The case study was originally aimed at setting up a strategy for a new innovation initiative within the company covering the area of logistics. As part of the logistics innovation strategy crafting efforts, a key company output was for the case study to unveil at least five possible specific future innovation projects. Such innovation projects should be aimed at providing improvement opportunities within the area of logistics, which the company could subsequently incubate and work on in collaboration with suppliers, academia and/or governments: a WP hypothesis being that LCoE reductions are one of the potential improvement opportunities innovation can bring.

We opted to be part of the case study because WP is uniquely positioned in the market as the largest global OWF developer and operator. We also thought the case to be interesting because DONG Energy itself is a Denmark-based, government-owned utility company going through a major strategic development as a result of the ascension of a new minority shareholder in the form of the United States of America (U.S.) investment bank, Goldman Sachs [6]. Finally, WP owns and operates a public-private partnership (PPP) joint-venture (JV) for logistics in the form of the subsidiary company, A2Sea. The ownership of A2Sea is in JV with the largest offshore wind turbine generator (WTG) original equipment manufacturer (OEM), as measured in market share for offshore wind [3,4], Siemens Wind Power (SWP).

Our case study is timely and highly relevant from different perspectives:

- *Policy*: Our case study indicates that a clear regulatory environment up to at least 2030 is critical for a conducive investment climate to exist. Such an investment climate is necessary in order to enable the needed logistics infrastructure, logistics assets and logistics personnel to be developed by government-owned and private organizations in order to support further offshore wind diffusion in an economical and safe/healthy manner.

- *Governance*: Our case study shows that necessary research and development (R&D) funding will need to be allocated by governments to proactively ensure logistics innovation support to the technological development of even larger offshore WTGs, yielding a greater nominal output as measured in mega-Watts (MW). This need is further amplified, as the diffusion of offshore wind is about to expand from North Europe to become a globally-applied technology, while OWFs are at the same time moving further out to sea, away from shore and into deeper waters.

- *Academic:* It is only after the term 'logistics' is defined that we may adequately start assembling, qualifying and measuring data and knowledge about this phenomenon. Our case study depicts that the definition of logistics itself may vary greatly depending on many factors, e.g., organizational vantage point and specific life-cycle phase [7] involvement of the individual person involved in offshore wind. For offshore wind, an all-encompassing definition of logistics is challenging to achieve mainly due to the complexity deriving from the many and distinctively different supply chains comprising a complete OWF life-cycle. Each supply chain provides unique frameworks for the respective logistics-related tasks.

- *Practitioner*: The strong empirical evidence from our case study suggests that logistics may be a somewhat overlooked frontier in the quest for lowering the LCoE of offshore wind. Our case study findings indicate that LCoE models and calculators do not separate out logistics as a stand-alone horizontal cost item throughout the entire OWF life-cycle, where clear levers can be used to impact LCoE in a simple and meaningful manner. Our case study also highlights how different offshore wind organizations do not seem yet to have dedicated logistics departments or competence centers, as in other industries. This prevents proper analysis horizontally across the life-cycle phases of an OWF, stopping synergies within a portfolio of many different OWFs within a single supply chain lead company to be realized. When we contrast this current state of logistical affairs within offshore wind to the latest Council of Supply Chain Management Professionals' (CSCMP) review [8], it becomes clear that having an organization and singular focus are key contributing factors that have helped drive down U.S. logistics cost across industries as a percentage of gross domestic product ("GDP").

After this Introduction, Section 2 will present our research objective, the key academic terms of reference (LCoE, logistics and logistics innovation) and the background of our case study. Section 3 will present the case study in more detail and focus on the findings of the analysis. In Section 4, we discuss the findings along the dimensions of the aforementioned policy, governance, academia and practitioner perspectives. Finally, Section 5 contains the conclusion, including our suggestions for further research efforts.

2. Research Objectives, Key Academic Terms and Case Study Introduction

Compared to other more mature energy sources, such as nuclear power, coal as well as oil and gas, wind energy still depends on government subsidies for production, diffusion and consumption [9,10]. Shafiee and Dinmohammadi [11] point out that offshore wind presents a greater maintenance risk compared to onshore wind. LCoE for offshore wind still needs to be dramatically reduced in order to be competitive in its own right with other energy sources and without government support. With OWFs representing publicly-subsidized Weberian ideal-type megaprojects, as defined by Flyvbjerg *et al.* [12], the four distinctively different life-cycle phases of wind farm projects [13] make these projects very hard to manage.

2.1. Research Objectives

From a supply chain perspective, this research offers an in-depth perspective on the different supply chains comprised within offshore wind farm megaprojects through the project life-cycle phases [13]. As such, wind energy tends to be a government-created market globally with the underlying industry fueled by government subsidies [9,14,15]. With geopolitical drivers to have Europe depend less on oil- and gas-rich nations, such as Russia and several Middle Eastern countries [16], DONG Energy has played an important role in the execution of the aggressive climate change mitigation strategy of the government of Denmark. DONG Energy's role in the Danish mitigation strategy is particularly noticeable when it comes to the diffusion of wind energy in the form of a showcase within Europe.

Our WP case study about logistics innovation within offshore wind is both timely and relevant due to our three initial propositions:

1. Logistics is a significant cost driver for offshore wind, as it is for other industries. For logistics in the U.S., as defined by CSCMP across all industries, costs were cut in half over a 20-year period from 15.8% of GDP in 1981 to 8.4% in 2014 [8]. Logistics therefore holds the promise and allure of cost savings due to its sheer relative share of offshore wind LCoE.
2. Innovation is generally a path towards the maturing of industries, for example through platform leadership [17]. Furthermore, innovation provides an opportunity for cost reductions in general. Logistics innovation within offshore wind therefore seems relevant to pursue in order to obtain cost savings and to reduce LCoE.
3. With a market share of 15.6% of the operating European OWFs by the end of 2015 [3] and a construction/engineering, procurement, construction, and installation (EPCi) track record of 26% of all OWFs built globally [5] (p. 27), WP is the recognized market leader within offshore wind globally. WP seems to be the most interesting case study company to investigate in terms of logistics innovation within offshore wind, as they have the largest portfolio of planned OWFs, OWFs under construction and OWFs already in operation. Only a large market constituency like WP with a correspondingly significant organization and big portfolio of OWFs seems to be able to take advantage of synergies and benefit from economies of scale generating cost savings and

LCoE reductions from logistics innovation. A strong organization with strong focus on logistics seems relevant in terms of being able to execute logistics cost savings for offshore wind.

2.2. Levelized Cost of Energy

Diffusion of different energy types can be compared in different ways [18], and from a financial perspective, LCoE is the most commonly-used metric. LCoE is defined by The Crown Estate [19] (p. VII) as "the lifetime cost of the project, per unit of energy generated". The International Energy Agency ("IEA") defines LCoE as "the ratio of total lifetime expenses *versus* total expected outputs, expressed in terms of the present value equivalent" [20]. Prognos and Fichtner Group [21] (p. 12) define LCoE as "the average cost for generating electricity over an operational time of 20 years". Heptonstall *et al.* [22] further explain how to calculate LCoE and define it as "levelised costs seek to capture the full lifetime costs of an electricity generating installation, and allocate these costs over the lifetime electrical output, with both future costs and outputs discounted to present values". Liu *et al.* [23] evaluate different frameworks and finally utilize the 'E3' methodology in their setting of LCoE for China. Megavind [24] defines LCoE as lifetime discounted cost in EUR divided by lifetime discounted production in MW-hours (MW/h). As these different definitions indicate, the overarching concept for calculating offshore wind LCoE would seem similar; however, different countries within Europe have adopted different interpretations on how to perform these calculations, and many attempts have been made to use the calculations when planning OWFs [25].

When reviewing the state-of-the-art within academia, the topic of LCoE from a macro and policy perspective is addressed, e.g., by Gross *et al.* [26], as they explain how the government policy setting in the United Kingdom (U.K.) concerns itself mainly with the cost side of LCoE and why policy makers ought to focus on the revenue implications also for offshore wind. Based on mainly industry reports from 2006 to 2007, Blanco [27] breaks the wind farm cost components down into upfront capital expenditure and reoccurring variable costs for operations and maintenance (O&M) to arrive at an estimated LCoE number for onshore, as well as offshore wind, reflecting a downward cost trajectory over time. Heptonstall *et al.* [22] describe how LCoE for offshore wind has unexpectedly increased in the U.K. and break down the different cost drivers to justify how they expect LCoE to decrease also beyond 2020.

When it comes to cost drivers specifically related to logistics within offshore wind, the topics researched are generally very specific and seem to focus mainly on vertical "slivers" of the logistics chain as opposed to a holistic perspective with a horizontal view across the entire life-cycle phase, let alone the entire life-cycle of an OWF. This is illustrated by a state-of-the-art review of the offshore wind O&M logistics [28], where an overview of all logistics literature for the O&M life-cycle phase of an OWF is presented. The literature review reveals that whereas some logistics research deals with LCoE reductions, none of the academic works analyzed research logistics across all life-cycle phases of an OWF, nor do they consider logistics synergies across a portfolio of operating OWFs.

When we contrast individual academic works with more extensive efforts to unify academia, industry and government representatives in larger groupings to work towards bringing down LCoE across the entire offshore wind industry of a country in a systemic manner, the potential of logistics becomes gradually more pronounced:

Denmark study: In their report for the Danish Ministry of Climate and Energy, Deloitte [29] breaks down key cost drivers of OWFs. The report points out that a key cost driver for capital expenditure is installation vessels, and the Germanischer Lloyd Garrad Hassan underlying wind turbine installation vessel (WTIV) database is used to document the role of the WTIVs. The report points to a rise in installation costs in general because OWFs move further away from shore and into deeper waters.

U.K. study: In the final report from the U.K. industry-wide Department of Energy and Climate Change (DECC) Cost Reduction Taskforce [19], a target to reduce LCoE from Great Britain Pounds (GBP) 140 per MW/h in 2011 to GBP 100 per MW/h in 2020 is presented based on a six-month effort organized with five separate analysis tracks involving a total of 120 companies, organizations and individuals. Here, four different scenarios are presented based on four predefined OWF sites located in different offshore conditions. The offshore conditions vary by site in terms of average water depth, distance to shore and wind speed assumptions. The scenarios and different sites make the calculations and results more detailed and credible than the previous Danish study. Logistics cost drivers now start to feature more prominently and across several phases of the OWF life-cycle. Examples of LCoE reduction opportunities identified include more extensive site surveys, early involvement of suppliers, front-end engineering and design (FEED), better procurement, construction of new vessels, more competition in terms of installation, optimization of installation methods and evolution of the overall offshore wind supply chain. Applying an even broader implied definition considering overall offshore wind project financing, logistics plays an important role, as the key financing risks are seen as installation costs and O&M costs. Financing risks are crucial: the U.K. study explains that a change of 1% in the cost of financing for an offshore wind project in the form of weighted average cost of capital has a 6% impact of total project LCoE.

Germany study: In their analysis of how to decrease the LCoE of offshore wind in Germany over the coming 10 years, Prognos and Fichtner Group [21] base their research on the U.K. DECC Cost Reduction Taskforce results as published by The Crown Estate [19]. Prognos and Fichtner produce two different scenarios for three predefined OWF sites located in different offshore conditions [21]. The scenarios and sites contain more granular assumptions that make the calculations even more credible and accurate compared to the U.K. study. As Prognos and Fichtner Group are consultancies hired on behalf of The German Offshore Wind Energy Foundation to produce the analysis, they seem to have prepared a larger part of the findings by themselves than the U.K. study. However, approximately 50 external interviewees have been involved in the Germany study for dialogue and validation purposes. Logistics considerations feature much more prominently in the German study, which even has a detailed calculation involving day-rate hire costing ranges for eight different vessel types within the installation phase, as well as two vessel types and helicopter rates for O&M. A large part of the LCoE reduction initiatives identified have to do with logistics. The examples cited include improved logistics infrastructure for installing wind power plants, installation logistics innovation, improved logistics for offshore substations/wind turbine installation, new installation methods for substations/foundations, changing vessel requirements, larger vessels for foundation installation, more competition in the area of installation vessels for substations/turbines/foundations/cables, weather risk considerations for vessel bookings, O&M logistics costs and costs for loading, as well as transporting dismantled OWFs back to port at the end of the life-cycle. The German study considers different scenarios for O&M based on the distance to port and assumes a *land-based* maintenance set-up *versus* that of a *sea-based* concept for OWFs at deeper waters further from shore. In addition, unforeseen events, especially pertaining to the logistics components of the installation risk, are set at some 15% of the total OWF LCoE in the German study. Last, but not least, logistics plays an important role in OWF portfolio synergies and synergies between different farm operators, because the German study considers LCoE savings generated from joint fleets of vessel, helicopters, ports, warehouses, *etc.*

It is important to note that *when comparing the different country LCoE studies* outlined above, a key difference in calculation methods with profound impact is found within the area of offshore transmission assets and connection to the onshore grid. The Denmark study [29] reveals that offshore transmission assets and onshore grid connection investments for wind farms in Danish waters are planned, constructed and operated by a state-owned enterprise called Energinet.dk. In the German study [21] (p. 21), the OWF developer is responsible for building the wind farm, including an offshore substation; however, the developer is not responsible for connecting the OWF to the onshore grid. The U.K. study [19] (p. 34) reveals that the developer must construct the offshore transmission assets and ensure grid connection to the onshore grid only to subsequently transfer these assets to a third party offshore transmission owner via a tender process by the U.K. government, the Office of Gas and Electricity Markets. In the U.K., the operator of the OWF must then later pay for use and balancing use of these transmission assets, which is included in the LCoE calculations [19] (p. 6). The differences in calculation methods allow for a significant variation in LCoE cost reduction impact calculations, as offshore transmission assets and onshore grid connection costs could be as high as 20% of CapEx, as was the case for the Anholt OWF in Denmark [30].

2.3. Logistics

As indicated from our LCoE review, logistics for offshore wind may be rather broadly defined and, as such, comprise a very extensive scope ranging from more traditional definitions involving operation of assets, such as trucks, ports and vessels, to more complex implications, such as the logistics component of installation and O&M risks involving both "unforeseen events" and changes in the life-cycle project financing/weighted average cost of capital.

As a term and word, "logistics" originates from the Greek word "*logisitki*" deriving from the verb "*logizomai*", which means to think deeply about something and to calculate the consequence of actions. Logistics can be dated back to the Roman Empire, ancient Greece and Byzantium, where military officers, referred to as "*logistikas*", were responsible for finance, distribution and supply already back then [31]. Academically speaking, "logistics" was coined in several contexts through time including how it relates to the physical distribution of agricultural products by Crowell back in 1901 and from a marketing perspective by Clark in 1922 [32]. The first academic accounts of logistics as a more technical and managerial discipline, including the notion of a flow, inventory control and optimum lot sizes, were coined by Magee [33]. Other scholars like Heskett [34,35] and Shapiro [36] also discussed logistics in terms of definitions, structure, composition, operations, as well as strategic implications.

When it comes to strategy alignment of the company, logistics can be part of the competitive business advantage within the overall value chain [37], and alignment between the strategic goals of the company with the logistics system of the company is discussed by Shapiro and Heskett [38]. Fisher [39] discusses the same topic from a supply chain structure perspective, and Chopra and Meindl [40] devote the entire second chapter of their book to discuss the benefits of strategic fit between a company's competitive strategy and the supply chain strategy.

Other academic scholars attempt to group various lines of thought into different overall theory streams. Hesse and Rodrigue [41] present what they call "the evolution of logistical integration" from 1960 to 2000: They state that theory streams relating to many concepts, such as materials handling (MH), inventory management (IM), materials management (MM) and physical distribution (PD), are all antecedents to "logistics" as a theory stream. Additionally, they continue to state that by scholars adding information technology, marketing and strategic planning disciplines to the logistics theory stream during the 1990s, supply chain management (SCM) has succeeded logistics as a more encompassing theory stream. In a later study, Hou *et al.* argue [42] that PD, logistics and SCM can be considered to be "under the umbrella of a new theory", called the materials flow (MF) theory.

2.4. Logistics Innovation

Within the arena of logistics innovation, competing theory streams are also found along with a number of broader theoretical frameworks that impact either innovation in general or logistics innovation specifically. Some of this ambiguity within academic definitions is a result of the evolution of the core term itself, *i.e.*, whether we are discussing innovation for logistics or innovation for MH, IM, MM, PD, SCM or MF. Competing with logistics innovation, theory streams with some degree of weight attached to them could be supply chain learning management [43] or supply chain/SCM innovation [44]. Broader theoretical frameworks that are of relevance to logistics innovation according to Grawe [45] include the knowledge-based view, the dynamic capabilities framework, the Schumpeterian innovation framework, the exploration/exploitation framework, the theory of S-curves, network theory and resource advantage theory.

Regarding the term "innovation" itself, it is used by practitioners in a very broad sense from the action of invention to the discipline of R&D to innovation as an outcome of a process or effort.

The innovation definition and innovation framework of Schumpeter [46,47] generally seem to be recognized as the original academic thought processes defining and dealing with innovation.

Through an extensive literature review of logistics innovation, Grawe [45] also points out that logistics innovation is based on a number of factors that either relate to the organization of a company or the societal context/environment of a company. Grawe [45] furthermore argues that a company perspective may be either that of the company creating the innovation or that of the company(ies) adopting the innovation. Flint *et al.* [43] argue that logistics managers may be considered successful in terms of innovation if they innovate within the area of logistics to create a competitive advantage for the company or if they generate logistics innovation in order support the company's core product innovation process. To support a product innovation, logistics managers need to be involved upfront in the product innovation process [43]. A good example of this is FEED for offshore wind [19]. Arlbjørn *et al.* [44] have performed a broad literature search and argue that logistics could equal SCM and in the presentation of their results, SCM innovation (SCMI) seems to equal supply chain innovation (SCI), prompting them to label the field of study "SCI". Whereas the convergence and evolution of the terms logistics and SCM have been covered above, some academic scholars and practitioners alike would disagree with Arlbjørn *et al.* [44] and argue that the supply chain is, however, not equal to the discipline of SCM.

2.5. DONG Energy Wind Power Case Study Introduction

The key topic of this case study is the role and relative importance logistics plays within offshore wind when it comes to LCoE reductions, as well as how logistics innovation may specifically be applied within the WP setting, also organizationally. Flint *et al.* [43], Grawe [45] and Arlbjørn *et al.* [44] agree that the theoretical frameworks of logistics innovation, respectively SCI, described need empirical testing in an empirical setting along several dimensions for the benefit of both academia and practitioners alike. It is with this goal of empirical dimensional testing that the following company case study was developed.

With an exclusive focus on offshore wind, WP presently counts in excess of 1600 full-time-equivalent (FTE) people in a matrix organization organized in a hierarchical tiered structure and along the OWF life-cycle phases (see Table 1). WP is a complex organization to navigate for people working inside the company, let alone for outside researchers. Within offshore wind logistics, WP has a fairly unique position inasmuch as it owns shipping and logistics subsidiary A2Sea in a 51% PPP partnership with conglomerate SWP [13]. In addition to being the minority owner of A2Sea in the PPP set-up with WP, SWP is also a "preferred supplier" of WP, as SWP holds large frame agreements with WP for WTG supply and related services, such as WTG installation, commissioning, servicing and warranty. The WP business model is unique in the market place because the company believes that it is the world leader at constructing and operating offshore wind farms. Unlike many other industries, shipping/logistics/SCM did, however, not seem to play a significant role within the company, and the goal of our project with WP was to develop an offshore wind logistics R&D strategy for the company going forward towards 2020, 2030 and 2050.

From an academic perspective, the key assumption at the start of the project was that WP would most likely not have a commonly-agreed definition of what "logistics" is. A secondary assumption was that WP would perhaps also not have a commonly-agreed definition of what R&D efforts are comprised of. It was known that WP did not have a logistics department or logistics competence center, and another assumption was therefore that the company could be faced with organizational challenges within the field of logistics skills and competencies. In order to explore this setting, to understand logistics innovation within WP and to gather information needed to craft the R&D strategy for logistics, the investigation method applied was the case study [48].

Table 1. Case study company organization in 2015: multidimensional employee count.

Organizational Layers	Management and Finance	Development & Consent (D&C) Life-Cycle Phase	Installation & Commissioning (D&C) Life-Cycle Phase	Operations & Maintenance (O&M) Life-Cycle Phase	De-Commissioning (De-Comms) and Site Abandonment Life-Cycle Phase	Full-Time Equivalent (FTE) Employee Count	% of Total
Management Board	4	2	4	2	0	12	0.76%
Top management	6	3	27	11	0	47	2.96%
Middle management	27	18	38	45	0	128	8.06%
Operations/execution/analytical	103	60	762	202	0	1127	70.93%
Site	0	0	0	275	0	275	17.31%
FTE count	140	83	831	535	0	1589	-
% of total	8.81%	5.22%	52.30%	33.67%	0%	-	100%

To explore the topic, a largely WP-driven selection process yielded a total of 15 company interviews comprising a total of 18 company interviewees. The interviewees were chosen in order to represent the entire WP business unit in the interview process. An extensive interview protocol was simultaneously designed by the research team in order to be able to cater to all of the different organizational constituencies selected for interview within WP. The interviewees were chosen along several different dimensions, as illustrated in Table 1: they had to represent different organizational layers of management within the company; they had to represent the different offshore wind farm life-cycle phases; and lastly, the interviewees had to have representative expertise within the key parts making up an offshore wind farm (for example, the WTG, the foundations, the underwater cables and the substations). It was also important that the interviewees had some knowledge of both logistics and R&D within the company or at least within the industry in general (see Figure 1).

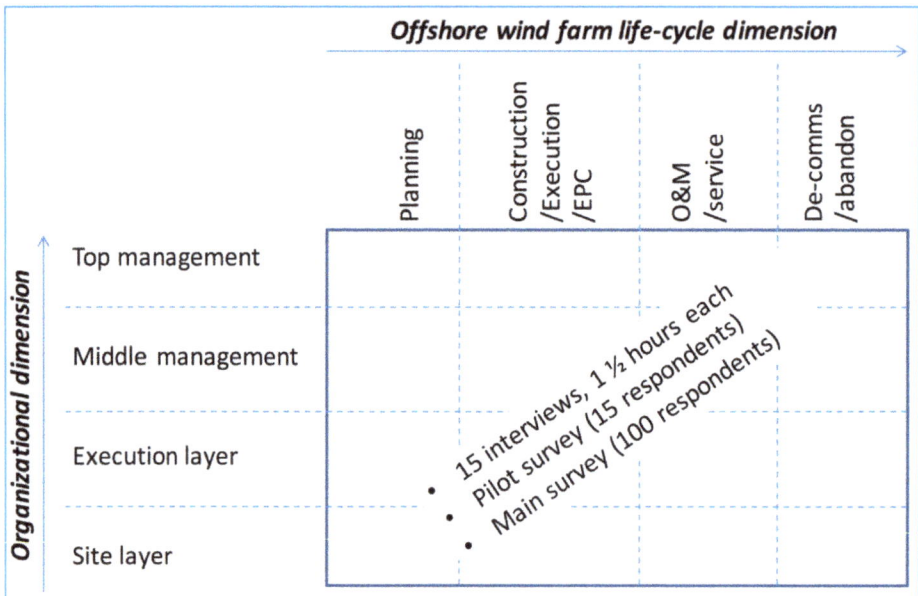

Figure 1. Interview and survey selection matrix.

The selection process for the interviewees and the interview protocol design efforts took from July–October 2014 to organize, and the 15 interviews were conducted from November 2014 through the middle of February 2015. Each interview lasted between 60 and 90 min, depending on availability. Two interviewers in the form of a company representative and an academic interviewer were present in all interviews, and in one of the interviews, a third interviewer participated as an observer. The company representative started off all interviews to set the scene and subsequently handed over the interview process to the academic interviewer.

The first phase of empirical data-gathering efforts in the form of the interviews was conducted in person, face-to-face, except two, which were conducted via video conference. Fourteen of the 15 interviews were, with due consent from the interviewees, audio taped for later transcription purposes, and 14 of the 15 interviews were conducted in English to enhance the scientific value to be derived from the subsequent academic team processing and interpretation. Each interview had an introductory section, which was aided by a hard-copy presentation for visualization purposes, and this was the same for all interviews in order to ensure that the background and purpose of the interview process was framed in the same way for all interviewees. The transcription was organized with the research team splitting and transcribing a number of interviews. Each transcript was subsequently reviewed and edited/completed by another research team member with an ultimate joint review conducted by the transcriber, the reviewer and the academic representative who was present within the interview itself. In nine cases, the transcribed interviews were sent back to the interviewee for validation/comments.

The second phase of the empirical data-gathering efforts reviewed the evidence gathered through the 15 interviews and used these findings to craft/issue a survey within the case study company. The survey was crafted in order for the research team to understand the topic of R&D within logistics as seen by a larger and randomly-selected, non-biased employee population. The survey was initially issued to 15 people in a pilot version. Subsequently, the survey was modified based on the pilot population input before being issued to a population of 100 employees within the case study company. A total of 38 useable survey responses were obtained from the survey effort. The objective of the survey was to test the overall understanding of logistics innovation topics within the company organization using general industry vocabulary as opposed to WP-specific vocabulary.

3. Results

According to the empirical findings of our case study, an important finding is that DONG Energy entered the market of offshore wind farms as a pioneer when no "traditional" EPCi companies had yet developed skills and competencies to move land-based WTGs offshore and build wind farms offshore. The senior manager responsible for the strategy of WP explained that " . . . the philosophy of course stems from the fact that we have been in the market when there had not been anybody available who could readily do what was needed. I mean, had it been started within the industry with a clear technique or something in order to be able to buy a full park fully installed, we probably would have taken that". Therefore, a strong set of in-house skills and competencies was developed by WP in what is portrayed as a vacuum of the market and where the company was an early mover. Still today, most competitors of WP in the offshore wind sector in Europe employ 5–50 employees to develop a wind farm where WP, in turn, now employs in excess of 1600 people: The case study company acts as both utility, offshore wind farm developer/EPCi and offshore wind farm operator with a multi-contracting governance structure "slicing" up the work tasks into small contract pieces. From a logistics perspective, this makes WP a very strong supply chain lead company with vast human resources available to plan, develop, monitor and manage many of the different sub-supply chains within each of the wind farm life-cycles. For almost all other wind farm developers and operators, the very low number of in-house employees results in single contracting set-ups, where typically 4–6 larger contracts are awarded to, for example large (and now capable) EPCi providers and WTG OEMs in the construction phase and, e.g., a WTG OEM and a service company in the operations phase.

Regarding the topic of logistics within WP, the interviewees were subjected to questions about the case study company's ownership of the major shipping and logistics company A2Sea. This PPP subsidiary company was first acquired directly by the Scandinavian state-owned utility case study company in the open market place, and subsequently, 49 percent of the shares were sold off to the dominant WTG OEM. The PPP subsidiary has increased its financial standing considerably and is now active both in the offshore wind farm construction and operations life-cycle phases with a much enhanced asset set-up and human resources infrastructure. The WP interviewees generally downplayed the importance of having such logistics, shipping and SCM skills available in-house and explained that it was operated at arm's length: the interviewees generally stated that at the time of the acquisition by the state-owned case study utility company, the market situation was such that a bottleneck surrounded key assets and competencies possessed by the subsidiary company, but that the situation has now changed to a supply/demand equilibrium. The interviewees generally did not seem to find the ownership of the PPP subsidiary to provide the case study utility company with an unfair advantage over both direct OWF developer/operator competitors nor shipping/logistics/SCM companies trying to serve the global wind energy sector. The interviewees generally stated that they also did not find the WTG OEM JV partner to be put in a more advantageous market position than its direct WTG OEM competitors, or its indirect EPCi competitors, or the shipping/logistics/SCM companies serving the global wind energy sector.

3.1. Definition of Logistics

It was clear from the interview process that WP does not have a logistics strategy as such. A member of the WP management board explained that " ... from the strategic perspective, we don't have a strategy on logistics, or what logistics is. Then I want to mention this because you ask 'What is the definition?' and there is none. There is none ... " This view was supported by other interviewees and another member of the WP management board said that " ... ok, when we now talk about logistics we have, either we have a definition, [or...] We don't have that! ... ".

As a leading practitioner association, CSCMP [8] defines logistics across multiple industries as: "*The part of supply chain management that plans, implements, and controls the efficient, effective forward and reverse flow and storage of goods, services and related information between the point of origin and the point of consumption in order to meet customers' requirements*". Within our case study, the logistics definition varied both across WP team member work scope within the OWF life-cycle phases, organizational layers of WP and depending on our methodology of obtaining the empirical data. In addition, we found that to a certain extent, WP has their own logistics terminology, which varies somewhat from the non-WP industry definitions. During the 15 *interviews*, the interview guide was designed in such a way that the interviewees were given an opportunity to freely discuss logistics issues, including how they would define logistics. Here, it became clear that their vantage point, definition and perspective were very much based on where in the OWF life-cycle they worked, as well as where they had prior experience from. The *surveys* were more structured in advance by the research team inasmuch as the logistics definition section gave a number of options for the respondents to tick, as well as a free text field option in terms of how they felt that logistics should be defined. The logistics definition options in the survey were based on industry definitions not specifically designed around the WP terminology (see Figure 2).

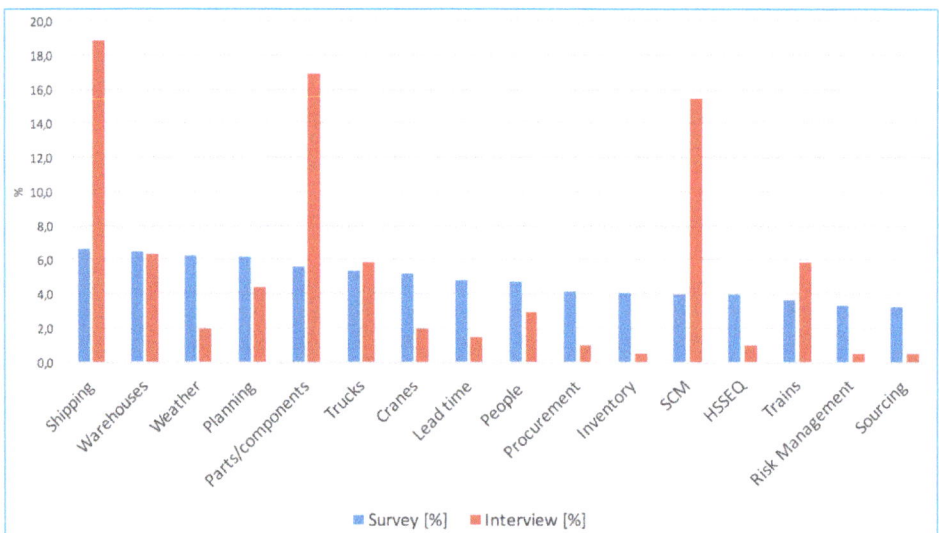

Figure 2. Frequency of terminology used (in %) during interviews and in the survey definition section.

The people interviewed at WP spoke much more about three of the keywords from the survey, *i.e.*, "shipping", "parts/components" and "SCM". When we disentangled these and other WP key terms during the interviews, we got into an underlying set of additional words associated with each of these keywords (see Table 2). These words we could further categorize along several dimensions, each forming part of the definition of offshore wind logistics:

- The term "shipping" could mean transportation by both vessel and helicopter (*mode of transport* being sea or air); different types of trucks/ships/boats/vessels/helicopters could be involved (*means of transport*); and different tasks could be performed (*activities* such as transporting personnel, performing surveys, preparation, loading, unloading).
- In terms of *what we ship*, different "parts and components" mentioned by the interviewees included both main WTG and BOP components, but also technicians including their tools, personal protection equipment (PPE), equipment, parts, as well as power to the grid.
- Just like we saw within academia, the definition of "supply chain management" was much wider during the interviews with the WP personnel. Here, the discussions ranged across a wide spectrum: from skills/knowledge (*competencies*), who is being served within which supply chains (who is the customer of either a *single or multiple supply chains*), the scale, scope and extent of the different supply chains (*beginning and ending points*) and the use of key performance indicators and computers (*IT and data management*).

Table 2. Words included in the interview dialogue about key survey terms.

Shipping	Parts/Components	Supply Chain Management (SCM)
Transport	Foundations	Delivery
Vessel	Turbine	Reduce delivery time
Crew transfer vessel (CTV)	Cable	Set-up around transportation
Helicopters	Goods/components	Preparation prior to execution
Transportation as part of installation	Towers	Coordinate logistics activities
Accommodation vessels	Building materials	Aligned flow of components
Survey vessels	Spare parts	Installation
Other vessels	Equipment	Logistics in operations & maintenance (O&M)
Offshore	Suppliers	Transport
Transportation with installation vessel	Survey equipment	Starts at production
Personnel logistics	Fixed platform	End-to-end (E2E)
Execution	Life vests	Between different countries
Installation vessel	Tools	Tier one customer
Unloading	Onshore activity	Idea to project hand-over
Prepare for shipping	Transition assets	Quay side
Sailing	Return of faulty component	Build an offshore wind farm (OWF)
-	Distribution	Supply
-	Technicians	Onshore projects
-	Logistics concepts	Knowledge regarding transportation process quality
-	Traffic	-

Both the discussions and survey reflected that *weather* considerations and health, safety, security, environmental and quality (*HSSEQ*) considerations play a very significant part in both OWF installation and O&M. Similarly, it was also clear that the context of logistics is very different if the *logistical focus* (unit of analysis) is that of an individual WTG (for example, break-down maintenance), an entire OWF (for example, during installation or in the event of a cable disruption during operations) or across a portfolio of OWFs (for example, survey vessel operations across more OWFs or synergies in terms of spare part storage for several OWFs). The risks and costs are much smaller for an individual WTG compared to an entire OWF or the synergies from portfolio asset management economies of scale.

When grouped along the definition category dimensions, the individual words used in the interviews and survey responses could be further sorted and contrasted, as seen in Figure 3, showing a difference in how the WP survey personnel responded differently from those interviewed because the surveys prompted industry terms rather than commonly-used WP in-house terminology. Our research resulted in a suggested and all-encompassing *definition for offshore wind logistics* as follows: "*Parts, modules, components, people and tools are responsibly stored and moved safely, weather permitting, onshore, as well as offshore by air/ocean/land using various transportation assets and transport equipment with a focus on an individual wind turbine generator, an offshore wind farm asset project or across a portfolio of projects by means of different in-house and outsourced logistics skills/capabilities/IT systems used across multiple supply chains spanning different starting and ending points*". This definition was a very important cornerstone in the efforts of the research team to come up with a tangible R&D strategy for logistics within WP.

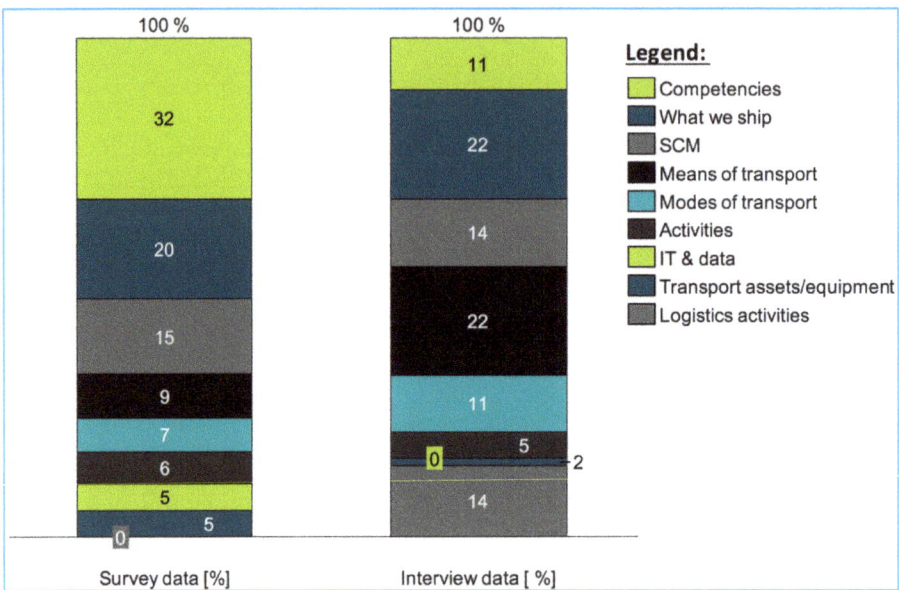

Figure 3. Logistics words frequency (in %) categorized along dimensions from surveys *vs.* interviews.

3.2. The cost of Logistics

Part of the interviews and a section of the surveys were dedicated to understanding the cost of logistics. Of 28 useful answers obtained regarding logistics costs from the interviews and surveys, eight answers had some degree of ambiguity in terms of whether the logistics costs portrayed could be directly associated with different life-cycle phases, for example installation and commissioning (CapEx), operations and maintenance (OpEx) or LCoE as measured in end-to-end (E2E) logistics costs. To resolve these ambiguity conflicts, the research team had to either review the overall context of the interview or the survey response submission in its entirety in order to determine the exact context for the logistics cost answer. The rest of the answers could be clearly categorized within CapEx, OpEx or E2E with one example being a senior DONG Energy Group finance manager who clearly had a full LCoE and E2E logistics scope in mind: " … I think that there is logistics all through the value chain from [when] you acquire the, the right to build wind turbines in a specific area until you take it down. But of course it's, it's different kind of logistic capabilities you need … ".

None of the respondents had a good sense of the size of the de-commissioning costs as a stand-alone cost component of LCoE, but many were discussing it. A member of the WP management board responsible for key component design and manufacturing: " ... if you have to remove a gravity foundation, what to do with that excess concrete afterwards? If you asked 10 years ago, we would say it could be used for pavements, *etc.* Looking into the future [now], perhaps it's going to be reused into a different form somewhere in a different way ... ". Furthermore, a WP manager with a leading role in the design and manufacturing process for WTGs said " ... and if at one point we do see a major failure in one of our turbines, we have to think about whether it is time for de-commissioning or how the business case is the best ... ". As can be derived from Figure 4, logistics costs form a relatively significant part of the overall costs irrespective of the vantage point within WP.

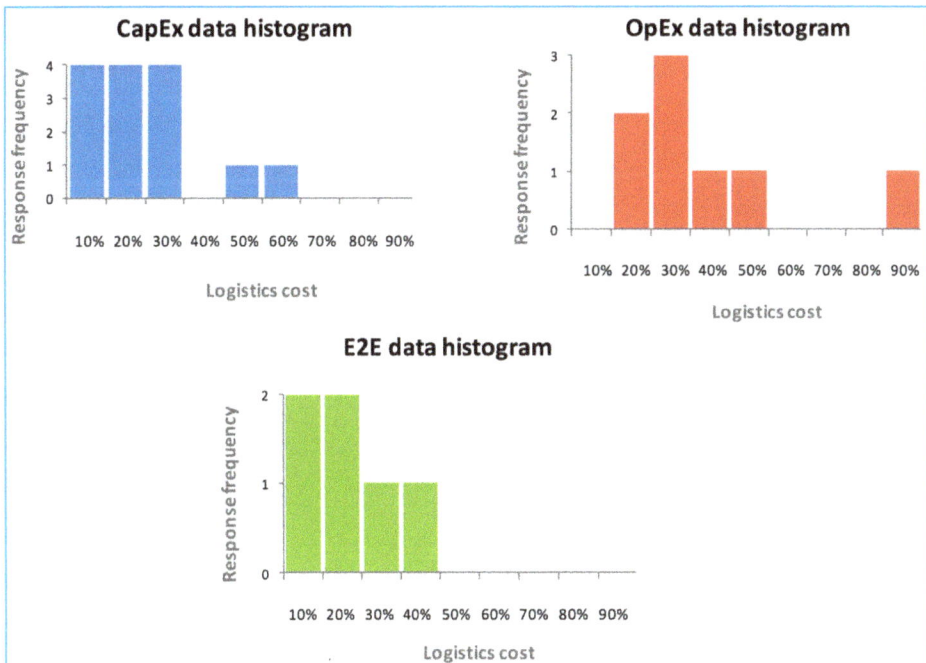

Figure 4. The 28 useful responses about logistics costs (in %) of CapEx, OpEx or LCoE (E2E).

Another LCoE initiative [24] practically substantiates that it is not possible to simply add CapEx and OpEx costs to get to the total costs within the LCoE calculation, because both the development and consent (project development expenditure, DevEx) costs prior to the OWF project final investment decision and the de-commissioning (site abandonment expenditure, AbEx) costs need to be included, as well. It was therefore only possible to review the useful WP logistics cost responses separately within their respective categories as depicted in Figure 5. In doing so, we can conclude that whereas 23% and 36% of CapEx and OpEx costs, respectively, are attributable to logistics, 18% of the E2E OWF project costs across life-cycles and equal to the cost equation of the LCoE can be attributed to logistics. Based on the ambiguity within both the country LCoE definitions themselves and the definition of logistics in its widest application (including the project risk from the U.K. [19] and German [21] LCoE studies), *logistics costs of 18% of LCoE* must be deemed to be a 'very conservative minimum level' according to our research.

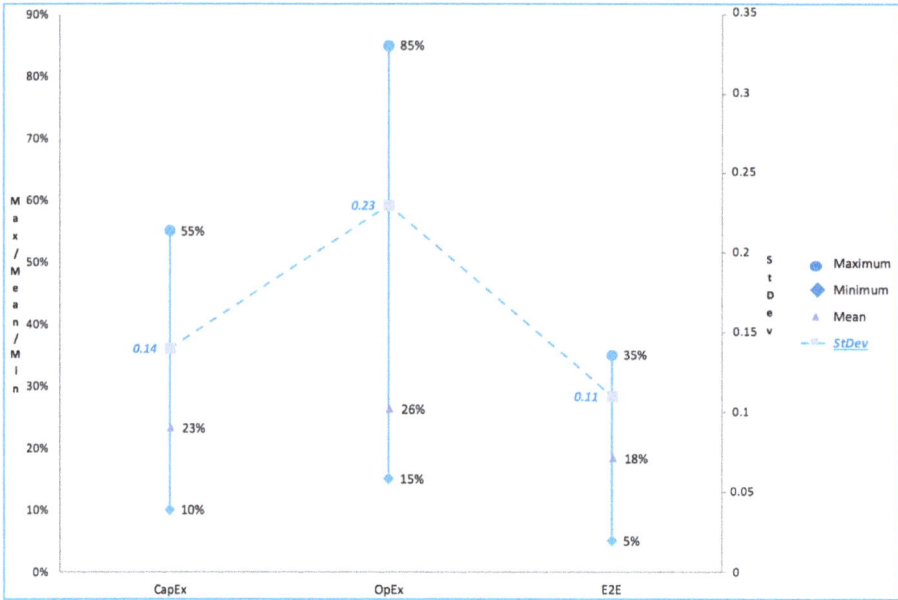

Figure 5. Distribution of responses about logistics costs as a share of total costs.

3.3. Logistics Innovation

The interpretation of logistics innovation within WP was clearly framed by a member of the WP management board, who said that " ... logical next step business issues ... " and " ... execution of the normal business strategy ... " should not be confused with logistics innovation. Another member of the management board said that logistics innovation within WP can be classified as " ... ideas that are known solutions but new to wind power in general, ideas that are known solutions but new to DONG Energy Wind Power, or new solutions ... ". A WP top manager within the area of procurement and LCoE defined the critical success factors ("CSFs") for logistics innovation as " ... sustainable improvements in cost of energy, health/safety/environment, or quality".

On this basis, the research team reviewed the interview transcripts and survey responses in order to come up with a gross list of potential logistics innovation ideas. A total of 159 quotes were identified and processed during three workshops involving the research team and case study company representatives. Several interviewees and survey respondents talked about the same or similar ideas, and some of the quotes from the interviews/survey responses needed further interpretation. This resulted in a gross list of 61 useful ideas generated from the case study process, and of these, eight were not related to logistics. Of the 53 remaining ideas in the catalogue, 38 could be considered a resourceful expansion of the daily work scope for different parts of the organization. When reviewing the remaining 19 idea catalogue items together with company representatives, these could be further consolidated into 12 innovative project ideas for WP to focus on. To focus on 12 projects is, however, not efficient, and a prioritization therefore took place both focusing on the aforementioned CSFs. The level of complexity, whether WP has the necessary personnel in-house to complete the task, and the estimated time required to implement the changes were factors also considered. Accordingly, the top five "must-win battles" were identified as depicted in Table 3. The goal to identify at least five tangible R&D projects for the new logistics R&D project organization to work on was achieved, which is in line with the original project charter to craft a logistics R&D strategy of the company.

Table 3. Top 5 "must-win battles 2016" for the WP R&D logistics project organization.

ID	2016 "Must-Win Battles"	CSF
1.	Establish preventive maintenance process for balance of plant (BOP) components including foundations/cables/offshore substation	LCoE
2.	Market analysis of future offshore accommodation options as offshore wind farms (OWF) move further from shore into deeper waters	LCoE
3.	Improve present and future crew transfer process to/from any offshore structure to reduce risk of accidents	HSSEQ
4.	Proactively support wind turbine generator (WTG) mega-Watt (MW) yield step-change in terms of logistics to cater for heavier and larger WTG and BOP components	LCoE
5.	Determine if present and future vessels can be used for multiple purposes (e.g., wind turbine installation vessels (WTIVs) for foundations, WTGs, cables, and OSS; crew transfer vessels (CTVs) for surveys)	LCoE

3.4. Organizational Implications

According to our research, expansion into the U.S. and Asian offshore wind markets is being contemplated at all levels of management of WP beyond 2020. Logistically, this means replicating the largely Scandinavian company culture, skills and competencies much further away from home than hitherto. This is recognized at the DONG Energy group level according to a manager in the Group finance organization: " ... the supplier relations and the culture change and I think today we are a very Scandinavian company ... " Now people, competencies, cultural integration, legislative understanding, WTG parts, wind components, ports, vessels and other transport assets/equipment will be needed in far-away markets where the rest of the case study company experiences little synergy. Within the WP finance team, a manager expressed it as " ... it's going to be a big challenge for DONG [WP] going really far abroad. I think culture wise it's going to be a massive change ... ". Today, logistics is not organized horizontally across the company in a centralized department, competence center or center of excellence. One member of the WP management board said a centralized function for logistics is needed in the future: " ... To be able to actually to build competence, to build culture, to build method, and build also the future... All that intelligence should be here. And, and why should it be in one department is, of course, that to be able to have that central expertise you need to gather these people who are working with this daily, to get the knowledge into, say, this center, so you can gather it ... ". With the rapid globalization of the WP offshore wind business model, the need for a centralized focus and attention to logistics becomes even more relevant.

Our findings indicate that an organizational shortcoming within logistics was confirmed through the interviews with both the interviewees and the survey respondents. A senior WP manager within the area of construction and EPCi explained that in terms of replicating a European offshore wind project in, for example, an Asian geography like China, Taiwan, South Korea or Japan, " ... there would be maybe a handful of those profiles where I would have that kind of trust that they would be able to develop this on, on their own ... " and he continued that " ... some of them are no longer in my organization and elsewhere in DONG [Energy] em, but still accessible ... ". He concluded that " ... it would generally be some of the quite senior, em, installation managers that I have". It is also a question of having the right skills and competencies available, both in the future as well as right now, as the portfolio of OWFs continues to expand. A member of the WP management board explained that tenure with the firm and industry experience is lacking within offshore wind, as the industry is still rather young: " ... if you look at the people working here, we have very experienced people that are on the ships and out in the projects. We don't have people ... with the 25 years in the business ... these guys are fact people ... [people who learned by doing]". In addition, the logisticians employed are considering mainly their own vertical area of responsibility and not horizontally across the project life-cycle or across multiple offshore wind projects. One WP middle management representative from the construction and execution arena explained that " ... there are very, very few that are, are good generalists. It is specialists that we have employed and I think that is the challenge. That many of these, they are so hardcore in their own discipline that they, they sometimes are difficult to lift up in a helicopter to give you the full perspective. So they would attempt to sub-optimize their own silo and that's some of the barriers that we would need to break down . . . ".

From a knowledge management point of view, it is difficult for the company to perform a hand-over of the experience gained by multiple people from multiple sources within an individual OWF project to future projects [49]. One WP management board member with R&D responsibilities said " . . . in the ideal world you would do the R&D work upfront before you have a problem. Or when you identify the problem on one wind farm then you would start an R&D project and once you have a solution, you could implement it on the next one. But with the timeframe we have [laughing] on our projects, often we have to develop almost as we built. . . . ". The challenge is great during individual life-cycle phases, such as the installation and commissioning process as, e.g., voiced by the senior manager in the construction and EPCi part of the WP organization, who said " . . . I think one of the challenges we have in DONG [WP] is that we are working in those [logistics] silos. We don't talk together, we have a lot of guys sitting over here, doing a lot of work—they don't talk with the end users out here. And we have seen it on a lot of our projects now that we have someone going that direction but we should have been in this direction and it costs us a lot of money because we didn't meet upfront to align this . . . ". Furthermore, between life-cycle phases, hand-overs present a logistical challenge, said a WP manager with full visibility of the WTG manufacturing process: " . . . one of the important things for us is to understand what abnormalities they [suppliers] see during construction. And that is actually logistics. When they are moving it on the harbor to do some tests, and then moving it into the sea and erecting them, that logistics part is also important for us to understand, because that is basically the baseline for the integrity. So if they have had some [damages] during this part of the logistics, which is important for us to know. Because when we do start to see some problems in the O&M phase that can be due to transportation or mishandling of the product during that erection period . . . ".

To conclude our case study findings, three macro factors were identified that seem to be going to make the offshore wind business more complex beyond 2020:

1. *OWFs will move further away from shore*. The near shore sites are becoming rarer, which means that OWFs are moving further offshore and into deeper waters. The individual OWFs will be GW-sized, which means that risk management efforts and focused contingency plans will be increased. Each WTG position must produce a greater yield in terms of MW/h, and this, in turn, requires more shore-based personnel to stay offshore for longer periods of time.
2. *WTG output yield will go through another step-change size increase*. The present WTGs yielding 4–8 MW will be replaced by WTGs yielding 10–15 MW by the early 2020s. Towards the end of the 2020s, WTGs yielding 20 MW will be introduced to the market along with floating WTG concepts.
3. *Offshore wind is rapidly going global*. The WTG supply chain is largely global already; however, the BOP supply chain is predominantly European. This means that new key markets, such as China, Japan, South Korea, Taiwan, India and the U.S., will largely depend on a European supply chain for BOP and a largely European experience base in terms of the process of moving land-based WTGs into the ocean.

4. Discussion

Our case study identified that these macro-level findings do have a profound impact on especially our overall case study policy and governance perspectives:

- *Policy-wise*, our work with WP shows that offshore wind is still a fairly young and immature industry with a large dependency on government subsidies to survive and expand diffusion. Up to 2020, the legislative environment is firm in key EU countries and especially the emerging Chinese offshore wind market. A stabile and long-term legislative environment also beyond 2020 is needed to ensure that the necessary investments can be made by shipping/logistics/SCM companies. This is needed to ensure that transportation assets and transport equipment of the necessary size, caliber and the right lifting abilities are in place for the expected advances in technology size and shape. Although downplayed in the interviews, the role of the case study firm's JV-owned PPP shipping/logistics/SCM subsidiary originally alleviated a significant supply bottleneck at the time of acquisition. Now, the PPP logistics subsidiary has, at a minimum, strengthened the relations between the case study company and the dominant WTG OEM, SWP, with whom the JV subsidiary is jointly owned. In addition, critical shipping/logistics/SCM skills and competencies are now available "in-house" via the JV PPP logistics subsidiary company. Although supposedly run at arm's length, the availability of both assets, people, competencies, skills and knowledge within the field of logistics seem to go hand-in-hand with the case study company's ambition to remain in the market leadership role for global offshore wind farm construction and operations. Additional players from the market are, however, needed in order for the industry sector of offshore wind to create the diffusion necessary to reach global renewable energy targets.

- *Governance-wise*, it is important that necessary government funding is allocated to the area of logistics innovation in order to support the core technological innovation of the WTG products. Only by ensuring proper alignment and due FEED several years in advance can new WTGs and supporting BOP structures be transported and installed to their offshore sites.

When it comes to the applicability to both practitioners and academicians alike, our case study findings are very useful:

- From an *academic perspective*, strategy alignment is necessary, as well as critical. The task of defining an R&D strategy for logistics within the case study company became more complex when the lack of a common logistics definition along with the inexistent logistics strategy became apparent early in the interviewing process. The strategy hierarchy seemed to be clear with company strategy placed squarely at the top and supported by business unit strategy; in this case, strategy within the offshore wind business unit. WP business unit strategy would ideally be comprised of different supporting pillars of which a logistics or supply chain strategy could expectedly be one such pillar. As defined by Chopra and Meindl [40], alignment of a company's supply chain strategy to the company strategy is critical to success and company survival. It follows from this argument that the strategy for R&D within the area of logistics should therefore be closely aligned with the overall strategy for logistics. The logistics strategy would be dependent on how logistics itself is defined. Our case study definition category shows that a proposed definition of offshore wind logistics across multiple dimensions should be a step in the right direction for the case study company and also for the offshore wind industry at large. With almost all other offshore wind farm developers and operators applying a single contracting business model, where large contracts are given to, e.g., EPCi companies and/or WTG OEMs, the market is not very transparent to the shipping/logistics/SCM companies trying to serve the global wind energy market. Who is actually the customer demanding the services to be rendered? When is the customer a competitor? Additionally, what alliances and allegiances exist between seemingly straight-forward companies with not so apparent links to sovereign nation states and their national agendas? These questions and the fact that the mere future existence of the wind energy market depends on continued government-sponsored subsidies are factors that may keep some shipping/logistics/SCM companies away from competing in the muddy waters of the global offshore wind industry; or perhaps causes some of the metaphorical blindness referred to by Mintzberg and Lampel [50] in their description of how both practitioners and scientists view this particular "elephant" in the safari of strategy. If the right companies do not enter the offshore wind logistics market place, the much needed professionalization of the supply chain may not happen. This lack of professionalization will be the beginning of a vicious circle that may lead to a lack of industrialization of the wind industry itself and inability to practically lower LCoE, a parameter that in itself is vital for offshore wind industry survival in the long-term without government subsidies; and an important factor for the OWFs already in operation as they start to move closer to their end of life service time [51].

- From a *practitioner perspective*, our case study findings indicating that logistics is at least 18% of LCoE should point towards the area of logistics being ripe to explore in terms of possible cost reduction exercises. Findings from the U.S. over an extensive period of time reveal that by making logistics a recognized and admirable focus area for a cross-section of all industries with support from academia had brought down logistics costs as a percentage of GDP from 15.8% in 1981 to 8.4% in 2014 [8]. Realizing a 50% reduction in cost is not easy and has taken in excess of 20 years in the U.S. Therefore, the offshore wind industry needs to get organized not only within project life-cycle phases, but also horizontally across the different OWF life-cycle phases and across a portfolio of more OWFs. As the LCoE calculations of respectively Denmark, the U.K. and Germany showed [19,21,29], it is always hard to determine exactly how to measure costs within offshore wind, as it needs to be made very clear from the context or questions asked what, for example, a percentage is related to. Here, the LCoE initiative [24] should be highlighted because it developed a LCoE calculator tool based on the company-specific LCoE calculation models of key offshore wind developers (DONG Energy Wind Power, E.On and Vattenfall), key offshore wind OEMs (Siemens Wind Power and MHI Vestas Offshore Wind) and with input to the initiative from an additional 15 organizations, including several academic institutions, such as Aalborg University and DTU Wind Energy. This LCoE calculator tool [24] takes all wind farm life-cycle stages into consideration, from project idea through site restoration at the end of service life, as it is organized along four main cost dimensions, DevEx, CapEx, OpEx and AbEx. The cost items to be included in the LCoE calculator tool are generic in nature and as such do not allow for a significant further itemized breakdown. However, this model offers a full scope regarding the different supply chains where logistics costs may be incurred throughout the entire OWF project life-cycle. The LCoE calculator tool also considers, for example, production in the construction phase, and as part of production, a large inbound logistics flow is required. None of the country studies accounted or allowed for such an inbound flow. As such, the LCoE calculator tool [24] comes closest to being able to establish a platform able to address the end-to-end logistics costs in a horizontal manner across an OWF project and, thus, also the opportunity to start optimizing across a portfolio or several portfolios of OWFs. The LCoE calculator [24] furthermore addresses the offshore grid connection challenges described earlier by establishing a "point of common coupling" between the onshore grid and the offshore transmission owner, which may be supported by the model. Finally, the terminology used within the Megavind LCoE calculator tool [24] matches almost identically the company-specific terminology we found within our case study company.

5. Conclusions

Our case study was comprised of 15 interviews and 38 usable survey responses out of a total of 115 possible responses within DONG Energy. This largely government-owned market share leader of the offshore wind market segment has positioned itself strongly within the field of logistics before a contemplated listing of the company on the stock market in Denmark [52]. When seen in conjunction with the large workforce employed in order to position the company as an offshore wind farm construction company and operator, the multi-contracting business model and on-going global market scaling efforts make the case study company a very serious player to be reckoned with in the market.

When analyzing the 28 useful qualitative responses about *logistics costs*, we conservatively identified that end-to-end offshore wind logistics across the four offshore wind farm life-cycle phases make up at least 18% of the offshore wind levelized cost of energy. Based on the fact that it took the United States in excess of 20 years to reduce logistics costs across all industries as a percentage of gross domestic product from 15.8% to 8.4% [8], our findings show that the offshore wind industry should focus on reducing logistics costs: It will take time; however, cost savings can be reaped.

Energies **2016**, *9*, 437

From the list of 12 specific *logistics innovation* ideas yielded by our case study for the case study company to focus on during 2016 and beyond, several of the "must-win battles" identified hold a lot of promise and potential, also for the offshore wind industry at large, in terms of cost reductions within the area of logistics. Efforts to create logistics innovation within the area of preventive maintenance for the balance of plant parts of offshore wind farms must be highlighted. Efforts should also be put into the idea to logistically innovate in terms of vessel types to be used for multiple purposes. Logistics innovation in the early stages of the technological product design process for larger wind turbines is critical for the industry in general due to the additional issues of them being placed further from shore in deeper waters.

Focus on the *organizational* set-up within offshore wind is of paramount importance, and our case study highlighted that economies of scale are required by optimizing across all assets across all wind farm life-cycles. These include logistics activities across a portfolio of offshore wind farms under development, under construction, as well as offshore wind farms already in operation. Being the market leader in terms of construction and operations of offshore wind farms, our case study company is a good example of the state of the industry. Our case study showed that the case study company is not yet ideally positioned organizationally to focus beyond vertical organizational silos, let alone replicate offshore wind logistics skills to markets outside Northern Europe. This implies that for the offshore wind industry in general, infusion of additional skilled logistics personnel trained from other industries with the required vertical specialist skills and strategic horizontal skills is a must to realize logistics cost savings.

We recommend that *further research* efforts be undertaken by other academic scholars and practitioners alike in order to ensure that the exact logistics cost components of offshore wind are unveiled and fully defined. We recommend that specific studies be completed regarding how the levelized cost of energy can be reduced and executed within logistics cost component groupings through specific cost-out initiatives. We also recommend that logistics be included as a vertical life-cycle phase cost component and that a horizontal logistics view be adopted and defined. This definition should be at a national level, a company-specific level and for use within academic levelized cost of energy models, calculators and initiatives. Finally, we recommend that our study be followed up by additional quantitative studies on what planned "ideal state" logistics costs are expected to attribute in terms of levelized cost of energy share compared to actual "realized" logistics costs for real offshore wind projects across the entire offshore wind farm project life-cycle, as well as across a portfolio of offshore wind farms.

Our research shows that at a level of at least 18% of the total life-time costs of offshore wind farms, logistics costs are considerable. Therefore, our overall conclusion is that logistics is an area that is expensive enough to be a major focus for innovation and that further work is essential in order to reduce cost for the offshore wind sector.

Acknowledgments: This research is sponsored by the Danish Maritime Foundation (Grant 2012-097) and Aalborg University. The research is published based on written consent obtained from DONG Energy. The authors would like to thank DONG Energy for case study access, including the interviewees, as well as case study respondents. The authors would like to thank the combined research team; particularly, Christina Aabo and Anders Greve Pihlkjær from DONG Energy Wind Power, as well as Aalborg University students Martins Paberzs, Alex Timar, Emel Zhao and Thomas Aabo. A special thanks is also extended to Thomas Poulsen's Ph.D. supervisor Lars Bo Henriksen of Aalborg University.

Author Contributions: Thomas Poulsen conceived of the research design. Thomas Poulsen performed the research with support from the research team. Thomas Poulsen and the research team analyzed the data. Thomas Poulsen wrote the paper with support from Charlotte Bay Hasager.

Conflicts of Interest: The authors declare no conflict of interest. The founding sponsors had no role in the design of the study; in the collection, analyses or interpretation of data; in the writing of the manuscript; nor in the decision to publish the results.

Abbreviations

The following abbreviations are used in this manuscript:

AAU	Aalborg University
AbEx	Abandonment expenditure
BOP	Balance of plant (cables, substations, wind turbine foundations)
CapEx	Capital expenditure
CSCMP	Council of Supply Chain Management Practitioners
CSF	Critical success factors
CTV	Crew transfer vessel
DECC	UK Department of Energy and Climate Change
De-comms	Decommissioning, site abandonment at the end of service life
DevEx	Development expenditure
DTU	Technical University of Denmark
E2E	End-to-end
EU	European Union
EPCi	Engineering, procurement, construction and installation companies
EWEA	European Wind Energy Association, now WindEurope
FEED	Front-end engineering and design
GBP	Great Britain Pounds
GW	Giga-Watt
GWEC	Global Wind Energy Organization
HSSEQ	Health, safety, security, environment and quality
I&C	The installation and commissioning life-cycle phase of an offshore wind farm
IEA	International Energy Agency
IM	Inventory management theory stream
IT	Information technology
JV	Joint-venture
LCoE	Levelized cost of energy
MF	Materials flow theory stream
MH	Materials handling theory stream
MM	Materials management theory stream
MW	Mega-Watt
MW/h	Mega-Watt hours
O&M	Operations and maintenance
OEM	Original equipment manufacturer
OpEx	Operational expenditure
OSS	Offshore (and onshore) sub-station
OWF	Offshore wind farm
PD	Physical distribution theory stream
PPP	Public-private partnership
R&D	Research and development
SCM	Supply chain management
SCI	Supply chain (management) innovation
SWP	Siemens Wind Power
U.K.	United Kingdom
U.S.	United States of America
WP	DONG Energy Wind Power
WTIV	Wind turbine installation vessel
WTG	Wind turbine generator

References

1. Global Wind Energy Council. Global Wind Energy Outlook. 2014. Available online: http://www.gwec.net/wp-content/uploads/2014/10/GWEO2014_WEB.pdf (accessed on 26 November 2015).
2. Milborrow, D. Windicator: Global Total Hits 400 GW as China Continues to Push Ahead. 2015. Available online: http://www.windpowermonthly.com/article/1365877/windicator-global-total-hits-400gw-china-continues-push-ahead (accessed on 2 February 2016).

3. European Wind Energy Association. The European offshore wind industry key trends and statistics 2015. Available online: http://www.ewea.org/fileadmin/files/library/publications/statistics/EWEA-European-Offshore-Statistics-2015.pdf (accessed on 2 February 2016).
4. Navigant Research. *A BTM Navigant Wind Report. World Wind Energy Market Update 2015. International Wind Energy Development: 2015–2019*; Navigant Research: Chicago, IL, USA, 2015.
5. DONG Energy. 2015 Annual Report Information (P. 27) about World Leading Construction and Operations of Offshore Wind. Available online: https://assets.dongenergy.com/DONGEnergyDocuments/com/Investor/Annual_Report/2015/dong_energy_annual_report_en.pdf (accessed on 14 February 2016).
6. DONG Energy. Shareholder Information. Available online: http://www.dongenergy.com/en/investors/shareholders (accessed on 14 February 2016).
7. Poulsen, T.; Rytter, N.G.M.; Chen, G. Global wind turbine shipping & logistics—A research area of the future? In Proceedings of the Conference Proceedings of International Conference on Logistics and Maritime Systems (LogMS) Conference, Singapore, 12–14 September 2013.
8. Council of Supply Chain Management Professionals. *CSCMP's Annual State of Logistics Report, Freight Moves the Economy in 2014*; National Press Club: Washington, DC, USA, 2015.
9. Mazzucato, M. *The Entrepreneurial State*, 2nd ed.; Anthem Press: London, UK, 2014.
10. Gosden, E. World's Biggest Offshore Wind Farm to Add £4.2 Billion to Energy Bills. Available online: http://www.telegraph.co.uk/news/earth/energy/windpower/12138194/Worlds-biggest-offshore-wind-farm-to-add-4.2-billion-to-energy-bills.html (accessed on 7 February 2016).
11. Shafiee, M.; Dinmohammadi, F. An FMEA-Based Risk Assessment Approach for Wind Turbine Systems: A Comparative Study of Onshore and Offshore. *Energies* **2014**, *7*, 619–642. [CrossRef]
12. Flyvbjerg, B.; Bruzelius, N.; Rothengatter, W. *Megaprojects and Risk. An Anatomy of Ambition*; Cambridge University Press: Cambridge, UK, 2003.
13. Poulsen, T. Changing strategies in global wind energy shipping, logistics, and supply chain management. In *Research in the Decision Sciences for Global Supply Chain Network Innovations*; Stentoft, J., Paulraj, A., Vastag, G., Eds.; Pearson Eduction: Old Tappan, NJ, USA, 2015; pp. 83–106.
14. Lacerda, J.S.; van den Bergh, J.C.J.M. International Diffusion of Renewable Energy Innovations: Lessons from the Lead Markets for Wind Power in China, Germany and USA. *Energies* **2014**, *7*, 8236–8263. [CrossRef]
15. Roehrich, J.; Lewis, M. Procuring complex performance: Implications for exchange governance complexity. *Int. J. Oper. Prod. Manag.* **2014**, *34*, 221–241.
16. Pregger, T.; Lavagno, E.; Labriet, M.; Seljom, P.; Biberacher, M.; Blesl, M.; Trieb, F.; O'Sullivan, M.; Gerboni, R.; Schranz, L.; et al. Resources, capacities and corridors for energy imports to Europe. *Int. J. Energy Sect. Manag.* **2011**, *5*, 125–156. [CrossRef]
17. Cusumano, M.A.; Gawer, A. The elements of platform leadership. *IEEE Eng. Manag. Rev.* **2003**, *31*, 8–15. [CrossRef]
18. Dale, M. A Comparative Analysis of Energy Costs of Photovoltaic, Solar Thermal, and Wind Electricity Generation Technologies. *Appl. Sci.* **2013**, *3*, 325–337. [CrossRef]
19. The Crown Estate. Offshore Wind Cost Reduction Pathways Study. 2012. Available online: http://www.thecrownestate.co.uk/media/5493/ei-offshore-wind-cost-reduction-pathways-study.pdf (accessed on 7 December 2015).
20. International Energy Association. *Projected Costs of Generating Electricity*; Organization for Economic Co-operation and Development: Paris, France, 2005.
21. Prognos & Ficthner Group. Cost Reduction Potentials of Offshore Wind Power in Germany, Long Version. 2013. Available online: http://www.offshore-stiftung.com/60005/Uploaded/SOW_Download%7cStudy_LongVersion_CostReductionPotentialsofOffshoreWindPowerinGermany.pdf (accessed on 6 December 2015).
22. Heptonstall, P.; Gross, R.; Greenacre, P.; Cockerill, T. The cost of offshore wind: Understanding the past and projecting the future. *Energy Policy* **2012**, *41*, 815–821. [CrossRef]
23. Liu, Z.; Zhang, W.; Zhao, C.; Yuan, J. The Economics of Wind Power in China and Policy Implications. *Energies* **2015**, *8*, 1529–1546. [CrossRef]
24. Megavind. LCoE Calculator Model. 2015. Available online: http://megavind.windpower.org/download/2452/1500318_documentation_and_guidelinespdf (accessed on 8 December 2015).

25. Hasager, C.B.; Madsen, P.H.; Giebel, G.; Réthoré, P.-E.; Hansen, K.S.; Badger, J.; Pena Diaz, A.; Volker, P.; Badger, M.; Karagali, I.; *et al.* Design tool for offshore wind farm cluster planning. In Proceedings of the EWEA Annual Event and Exhibition, 2015, European Wind Energy Association (EWEA), Paris, France, 17–20 November 2015.

26. Gross, R.; Blyth, W.; Heptonstall, P. Risks, revenues and investment in electricity generation: Why policy needs to look beyond costs. *Energy Econ.* **2010**, *32*, 796–804. [CrossRef]

27. Blanco, M.I. The economics of wind energy. *Renew. Sustain. Energy Rev.* **2009**, *13*, 1372–1382. [CrossRef]

28. Shafiee, M. Maintenance logistics organization for offshore wind energy: Current progress and future perspectives. *Renew. Energy* **2015**, *77*, 182–193. [CrossRef]

29. Deloitte. Analysis on The Furthering of Competition in Relation to the Establishment of Large Offshore Wind Farms in Denmark. 2011. Available online: http://www.ens.dk/sites/ens.dk/files/info/news-danish-energy-agency/cheaper-offshore-wind-farms-sight/deloitte_background_report_2_-_analysis_of_competitive_conditions_within_the_offshore_wind_sector.pdf (accessed on 8 December 2015).

30. Poulsen, T.; Rytter, N.G.M.; Chen, G. Offshore windfarm shipping and logistics—The Danish Anholt offshore windfarm as a case study. In Proceedings of the 9th EAWE PhD Seminar on Wind Energy in Europe, Uppsala, Sweden, 18–20 September 2013.

31. Tudor, F. Historical Evolution of Logistics. *Revue Sci. Politiques* **2012**, *36*, 22–32.

32. Stock, J.R.; Lambert, D.M. *Strategic Logistics Management*, 4th ed.; Irwin/McGraw-Hill: Chicago, IL, USA, 2001.

33. Magee, J.F. Guides to Inventory Policy: Functions and Lot Sizes. *Harvard Business Rev.* **1956**, *34*, 49–60.

34. Heskett, J.L.; Glaskowsky, N.A., Jr.; Ivie, R.M. *Business Logistics: Physical Distribution and Materials Management*; Ronald Press: New York, NY, USA, 1973.

35. Heskett, J.L. Logistics—Essential to strategy. *Harvard Business Rev.* **1977**, *55*, 85–96.

36. Shapiro, R.D. Get Leverage from Logistics. *Harvard Business Rev.* **1984**, *62*, 119–126.

37. Porter, M.E. *Competitive Advantage*; Free Press: New York, NY, USA, 1985; Chapter 2.

38. Shapiro, R.D.; Heskett, J.L. *Logistics Strategy*; West Publishing: St. Paul, MN, USA, 1985.

39. Fisher, M.L. What is the right supply chain for your product? *Harvard Business Rev.* **1997**, *75*, 105–116.

40. Chopra, S.; Meindl, P. *Supply Chain Management: Strategy, Planning, and Operation*, 5th ed.; Pearson Education Limited: Harlow, Essex, UK, 2013.

41. Hesse, M.; Rodrigue, J.-P. The transport geography of logistics and freight distribution. *J. Transp. Geogr.* **2004**, *12*, 171–184. [CrossRef]

42. Hou, H.; Kataev, M.Y.; Zhang, Z.; Chaudhry, S.; Zhu, H.; Fu, L.; Yu, M. An evolving trajectory—From PD, logistics, SCM to the theory of material flow. *J. Manag. Anal.* **2015**, *2*, 138–153. [CrossRef]

43. Flint, D.J.; Larsson, E. Exploring processes for customer value insights, supply chain learning and innovation: An international study. *J. Business Logist.* **2008**, *29*, 257–281. [CrossRef]

44. Arlbjørn, J.S.; de Haas, H.; Munksgaard, K.B. Exploring supply chain innovation. *Logist. Res.* **2011**, *3*, 3–18.

45. Grawe, S.J. Logistics innovation: A literature-based conceptual framework. *Int. J. Logist. Manag.* **2009**, *20*, 360–377. [CrossRef]

46. Schumpeter, J.A. *The Theory of Economic Development*; Harvard University Press: Boston, MA, USA, 1934.

47. Schumpeter, J.A. *Capitalism, Socialism, and Democracy*; Harper and Brothers: New York, NY, USA, 1942.

48. Flyvbjerg, B. Five Misunderstandings about Case-Study Research. *Qual. Inq.* **2006**, *12*, 219–245. [CrossRef]

49. Henriksen, L.B. Knowledge management and engineering practices: The case of knowledge management, problem solving and engineering practices. *Technovation* **2001**, *21*, 595–603. [CrossRef]

50. Mintzberg, H.; Lampel, J. Reflecting on the Strategy Process. *Sloan Manag. Rev.* **1999**, *40*, 21–30.

51. Luengo, M.M.; Kolios, A. Failure Mode Identification and End of Life Scenarios of Offshore Wind Turbines: A Review. *Energies* **2015**, *8*, 8339–8354. [CrossRef]

52. Reuters. Goldman Sachs Likely to Keep Stake in DONG After Float—Borsen. 2015. Available online: http://www.reuters.com/article/dongenergy-ipo-goldman-idUSL5N11S0I120150922#sCvzKEwT9OjFV672.97 (accessed on 20 December 2015).

energies

MDPI

Article

Enhanced Forecasting Approach for Electricity Market Prices and Wind Power Data Series in the Short-Term

Gerardo J. Osório [1], Jorge N. D. L. Gonçalves [2], Juan M. Lujano-Rojas [1,3] and João P. S. Catalão [1,2,3,]*

[1] C-MAST, University of Beira Interior, Covilhã 6201-001, Portugal; gjosilva@gmail.com (G.J.O.); lujano.juan@gmail.com (J.M.L.-R.)
[2] INESC TEC and the Faculty of Engineering of the University of Porto, Porto 4200-465, Portugal; ee10191@fe.up.pt
[3] INESC-ID, Instituto Superior Técnico, University of Lisbon, Lisbon 1049-001, Portugal
[*] Correspondence: catalao@fe.up.pt; Tel.: +351-220-413-295

Academic Editor: Frede Blaabjerg
Received: 31 July 2016; Accepted: 25 August 2016; Published: 31 August 2016

Abstract: The uncertainty and variability in electricity market price (EMP) signals and players' behavior, as well as in renewable power generation, especially wind power, pose considerable challenges. Hence, enhancement of forecasting approaches is required for all electricity market players to deal with the non-stationary and stochastic nature of such time series, making it possible to accurately support their decisions in a competitive environment with lower forecasting error and with an acceptable computational time. As previously published methodologies have shown, hybrid approaches are good candidates to overcome most of the previous concerns about time-series forecasting. In this sense, this paper proposes an enhanced hybrid approach composed of an innovative combination of wavelet transform (WT), differential evolutionary particle swarm optimization (DEEPSO), and an adaptive neuro-fuzzy inference system (ANFIS) to forecast EMP signals in different electricity markets and wind power in Portugal, in the short-term, considering only historical data. Test results are provided by comparing with other reported studies, demonstrating the proficiency of the proposed hybrid approach in a real environment.

Keywords: adaptive neuro-fuzzy inference system (ANFIS); differential evolutionary particle swarm optimization (DEEPSO); electricity market prices (EMP); forecasting; short-term; time series; wavelet transform (WT); wind power

1. Introduction

In competitive and deregulated electricity markets, potential integration of renewables, especially wind power, which naturally introduces its stochastic, volatile, and uncertain behaviour, is totally reflected in the market players' strategies and presents more difficulties for a sustainable and robust management of the power framework. Even more when the renewable potential is introduced very widely, yielding higher production costs, inflexibility, and unnecessary penalties due to wrong strategies by players or an increment in emissions caused by conventional producers filling the gaps, especially when the renewable resources suddenly fail or do not cover the required demand [1]. Moreover, with the growing need for smart grids, for example to meet the growing interest in electric vehicles and their integration, the above concerns may be even more pronounced without the use of innovative tools or mechanisms to ensure the quality, safety, and robustness of the electrical system [2].

One of the approaches discussed nowadays in the scientific domain to mitigate some of the problems described above and to achieve a profitable and sustainable management of the electrical framework involves the integration of energy storage systems, which makes the electrical system more flexible due to the increased exploitation of potential usage of renewables, especially under peak loads, reducing the operational cost or curtailment events; however their implementation is still highly costly and in experimental phases in some cases [3].

An alternative way to tackle the aforementioned concerns in power systems and in competitive electricity markets, which are by nature more economical and useful for all agent players, is through the use of innovative forecasting tools to determine the future behaviour of the renewable potential or electricity market price (EMP) signals; making the creation of sets of possible market strategies suitable, considering other important indicators such as social behaviour, environmental factors, electrical constraints, and the behaviours of other electricity agents; in other words, the forecasting tools may be used as a first stage of defense for all market players [4]. In the last years, massive efforts, supported by the scientific community, have been made to propose more viable and reliable solutions, allowing mitigation of the countless concerns regarding power systems, which are reflected in widespread techniques and forecasting approaches for EMPs or wind power behaviour, considering statistical or physical models in soft or hard computing, as shown for instance in [5–7], considering very short-, short-, and long-term horizon forecasting [8,9].

Regarding EMP forecasting tools, since 2005, models such as autoregressive integrated moving average (ARIMA) combined with wavelet transform (WT) [10] can be found. This model belongs to the family of hard computing tools, which require a large amount of physical information and an exact modelling of the system, resulting in high computational complexity, and in this sense, will not be considered in this review of the state of the art. However, soft computing models, such as fuzzy neural network (FNN) [11] or hybrid intelligent system (HIS) [12], are among the soft computing models, which require the usage of any auto learning process from historical sets to identify future patterns and therefore require less computational complexity or information to model the problem. In this regard, several examples can be found such as neural network (NN) models [13], adaptive wavelet NN (AWNN) [14], cascaded neuro-evolutionary algorithm (CNEA) [15], cascaded NN (CNN) [16], the hybrid neuro-evolutionary system (HNES) [17], and some hybrid forecasting models, such as those presented in [18], or a combination of WT with particle swarm optimization (PSO) and the adaptive neuro-fuzzy inference system (ANFIS) (WPA) [19] and other hybrids [20], the hybrid fundamental-econometric model [21], or two-stage approaches such as those reported in [22,23]. Furthermore, more approaches considering singular spectrum analysis [24], informative vector machine [25], or even new genetic algorithms such as Levenberg-Marquardt and cuckoo search algorithms [26] and genetic regression of relevance vector machines [27] can be found for different EMP prices analyses, considering the Spanish, Pennsylvania-New Jersey-Maryland (PJM), Australian National Electricity Market (ANEM), and other liberalized electricity markets around the world as real case studies.

In wind power forecasting, widespread use of forecasting models for the very short and short term can be found in specialized literature considering soft computing and statistical models. In this sense, several examples are usually found, such as an evolutionary algorithm using an artificial intelligence model [28], NN [29,30], ridgelet NN [31], hybrid approaches composed of WT and a neuro-fuzzy network (NF) [32], WT with NN [33], WT with ANFIS (WNF) [34], or WPA [35]. Also, wind power forecasting can be tackled by considering a combination of WT with support vector machine (SVM) and statistical analysis [36], adaptive WT combined with feed-forward NN (AWNN) [37], WT combined with ARTMAP [38], and optimized SVM using a genetic algorithm [39]. More recently some proposals have considered a principal component analysis algorithm [40], hybrid WT, PSO, and NN [41], multi-layer artificial NN improved with simplified swarm optimization [42], and WT combined with NN, trained by an improved clonal selection algorithm [43]. All of the aforementioned models were run considering real cases with data from wind farms or historical data collected from the public domain in different locations around the world.

In this paper, in accordance with the features demonstrated by hybrid forecasting models briefly presented above, a new approach to forecast the EMP or wind power performance in the short term (from a few to 168 h ahead) is proposed.

Specifically, in the case of EMP forecasting, the proposed approach will perform a forecast for the next 168 h ahead with a time step of 1 h, considering only historical data available from the public domain, without considering the inclusion of exogenous data such as load and other energy prices, among others, to allow a fair and clean comparison with other already published methodologies. In the case of wind power forecasting, the proposed forecasting approach will perform the forecasting for a range of 3 h ahead with a time-step of 15 min, refreshing the system (input data and forecast results) until completion of the forecasting results for 24 h ahead. As in the previous case study, in wind power forecasting the proposed approach does not consider the inclusion of exogenous data such as wind profile and atmospheric data, among others, in order to make a fair and clean comparison with previously published approaches.

Furthermore, the proposed approach is composed of an innovative combination of WT as the pre-processing tool, which provides a smoothing effect of all inputs, providing more flexibility and more convergence to forecast the future behaviour, differential evolutionary particle swarm optimization (DEEPSO), which is itself a hybrid method and will be responsible for augmenting the performance of ANFIS (which is by nature a hybrid tool) by tuning the ANFIS membership functions to attain a lower forecasting error. Finally, the inverse WT will be used to introduce again the smoothing information collected at the beginning, providing the final forecasting signal. In this sense, hereafter the proposed approach will be called the hybrid WT+DEEPSO+ANFIS (HWDA) approach. In all case studies, the real historical data used will be comparable to those data used in reported and published models [44,45]. The remainder of the manuscript is organized as follows: Section 2 describes the concepts used to create the HWDA approach, the algorithm used for EMP or wind power forecasting, and the criteria used to validate and compare the capabilities of the proposed HWDA approach with previous and published methodologies. Section 3 describes the historical data used to carry out the forecasting considering the EMP or wind power, the detailed results, and the comparison carried out; finally, Section 4 presents the main conclusions drawn in this paper.

2. Proposed Approach

The HWDA approach results from the successful combination of WT, DEEPSO, and ANFIS. The WT is employed as a pre-processing step to decompose the historical sets of EMP or wind power into new constitutive sets with better behaviour. Then, the forthcoming values of those constitutive sets are the feeding sets of ANFIS responsible for creating the forecast results. DEEPSO augments the performance of ANFIS by tuning the ANFIS membership functions, resulting in lower forecasting error. In comparison with its ancestor, evolutionary particle swarm optimization (EPSO), the underlying evolutionary and differential concepts make real differences in terms of robustness, convergence, and computational time. So the combination of DEEPSO features with the adaptive characteristics of ANFIS means that they complement each other in positive way. Finally, the inverse WT is used to reconstruct the forecasting signal, and thus the final forecasting results are obtained.

2.1. Wavelet Transform

As reported in most of the previously described works on the state of the art, the application of WT in forecasting approaches is important for overcoming the limitations of non-stationary time series such as EMP or wind power; however, it may be applied in other engineering fields, since it enables the analysis of time series in their natural state. WT is used as a pre-processing tool for understanding non-stationary or time varying data [46], with sensibility to the irregularities of input data. In this sense, WT is especially useful for showing different aspects that constitute the data without losing the real signal content [47]. Despite the problems related with continuous WT (CWT) analysis, discrete WT (DWT) was created to give, in an effective way, a description relative to CWT, which is widely used to decompose the time series under study:

$$DWT\left(m_{wt}, n_{wt}\right) = 2^{-(m_{wt}/2)} \sum_{h=0}^{H} p\left(t_{wt}\right) \varphi\left(\frac{t_{wt} - b}{a}\right) \qquad (1)$$

where H represents the length $p\left(t_{wt}\right)$, and the parameters of scaling (a) and translation (b) are changed to integer variables $a_{wt} = 2^{m_{wt}}$ and $b_{wt} = n_{wt}\,2^{m_{wt}}$ respectively, with a time-step t_{wt}, i.e.,:

$$DWT\left(m_{wt}, n_{wt}\right) = 2^{-(m_{wt}/2)} \sum_{h=0}^{H} p\left(t_{wt}\right) \varphi\left(\frac{t_{wt} - n_{wt}2^{m_{wt}}}{2^{m_{wt}}}\right) \qquad (2)$$

The DWT is performed by multi-resolution analysis, where a "father wavelet", responsible for the low-frequency series, is used with a complementary "mother wavelet", which is responsible for the high-frequency series components [38]. In this paper and following the description cited in [44,45] the Daubechies of fourth order, or Db4, was used as the mother-wavelet function. The Db4 has asymmetrical and continuous proprieties, where a higher order level will create a higher level oscillation, which is desirable in forecasting [38,47]. The coefficients of approximations A_n and details D_n are expressed as:

$$A_n = \sum_{n} DWT\left(m_{wt}, n_{wt}\right) \varphi_{mn}\left(t\right) \qquad (3)$$

$$D_n = \sum_{n} DWT\left(m_{wt}, n_{wt}\right) \psi_{mn}\left(t\right) \qquad (4)$$

where $\varphi_{mn}\left(t_{wt}\right)$ is the father-wavelet and $\psi_{mn}\left(t_{wt}\right)$ is the mother-wavelet, and $DWT\left(m_{wt}, n_{wt}\right)$ are the coefficients obtained from Equation (2) [33]. Furthermore, the Db4 is chosen as the mother-wavelet function due to a better trade-off between smoothness and length [19]. Also, the DWT used in this paper was created on four filters divided into two groups: the decomposition group, composed of low-pass and high-pass filters, and the reconstruction group, composed of low-pass and high-pass filters as described in [44,45]. Figure 1 shows a general decomposition model of WT, where approximation steps A_n are able to analyse the universal information of original sets; that is, the low-frequency representation and description of the high-frequency component and the detailed steps D_n are able to describe the difference between the successive approximations.

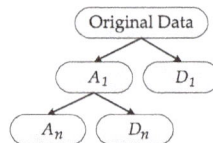

Figure 1. Universal n level decomposition model of WT.

2.2. Differential Evolutionary Particle Swarm Optimization

DEEPSO is a successful hybrid combination of the EPSO model [44,45], which is itself a hybrid combination of its ancestor model, namely PSO, where weight factors have self-adaptive features, with evolutionary programming, which brings self-adaptive operators [48], and a differential evolution algorithm, which provides a new solution from the current particle of the swarm by adding a fraction difference between two other points found from the previously evaluated swarm [49]. The DEEPSO schema is similar to EPSO [50]; however, the movement rule Equation (6) has new notation:

$$X_i^{new} = X_i + V_i^{new} \qquad (5)$$

$$V_i^{new} = w_{i0}^* V_i + w_{i1}^* \left(X_{r1}^i - X_{r2}^i\right) + P\,w_{i2}^* \left(b_g^* - X_i\right) \qquad (6)$$

where the weights w^*_{in-1} (inertia, memory and cooperation) are defined as:

$$w^*_{ik} = w_{ik} + \tau N\,(0,1) \tag{7}$$

and the global position is defined as:

$$b^*_g = b_g\,(1 + w_g N\,(0,1)) \tag{8}$$

From Equation (6), components X^i_r should be any pair of different particle already tested from the swarm, and ordered to minimize at the end of respective iteration, i.e.,:

$$f\left(X^i_{r1}\right) < f\left(X^i_{r2}\right) \tag{9}$$

From Equations (5)–(9), X^{new}_i is the new position of the particle, V^{new}_i is the new velocity found, P is a diagonal binary matrix with a value of 1 when the probability is p and 0 when the probability is $\{1 - p\}$, w^*_{ik} are the mutated weights of inertia, memory, and cooperation of the swarm, given by a learning parameter τ (fixed or mutated), and $N\,(0,1)$ is a random Gaussian variable with 0 mean and variance 1.

Also, b^*_g is the global position provided by the new weight w_g, which is collected from a diagonal matrix, having a self-adaptive feature, and in this sense, it is a mutated element [48,49]. Components X^i_{r1} and X^i_{r2} guarantee that a suitable extraction really happens, considering macro-gradient points in a descending direction depending on the structured comparison of $f\left(X^i_{r1}\right)$ and $f\left(X^i_{r2}\right)$. In this sense, component X^i_{r2} is assumed to be as $X^i_{r2} = X_i$, and component X^i_{r1} is sampled from the set of best ancestors from the swarm of n particles, that is, $S_{bA} = \{b_1, b_2, \ldots, b_n\}$ [50–52]. The main idea underlying DEEPSO movement is briefly illustrated in Figure 2.

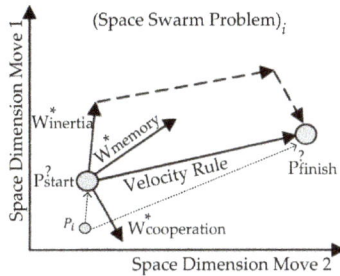

Figure 2. Brief illustration of DEEPSO (differential evolutionary particle swarm optimization) particle movement rule.

2.3. Adaptive Neuro-Fuzzy Inference System

ANFIS is a well-known hybrid combination of NN and fuzzy algorithms combining useful features such as low computational requirements, the possibility of dealing with a large number of data, and high response features. Furthermore, it has self-learning capabilities provided by the NN, which help it to self-adjust its parameters due to fuzzy capabilities [19,45]. The general ANFIS structure is based on several layers, which provide the fuzzification, rules, normalization data, desfuzzification, and data reconstruction process as described in [35,44]. Figure 3 briefly describes the multi-layer feed-forward network ANFIS structure. Mathematically, each of the five layers l_{n_k} used is:

$$\begin{cases} l_{1_k} = \mu A_i\,(x), & k = 1,2 \\ l_{1_k} = \mu B_{i-2}\,(y), & k = 3,4 \end{cases} \tag{10}$$

$$\mu A_k (x) = \frac{1}{1 + \left| \frac{x - r_k}{p_{ik}} \right|^{2q_k}} \tag{11}$$

$$l_{2_k} = w_k = \mu A_k (x) \, \mu B_k (y), \quad k = 1, 2 \tag{12}$$

$$l_{3_k} = \overline{w}_i = \frac{w_k}{w_1 + w_2}, \quad k = 1, 2 \tag{13}$$

$$l_{4_k} = \overline{w}_k z_k = \overline{w}_k (a_k x + b_k y + c_k), \quad k = 1, 2 \tag{14}$$

$$l_{5_k} = \sum_k \overline{w}_k z_k = \frac{\sum_k w_k z_k}{\sum_k w_k} \tag{15}$$

From Equation (10), all nodes k are adaptive nodes with node function l_{1_k}, where x and y are the input of the k_{th} node and A_k and B_{k-2} are the membership function, also called the linguistic label, associated with these nodes. In this paper, a triangular membership function is normally used [44,45], where $\{p_k, q_k, r_k\}$ are parameter sets, because it is a continuous and piecewise differentiable function, described in Equation (11), which represents the first layer. In Equation (12), all output nodes represent the firing strength of the rule w_k, where each node signal is multiplied by the previous inputs signals, representing the second layer. In Equation (13), the third layer, every node computes the ratio of firing strength rules k_{th} to the sum of all firing strength rules. Equation (14) represents the computation of all nodes' contribution to k_{th} rule with global output, where $\{a_k, b_k, c_k\}$ are parameter sets, and $\overline{w}_k$ is the layer output (fourth layer). Finally, Equation (15) defines the ANFIS output node, that is, the fifth layer where the summation Σ is made. As reported in [19,35], in this paper, the ANFIS structure follows the least-squares and back-propagation gradient descent method, considering the Takagi-Sugeno approach.

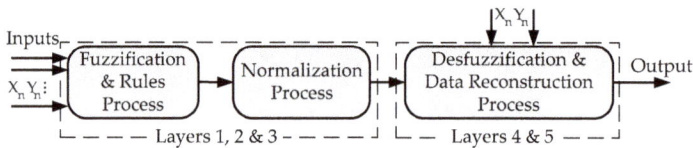

Figure 3. Brief illustration of ANFIS (adaptive neuro-fuzzy inference system) structure.

2.4. Hybrid Proposed Approach

As stated before, the HWDA approach results from a combination of WT, DEEPSO, and ANFIS. The WT is employed as a pre-processing step to decompose the historical sets. The DEEPSO augments the ANFIS performance by tuning the ANFIS membership functions. Finally, the inverse WT is used to reconstruct the forecasting signal, and then the final forecasting results are obtained. Figure 4 shows the HWDA flowchart. In detail, HWDA follows the following steps:

- Step 1: Initialize the HWDA approach with a historical data matrix of EMP or wind power, respectively, considering the forecasting time-scale of each forecast field;
- Step 2: Choose a set of historical data of the previous step to run the pre-processing process carried out by the WT tool. This step is performed by a backtracking process, in order to attain a smaller error at the end by choosing the best set of candidates. Also, the approach considered in this paper uses A_3, D_3, and D_1 steps as inputs for the next step;
- Step 3: Train the ANFIS tool with the previous sets of constitutive historical data obtained from WT. The optimization process of the ANFIS membership function parameters will be achieved with the DEEPSO method. All parameters considered from all methods are summarized in Table 1.

As in [44,45], the ANFIS inference rules are obtained by considering the automatic ANFIS mode, due to the nature of the data, which requires a large number of inference rules, and thus additional improvement is achieved.

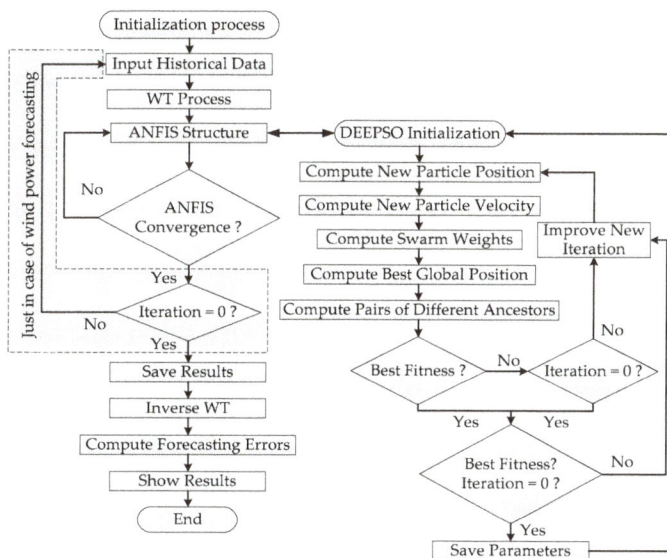

Figure 4. HWDA (hybrid WT+DEEPSO+ANFIS) forecasting approach flowchart.

Table 1. DEEPSO (differential evolutionary particle swarm optimization) and ANFIS (adaptive neuro-fuzzy inference system) parameters used for EMP (electricity market price) and wind power forecasting.

Methods	Parameters	Type or Size
WT	Decomposition Direction	Row
	Level of Decomposition	3
	Mother-Wavelet Function	Db4
	Denoising Methods	"sqtwolog"–"minimaxi"
	Multiplicative Thresholds Rescaling	"one"–"sln"
DEEPSO	Communication Probability	0.10
	Final Inertia Wight	0.01–0.15
	Initial Inertia Weight	0.50–0.90
	Initial Population Size	100
	Initial Sharing Acceleration	0.50–2.00
	Initial Swarm Learning Process	1.00–2.00
	Initial Swarm Sharing Process	2.00
	Learning Parameter	1
	Maximum Value of New Position	Set of Max. Inputs
	Minimum Value of New Position	Set of Min. Inputs
	Necessary iterations	100–1000
ANFIS	Structure Type	Takagi-Sugeno
	Style of Membership Function	Triangular
	Number of Inference Rules	Automatic
	Membership Functions	2–15
	Number of Epochs	2–50
	Number of Nodes	3–9
	Number of Inputs / Outputs	2–5/1

- Step 4: until the best results are obtained or convergence is reached:

  ○ Step 4.1: Jump to Step 4 in the case of EMP if convergence is not reached;
  ○ Step 4.2: Jump to Step 2 in the case of wind power forecasting, refreshing the historical data matrix.

When the best result is found or convergence is reached, the wind power data are forecasted for the next 3 h until the forecast for the next 24 h ahead is complete.

- Step 5: Apply the inverse WT. The output of the proposed HWDA approach is attained; that is, the forecasted EMP or wind power results are ready to be presented;
- Step 6: Compute the forecasting errors of EMP or wind power results with different criteria to validate the proposed HWDA approach and show the results.

2.5. Forecasting Error Evaluation

To compare the proposed approach with other methodologies for EMP or wind power forecasting previously published in the specialized literature, the mean absolute percentage error (MAPE) criterion is used. This criterion is given as [44,45]:

$$MAPE = \frac{100}{N} \sum_{n=1}^{N} \frac{|\hat{p}_n - p_n|}{\bar{p}} \tag{16}$$

$$\bar{p} = \frac{1}{N} \sum_{n=1}^{N} p_n \tag{17}$$

where $\hat{p}_n$ is the data forecasted at hour n, p_n is the real data at hour n, $\bar{p}$ is the average value for the forecasting time horizon, and N has the length value of observed points. Following the same concept from the MAPE criterion, the uncertainty of the HWDA model is evaluated using the error variance, described as [19,35]:

$$\sigma_{e,n}^2 = \frac{1}{N} \sum_{n=1}^{N} \left(\frac{|\hat{p}_n - p_n|}{\bar{p}} - e_n \right)^2 \tag{18}$$

$$e_n = \frac{1}{N} \sum_{n=1}^{N} \frac{|\hat{p}_n - p_n|}{\bar{p}} \tag{19}$$

Moreover, for wind power forecasting, the normalized mean absolute error (NMAE) criterion is used [35,45]:

$$NMAE = \frac{100}{N} \sum_{n=1}^{N} \frac{|\hat{p}_n - p_n|}{P_{installed}} \tag{20}$$

where $P_{installed} = 2700\,MW$, which corresponds to the total wind power capacity installed in accordingly to [53]. Furthermore, the normalized root mean square error (NRMSE) is also used and is described as [45]:

$$NRMSE = \sqrt{\frac{1}{N} \sum_{n=1}^{N} \left(\frac{\hat{p}_n - p_n}{P_{intalled}} \right)^2} \times 100 \tag{21}$$

3. Case Studies and Results

3.1. Electricity Market Prices Forecasting

As briefly stated before, the HWDA approach is used first to forecast EMP for the next 168 and 24 h considering the historical data from the Spanish market available in [54].

As mentioned in [10,21], this market has features that are difficult to forecast due to influences from dominant players, which are reflected in historical data. The EMP historical data used for the Spanish market date back to the year 2002, allowing a clear and fair comparison with the already published results from other proposed methodologies, considering the same four test weeks of the year 2002, which are consistent with the four seasons. As stated before, only EMP historical data sets were used, for the reasons stated above, otherwise a correct comparative study would not be possible.

The HWDA approach forecasts the next 168 h of EMP considering the previous 1008 h (six weeks), which are used as input sets. In order to avoid over-training during the learning process, very large training sets are not used. The output of the HWDA approach results in a set of 168 points representing the forecasting horizon. For day-ahead forecasting, the same idea may be followed; that is, the HWDA approach has as its input the previous six days, considering the historical data from the same market for the year 2006, which were analysed by the case studies reported in [44].

Furthermore, the HWDA approach is tested for the PJM market, forecasting the EMP for the next 24 and 168 h ahead. The historical data of electricity prices are available in [55]. Similarly to the Spanish market, no exogenous data were considered for the same reason as described above.

3.1.1. Spanish Market Results

The results obtained with the HWDA approach are provided in Figures 5–8 for the four test weeks (168 h ahead) of 2002, where the solid and dash-dot black lines represent the actual and forecasted EMP, respectively, while the blue line at the bottom of each figure represents the resulting errors as absolute values. Tables 2 and 3 shows the comparative MAPE criterion and weekly error variance criterion results, respectively, between the HWDA approach and ten previous published methodologies, namely NN [13], FNN [11], AWNN [14], HIS [12], CNEA [15], CNN [16], WPA [19], mutual information with composite NN (MI+CNN) [22], and hybrid evolutionary algorithm (HEA) [44], indicating the enhancements as the percentage evolution between the HWDA approach and the respective comparative methodology under analysis.

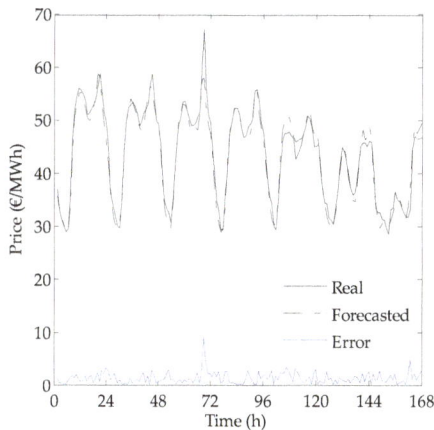

Figure 5. Winter week 2002 results for the Spanish market.

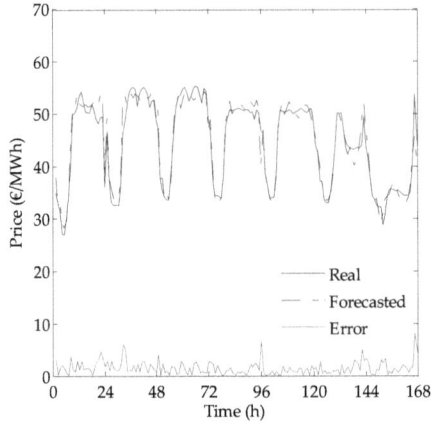

Figure 6. Spring week 2002 results for the Spanish market.

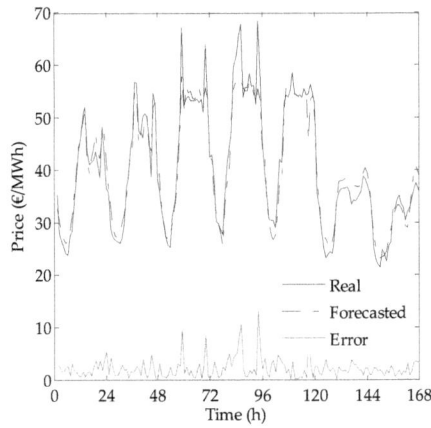

Figure 7. Summer week 2002 results for the Spanish market.

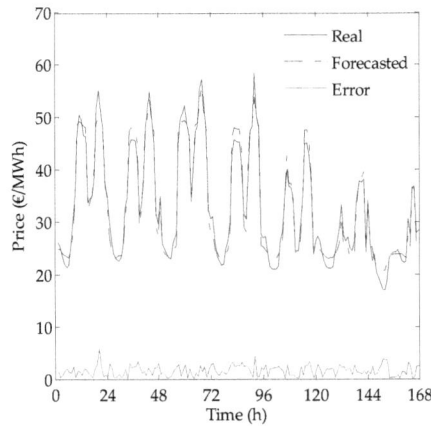

Figure 8. Autumn week 2002 results for the Spanish market.

Table 2. MAPE (the mean absolute percentage error) comparison considering the year 2002 Spanish market case study for 168 h ahead.

Methods	Winter	Spring	Summer	Fall	Average	Enhancement
NN [13], 2007	5.23	5.36	11.40	13.65	8.91	54.66%
FNN [11], 2006	4.62	5.30	9.84	10.32	7.52	46.28%
HIS [12], 2009	6.06	7.07	7.47	7.30	6.97	42.04%
AWNN [14], 2008	3.43	4.67	9.64	9.29	6.75	40.15%
CNEA [15], 2009	4.88	4.65	5.79	5.96	5.32	24.06%
CNN [16], 2009	4.21	4.76	6.01	5.88	5.22	22.61%
HNES [17], 2010	4.28	4.39	6.53	5.37	5.14	21.40%
MI+CNN [22], 2012	4.51	4.28	6.47	5.27	5.13	21.25%
WPA [19], 2011	3.37	3.91	6.50	6.51	5.07	20.32%
HEA [44], 2014	3.04	3.33	5.38	4.97	4.18	3.35%
HWDA	3.00	3.16	5.23	4.76	4.04	-

Table 3. Weekly error variance comparison considering the year 2002 Spanish market case study for 168 h ahead.

Methods	Winter	Spring	Summer	Fall	Average	Enhancement
NN [13], 2007	0.0017	0.0018	0.0109	0.0136	0.0070	82.86%
FNN [11], 2006	0.0018	0.0019	0.0092	0.0088	0.0054	77.78%
AWNN [14], 2008	0.0012	0.0031	0.0074	0.0075	0.0048	75.00%
HIS [12], 2009	0.0034	0.0049	0.0029	0.0031	0.0036	66.67%
CNEA [15], 2009	0.0036	0.0027	0.0043	0.0039	0.0036	66.67%
CNN [16], 2009	0.0014	0.0033	0.0045	0.0048	0.0035	65.71%
WPA [19], 2011	0.0008	0.0013	0.0056	0.0033	0.0027	55.56%
MI+CNN [22], 2012	0.0014	0.0014	0.0033	0.0022	0.0021	42.86%
HNES [17], 2010	0.0013	0.0015	0.0033	0.0022	0.0021	42.86%
HEA [44], 2014	0.0008	0.0011	0.0026	0.0014	0.0015	20.00%
HWDA	0.0007	0.0008	0.0022	0.0010	0.0012	-

When the HWDA approach was used, the MAPE criterion reached an average value of 4.04%, which is significant, even when it is compared for each week independently or considering the improvements over all comparative methodologies. The weekly error variance criterion results obtained using the HWDA approach reached an average value of 0.0012, showing a notable accuracy compared with the other methodologies described and reported, even when its improvements are analysed independently.

3.1.2. PJM (Pennsylvania-New Jersey-Mary) Land Market Results

The HWDA approach was also used to forecast the EMP considering the historical data from the PJM market, available in [55], providing results for the next 24 and 168 h ahead. As in the previous case study, no exogenous data are taken into account. Figures 9–11 illustrate some results for some days and weeks tested considering the historical data of 2006 for the PJM market, and the same condition as described in [44] is applied to give a clear and fair comparison with other published methodologies. Moreover, in all figures, the solid and dash-dot black lines represent the actual and forecasted EMP, respectively, while the blue line at the bottom of each figure represents the resulting errors as absolute values.

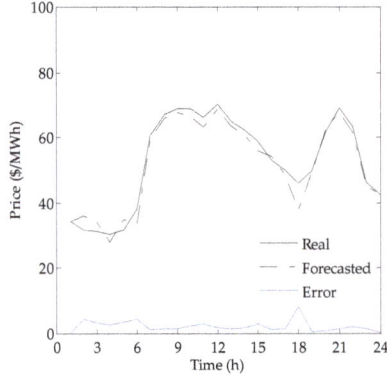

Figure 9. 7 April 2006 results for the PJM (Pennsylvania-New Jersey-Mary) market.

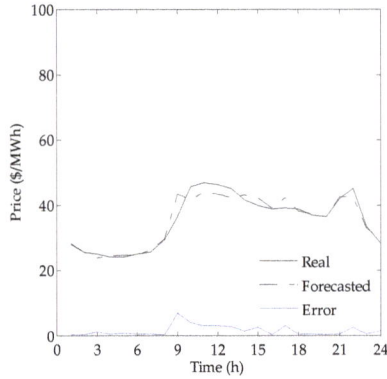

Figure 10. 13 May 2006 results for the PJM (Pennsylvania-New Jersey-Mary) land market.

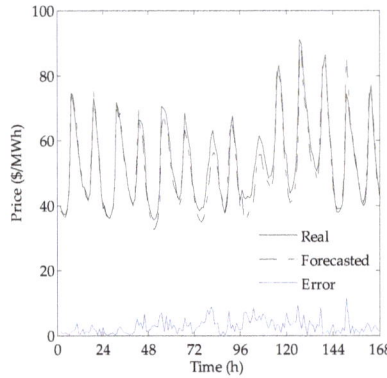

Figure 11. 22–28 February 2006 results for the PJM (Pennsylvania-New Jersey-Mary) land market.

Tables 4 and 5 shows the MAPE and error variance results, respectively, for the HWDA approach and four other methodologies. When using the HWDA approach, the MAPE criterion reached an average value of 3.16% and the error variance reached an average of 0.0011, which is notable for this competitive market.

Table 4. MAPE (the mean absolute percentage error) comparison considering the year 2006 PJM (Pennsylvania-New Jersey-Mary) land market case study for 24/168 h ahead.

	HNES [17], 2010	Hybrid [44], 2010	CNEA [15], 2009	HEA [44], 2014	HWDA
Jan. 20	4.98	3.71	4.73	3.29	3.22
Feb. 10	4.10	2.85	4.50	2.80	2.71
Mar. 5	4.45	5.48	4.92	3.32	3.27
Apr. 7	4.67	4.17	4.22	3.55	3.42
May 13	4.05	4.06	3.96	3.43	3.40
Feb. 1–7	4.62	5.27	4.02	3.11	3.09
Feb. 22–28	4.66	5.01	4.13	3.08	3.02
Average	4.50	4.36	4.35	3.23	3.16
Enhancement	29.78%	27.52%	27.36%	2.17%	-

Table 5. Error variance comparison considering the year 2006 PJM (Pennsylvania-New Jersey-Mary) land market case study for 24/168 h ahead.

	CNEA [15], 2009	Hybrid [44], 2010	HNES [17], 2010	HEA [44], 2013	HWDA
Jan. 20	0.0031	0.0010	0.0020	0.0010	0.0010
Feb. 10	0.0036	0.0015	0.0012	0.0009	0.0008
Mar. 5	0.0042	0.0033	0.0015	0.0011	0.0010
Apr. 7	0.0022	0.0013	0.0018	0.0011	0.0011
May 13	0.0027	0.0015	0.0013	0.0012	0.0012
Feb. 1–7	0.0044	0.0037	0.0016	0.0012	0.0011
Feb. 22–28	0.0035	0.0025	0.0017	0.0017	0.0016
Average	0.0034	0.0021	0.0016	0.0012	0.0011
Enhancement	67.65%	47.62%	45.45%	8.33%	-

3.2. Wind Power Forecasting

The HWDA approach was used to forecast the wind power for 3 h ahead with a time-step of 15 min until the forecast for the whole 24 h ahead was complete, considering the historical data of wind power in Portugal between 2007 and 2008 as described in [45,53] and considering the different seasons of the year. Also, as in the previous case studies, to allow a fair and clean comparison, only historical wind power data are considered, for the same reason as described above. Figures 12–15 show the numerical wind power results for winter, spring, summer, and autumn days, respectively, where solid and dash-dot black lines represent the actual and forecasted wind power, respectively, while the blue line in the bottom figures represents the errors as absolute values. For all results, it is possible to observe how the HWDA approach correctly forecasts the unexpected and abrupt changes of the wind power profile, that is, its uncertainty behaviour during the whole day of forecasting.

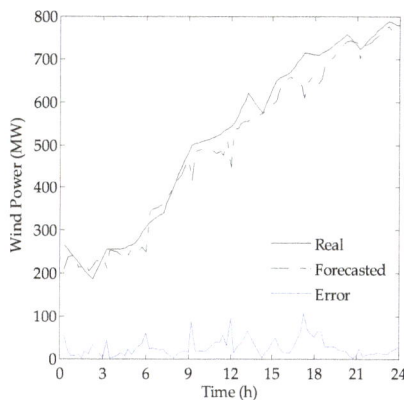

Figure 12. Real and forecasted wind power results (15 min intervals) for the Winter day.

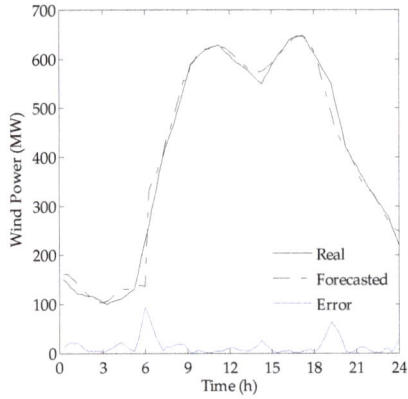

Figure 13. Real and forecasted wind power results (15-min intervals) for the Spring day.

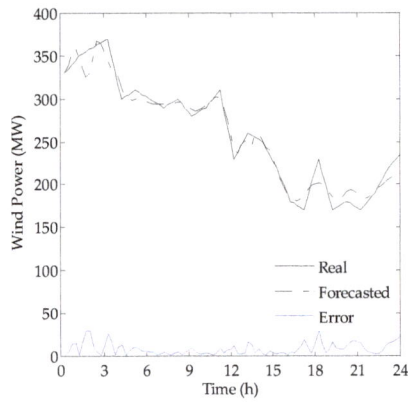

Figure 14. Real and forecasted wind power results (15-min intervals) for the Summer day.

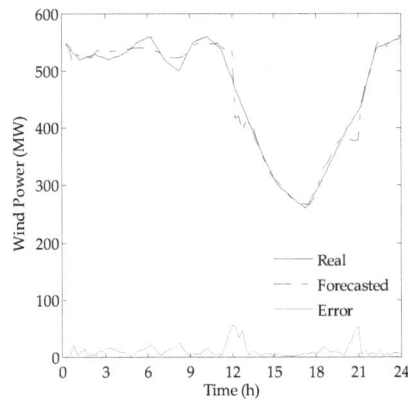

Figure 15. Real and forecasted wind power results (15-min intervals) for the Autumn day.

Tables 6 and 7 provide a comparative study between the HWDA approach using MAPE and the daily error variance criterion and five other previously published methodologies, namely NN [29],

NF [32], WNF [34], WPA [35], and HEA [45], respectively. When the HWDA approach is used, the MAPE criterion has an average value of 3.37%, representing an enhancement of 11.28% compared to the HEA methodology, which is again significant.

Table 6. MAPE (the mean absolute percentage error) comparison for wind power forecasting.

	Winter	Spring	Summer	Fall	Average	Enhancement
NN [29]	9.51	9.92	6.34	3.26	7.26	53.58%
NF [32]	8.85	8.96	5.63	3.11	6.64	49.25%
WNF [34]	8.34	7.71	4.81	3.08	5.99	43.74%
WPA [35]	6.47	6.08	4.31	3.07	4.98	32.33%
HEA [45]	5.74	3.49	3.13	2.62	3.75	11.28%
HWDA	5.08	3.19	2.96	2.27	3.37	-

Table 7. Daily error variance comparison for wind power forecasting.

	Winter	Spring	Summer	Fall	Average	Enhancement
NN [29]	0.0044	0.0106	0.0043	0.0010	0.0051	76.47%
NF [32]	0.0041	0.0086	0.0038	0.0008	0.0043	72.09%
WNF [34]	0.0046	0.0051	0.0021	0.0011	0.0032	62.50%
WPA [35]	0.0021	0.0035	0.0016	0.0011	0.0021	42.86%
HEA [45]	0.0019	0.0015	0.0010	0.0008	0.0013	7.69%
HWDA	0.0017	0.0016	0.0007	0.0006	0.0012	-

Furthermore, the daily error variance obtained using the HWDA approach has an average value of 0.0013%, presenting lower uncertainty in the forecasts done, and again, in all results the HWDA approach shows better accuracy in comparison with analyses of the same real case by all other previously published methodologies.

Finally, Tables 8 and 9 show a comparison of the results obtained with the HWDA approach according to the NMAE and NRMSE criteria, respectively. In all cases analysed, it is possible to observe that the HWDA approach gave better results than the other published methodologies considering the same cases studies. The proposed HWDA approach was performed on a standard PC equipped with an Intel Core i7-3537U, 2 GHz CPU and 4 GB of RAM with Windows 10 and the MATLAB®2016a platform. The authors used the ANFIS and WT structure functions available in MATLAB toolboxes, while DEEPSO was programmed from scratch in MATLAB considering the information available in [49–52].

Table 8. NMAE (the normalized root mean square error) comparison for wind power forecasting.

	Winter	Spring	Summer	Fall	Average	Enhancement
NN [29]	5.22	3.72	2.35	2.15	3.36	84.23%
NF [32]	4.86	3.36	2.09	2.05	3.09	82.85%
WNF [34]	4.58	2.89	1.78	2.03	2.82	81.21%
WPA [35]	3.56	2.28	1.60	2.02	2.37	77.64%
HEA [45]	2.73	1.48	0.74	1.10	1.51	64.90%
HWDA	0.94	0.49	0.28	0.39	0.53	-

Table 9. NRMSE (the normalized root mean square error) comparison for wind power forecasting.

	Winter	Spring	Summer	Fall	Average	Enhancement
HEA [45]	3.60	3.18	1.78	2.07	2.66	39.47%
HWDA	2.19	1.27	1.81	1.18	1.61	-

4. Conclusions

An enhanced HWDA approach was proposed in this paper for short-term EMP and wind power forecasting considering real cases studies, specifically the analyses from Spanish and PJM markets, as well as the wind power behavior in Portugal. The innovative and successful combination of WT, DEEPSO and ANFIS provided interesting and valuable results. The main findings resulting from this study are related to the lower forecasting errors attained while providing an acceptable computational time. The MAPE criterion reached an average value of 4.04% for the Spanish Market, surpassing all other methodologies, and for the PJM market reached an average value of 3.16%. Regarding the wind power forecasting results, the MAPE criterion had an average value of 3.37%. Lower error variances were also obtained in all cases. Moreover, the computational time required for HWDA approach was less than two min, on average, for the EMP results, and for wind power forecasting took less than one min per iteration. Hence, the overall results obtained with the HWDA approach provided an excellent trade-off between computational time and accuracy, which is crucial for real-life and real-time applications.

Acknowledgments: João Catalão and Juan Lujano-Rojas thank the EU Seventh Framework Programme FP7/2007–2013 under grant agreement No. 309048, FEDER through COMPETE and FCT, under FCOMP-01-0124-FEDER-020282 (Ref. PTDC/EEA-EEL/118519/2010), UID/CEC/50021/2013 and SFRH/BPD/103079/2014.

Author Contributions: All authors have worked on this manuscript together and all authors have read and approved the final manuscript.

Conflicts of Interest: The authors declare no conflict of interest.

Nomenclature

a	WT scaling integer variable
A_k	ANFIS linguistic label
a_k	ANFIS contribution parameter set
A_n	WT approximation coefficient
b	WT translation integer variable
b_g	DEEPSO actual global position
b_g^*	DEEPSO global position provided by a new weight w_g
B_k	ANFIS linguistic label
b_k	ANFIS contribution parameter set
c_k	ANFIS contribution parameter set
D_n	WT detail coefficient
DWT	Discrete wavelet transform set
e_n	Error at hour n
φ_{mn}	WT father-wavelet function
H	WT length of set $p\,(t_{wt})$
i	DEEPSO integer time-step from global search space
k	ANFIS number of nodes
k_{th}	ANFIS output node
l_{n_k}	ANFIS layer
$MAPE$	Mean absolute percentage error
m_{wt}	WT integer scaling parameter
N	Length of observed values points
$N\,(0,1)$	DEEPSO random Gaussian variable with 0 mean and variance 1
$NMAE$	Normalized mean absolute error
$NRMSE$	Normalized root mean square error
n_{wt}	WT integer translation parameter

$\bar{p}$	Average value for the forecasting horizon
P	DEEPSO probabilistic diagonal binary matrix
$\hat{p}_n$	Data forecasted at hour n
$P_{installed}$	Total wind power capacity installed
p_k	ANFIS parameter set of membership function
p_n	Real data at hour n
ψ_{mn}	WT mother-wavelet function
$p(t_{wt})$	WT signal input
q_k	ANFIS parameter set of membership function
r_k	ANFIS parameter set of membership function
$\sigma_{e,n}^2$	Error variance from the forecasting horizon
τ	DEEPSO learning parameter
t_{wt}	WT time-step
V_i	DEEPSO actual velocity
V_i^{new}	DEEPSO new velocity of the particle
w_g	DEEPSO new weight with self-adaptive features
w_{ik}^*	DEEPSO mutated weights of inertia, memory and cooperation
w_k	ANFIS firing strength
$\bar{w}_k$	ANFIS output firing strength
x	ANFIS input data
X_i	DEEPSO actual position
X_i^{new}	DEEPSO new position of the particle
X_{r1}^i	DEEPSO set of best ancestors from the swarm
X_{r2}^i	DEEPSO set of recorded positions of the swarm
y	ANFIS input data
z_k	ANFIS defuzzification parameters data

References

1. Dufo-López, R.; Bernal-Agustín, J.L.; Monteiro, C. New methodology for the optimization of the management of wind farms, including energy storage. *Appl. Mech. Mater.* **2013**, *330*, 183–187. [CrossRef]
2. Catalão, J.P.S. *Smart and Sustainable Power Systems: Operations, Planning, and Economics of Insular Electricity Grids*, 1st ed.; CRC Press, Taylor and Francis Group: Boca Raton, FL, USA, 2015.
3. Rodrigues, E.M.G.; Osório, G.J.; Godina, R.; Bizuayehu, A.W.; Lujano-Rojas, J.M.; Matias, J.C.O.; Catalão, J.P.S. Modelling and sizing of NaS (sodium sulfur) battery energy storage system for extending wind power performance in Crete Island. *Energy* **2015**, *90*, 1606–1617. [CrossRef]
4. Li, L.; Wang, J. Sustainable energy development scenario forecasting and energy saving policy analysis of China. *Renew. Sust. Energy Rev.* **2016**, *58*, 718–724.
5. Weron, R. Electricity price forecasting: A review of the state-of-the-art with a look into the future. *Int. J. Forecas.* **2014**, *30*, 1030–1081. [CrossRef]
6. Chang, W.Y. A literature review of wind forecasting methods. *J. Power Energy Eng.* **2014**, *2*, 161–168. [CrossRef]
7. Ren, Y.; Suganthan, P.N.; Srikanth, N. Ensemble methods for wind and solar power forecasting—A state of the art review. *Renew. Sust. Energy Rev.* **2015**, *50*, 82–91. [CrossRef]
8. Okumus, I.; Dinler, A. Current status of wind energy forecasting and a hybrid method for hourly predictions. *Energy Conv. Manag.* **2016**, *123*, 362–371. [CrossRef]
9. Wang, X.; Guo, P.; Huang, X. A review of wind power forecasting models. *Energy Proc.* **2011**, *12*, 770–778. [CrossRef]
10. Conejo, A.J.; Plazas, M.A.; Espínola, R.; Molina, A.B. Day-ahead electricity price forecasting using wavelet transform and ARIMA models. *IEEE Trans. Power Syst.* **2005**, *20*, 1035–1042. [CrossRef]
11. Amjady, N. Day-ahead price forecasting of electricity markets by a new fuzzy neural network. *IEEE Trans. Power Syst.* **2006**, *21*, 887–896. [CrossRef]

12. Amjady, N.; Hemmati, H. Day-ahead price forecasting of electricity markets by a hybrid intelligent system. *Euro Trans. Electron. Power* **2006**, *19*, 89–102. [CrossRef]
13. Catalão, J.P.S.; Mariano, S.J.P.S.; Mendes, V.M.F.; Ferreira, L.A.F.M. Short-term electricity prices forecasting in a competitive market: A neural network approach. *Electron Power Syst. Res.* **2007**, *77*, 1297–1304. [CrossRef]
14. Pindoriya, N.M.; Singh, S.N.; Singh, S.K. An adaptive wavelet neural network-based energy price forecasting, in electricity market. *IEEE Trans. Power Syst.* **2008**, *23*, 1423–1432. [CrossRef]
15. Amjady, N.; Keynia, F. Day-ahead price forecasting of electricity markets by mutual information technique and cascaded neuro-evolutionary algorithm. *IEEE Trans. Power Syst.* **2009**, *24*, 306–318. [CrossRef]
16. Amjady, N.; Daraeepour, A. Design of input vector for day-ahead price forecasting of electricity markets. *Exp. Syst. Appl.* **2009**, *36*, 12281–12294. [CrossRef]
17. Amjady, N.; Keynia, F. Application of a new hybrid neuro-evolutionary system for day-ahead price forecasting of electricity markets. *Appl. Soft Comput.* **2010**, *10*, 784–792. [CrossRef]
18. Wu, L.; Shahidehpour, M. A hybrid model for day-ahead price forecasting. *IEEE Trans. Power Syst.* **2010**, *25*, 1519–1530.
19. Catalão, J.P.S.; Pousinho, H.M.I.; Mendes, V.M.F. Hybrid wavelet-PSO-ANFIS approach for short-term electricity prices forecasting. *IEEE Trans. Power Syst.* **2011**, *26*, 137–144. [CrossRef]
20. Shafie-khah, M.; Moghaddam, M.P.; Sheikh-El-Eslami, M.K. Price forecasting of day-ahead electricity markets using a hybrid forecast method. *Energy Convers. Manag.* **2011**, *52*, 2165–2169. [CrossRef]
21. González, V.; Contreras, J.; Bunn, D.W. Forecasting power prices using a hybrid fundamental-econometric model. *IEEE Trans. Power Syst.* **2012**, *27*, 363–372. [CrossRef]
22. Keynia, F. A new feature selection algorithm and composite neural network for electricity price forecasting. *Eng. Appl. Artif. Intell.* **2012**, *25*, 1687–1697. [CrossRef]
23. Shayeghi, H.; Ghasemi, A. Day-ahead electricity prices forecasting by a modified CGSA technique and hybrid WT in LSSVM based scheme. *Energy Convers. Manag.* **2013**, *74*, 482–491. [CrossRef]
24. Miranian, A.; Abdollahzade, M.; Hassani, H. Day-ahead electricity price analysis and forecasting by singular spectrum analysis. *IET Gener. Trans. Distrib.* **2013**, *7*, 337–346. [CrossRef]
25. Elattar, E.; Shebin, E.K. Day-ahead price forecasting of electricity markets based on local informative vector machine. *IET Gener. Transm. Distrib.* **2013**, *7*, 1063–1071. [CrossRef]
26. Kim, M.K. Short-term price forecasting of Nordic power market by combination Levenberg-Marquardt and cuckoo search algorithms. *IET Gen. Trans. Distrib.* **2015**, *9*, 1553–1563. [CrossRef]
27. Alamaniotis, M.; Bargiotas, D.; Bourbakis, N.G.; Tsoulalas, L.H. Genetic optimal regression of relevance vector machines for electricity pricing signal forecasting in smart grids. *IEEE Trans. Smart Grid* **2015**, *6*, 2997–3005. [CrossRef]
28. Jursa, R.; Rohrig, K. Short-term wind power forecasting using evolutionary algorithms for the automated specification of artificial intelligence models. *Int. J. Forecast.* **2008**, *24*, 694–709. [CrossRef]
29. Catalao, J.P.S.; Pousinho, H.M.I.; Mendes, V.M.F. An artificial neural network approach for short-term wind power forecasting in Portugal. *Eng. Intell. Syst. Electron. Eng. Commun.* **2009**, *17*, 5–11.
30. Rosado, I.J.R.; Jimenez, L.A.F.; Monteiro, C.; Sousa, J.; Bessa, R. Comparison of two new short-term wind-power forecasting systems. *Renew. Energy* **2009**, *34*, 1848–1854. [CrossRef]
31. Amjady, N.; Keynia, F.; Zareipour, H. Short-term wind power forecasting using ridgelet neural network. *Electr. Power Syst. Res.* **2011**, *81*, 2099–2107. [CrossRef]
32. Catalão, J.P.S.; Pousinho, H.M.I.; Mendes, V.M.F. Hybrid intelligent approach for short-term wind power forecasting in Portugal. *IET Renew. Power Gener.* **2011**, *5*, 251–257. [CrossRef]
33. Catalão, J.P.S.; Pousinho, H.M.I.; Mendes, V.M.F. Short-term wind power forecasting in Portugal by neural network and wavelet transform. *Renew. Energy* **2011**, *36*, 1245–1251. [CrossRef]
34. Pousinho, H.M.I.; Mendes, V.M.F.; Catalão, J.P.S. Application of adaptive neuro-fuzzy inference for wind power short-term forecasting. *IEEJ Trans. Electr. Electron. Eng.* **2011**, *6*, 571–576. [CrossRef]
35. Catalão, J.P.S.; Pousinho, H.M.I.; Mendes, V.M.F. Hybrid wavelet-PSO-ANFIS approach for short-term wind power forecasting in Portugal. *IEEE Trans. Sustain. Energy* **2011**, *2*, 50–59.
36. Liu, Y.; Shi, J.; Yang, Y.; Lee, W.J. Short-term wind-power prediction based on wavelet transform-support vector machine and statistic-characteristics analysis. *IEEE Trans. Ind. Appl.* **2012**, *48*, 1136–1141. [CrossRef]
37. Bhaskar, K.; Singh, S. AWNN-assisted wind power forecasting using feedforward neural network. *IEEE Trans. Sustain. Energy* **2012**, *3*, 306–315. [CrossRef]

38. Haque, A.U.; Mandal, P.; Meng, J.; Srivastava, A.K.; Tseng, T.L.; Senjyu, T. A novel hybrid approach based on wavelet transform and fuzzy ARTMAP networks for predicting wind farm power production. *IEEE Trans. Ind. Appl.* **2013**, *49*, 2253–2261. [CrossRef]

39. Liu, D.; Niu, D.; Wang, H.; Fan, L. Short-term wind speed forecasting using wavelet transform and support vector machines optimized by genetic algorithm. *Renew. Energy* **2013**, *62*, 592–597. [CrossRef]

40. Skittides, C.; Früh, W.G. Wind forecasting using principal component analysis. *Renew. Energy* **2014**, *69*, 365–374. [CrossRef]

41. Mandal, P.; Zareipour, H.; Rosehart, W.D. Forecasting aggregated wind power production of multiple wind farms using hybrid wavelet-PSO-NNs. *Int. J. Energy Res.* **2014**, *38*, 1654–1666. [CrossRef]

42. Yeh, W.C.; Yeh, Y.M.; Chang, P.C.; Ke, Y.C. Forecasting wind power in the Mai Liao wind farm based on the multi-layer perceptron artificial neural network model with improved simplified swarm optimization. *Elect. Power Energy Syst.* **2014**, *55*, 741–748. [CrossRef]

43. Chitsaz, H.; Amjady, N.; Zareipour, H. Wind power forecast using wavelet neural network trained by improved clonal selection algorithm. *Energy Conv. Manag.* **2015**, *89*, 588–598. [CrossRef]

44. Osório, G.J.; Matias, J.C.O.; Catalão, J.P.S. Electricity prices forecasting by a hybrid evolutionary-adaptive methodology. *Energy Conv. Mang.* **2014**, *80*, 363–373. [CrossRef]

45. Osório, G.J.; Matias, J.C.O.; Catalão, J.P.S. Short-term wind power forecasting using adaptive neuro-fuzzy inference system combined with evolutionary particle swarm optimization, wavelet transform and mutual information. *Renew. Energy* **2015**, *75*, 301–307. [CrossRef]

46. Eynard, J.; Grieu, S.; Polit, M. Wavelet-based multi-resolution analysis and artificial neural networks, for forecasting temperature and thermal power consumption. *Eng. App. Art. Intell.* **2011**, *24*, 501–516. [CrossRef]

47. Amjady, N.; Keynia, F. Short-term loads forecasting of power systems by combining wavelet transform and neuro-evolutionary algorithm. *Energy* **2009**, *34*, 46–57. [CrossRef]

48. Miranda, V.; Carvalho, L.M.; Rosa, M.A.; Silva, A.M.L.; Singh, C. Improving power system reliability calculation efficiency with EPSO variants. *IEEE Trans. Power Syst.* **2009**, *24*, 1772–1779. [CrossRef]

49. Miranda, V.; Alves, R. Differential evolutionary particle swarm optimization (DEEPSO): A successful hybrid. In Proceedings of the 1st BRICS Congress on Computational Intelligence and 11th Brazilian Congress on Computational Intelligence (BRICS-CCI and CBIC), Recife, Brazil, 8–11 September 2013.

50. Carvalho, L.M.; Loureiro, F.; Sumaili, J.; Keko, H.; Miranda, V.; Marcelino, C.G.; Wanner, E.F. Statistical tuning of DEEPSO soft constraints in the security constrained optimal power flow problem. In Proceedings of the 2015 18th International Conference on Intelligent System Application to Power Systems (ISAP), Porto, Portugal, 11–17 September 2015.

51. Pinto, P.; Carvalho, L.M.; Sumaili, J.; Pinto, M.S.S.; Miranda, V. Coping with wind power uncertainty in unit commitment: A robust approach using the new hybrid metaheuristic DEEPSO. In Proceedings of the Towards Future Power Systems and Emerging Technologies, Powertech Eindhoven, Eindhoven, The Netherlands, 29 June–2 July 2015.

52. Differential Evolutionary Particle Swarm Optimization (DEEPSO). Available online: http://epso.inescporto.pt/deepso/deepso-basics (accessed on 10 February 2016).

53. Portuguese Transmission System Operator—REN. Available online: http://www.centrodeinformacao.ren.pt/ (accessed on 13 June 2016).

54. Electricity Market Operator—OMEL. Available online: http://www.omelholding.es/omel-holding/ (accessed on 10 February 2016).

55. Pennsylvania-New Jersey-Maryland (PJM) Electricity Markets. Available online: http://www.pjm.com (accessed on 20 June 2016).

energies

MDPI

Article

Effects of Increased Wind Power Generation on Mid-Norway's Energy Balance under Climate Change: A Market Based Approach

Baptiste François [1,*,†]**, Sara Martino** [2]**, Lena S. Tøfte** [2]**, Benoit Hingray** [1]**, Birger Mo** [2]
and Jean-Dominique Creutin [1]

[1]   Université Grenoble Alpes, CNRS, IGE, F-38000 Grenoble, France;
       benoit.hingray@univ-grenoble-alpes.fr (B.H.); jean-dominique.creutin@univ-grenoble-alpes.fr (J.-D.C.)
[2]   SINTEF Energy Research, 7465 Trondheim, Norway; sara.martino@sintef.no (S.M.);
       lena.s.tofte@sintef.no (L.S.T.); birger.mo@sintef.no (B.M.)
*   Correspondence: bfrancois@umass.edu
†   Current address: Department of Civil and Environmental Engineering, University of Massachusetts
       Amherst, Amherst, MA 01003-9303, USA.

Academic Editor: Robert Lundmark
Received: 25 November 2016; Accepted: 8 February 2017; Published: 15 February 2017

Abstract: Thanks to its huge water storage capacity, Norway has an excess of energy generation at annual scale, although significant regional disparity exists. On average, the Mid-Norway region has an energy deficit and needs to import more electricity than it exports. We show that this energy deficit can be reduced with an increase in wind generation and transmission line capacity, even in future climate scenarios where both mean annual temperature and precipitation are changed. For the considered scenarios, the deficit observed in winter disappears, i.e., when electricity consumption and prices are high. At the annual scale, the deficit behaviour depends more on future changes in precipitation. Another consequence of changes in wind production and transmission capacity is the modification of electricity exchanges with neighbouring regions which are also modified both in terms of average, variability and seasonality.

Keywords: variable renewable energy; wind; hydro; energy balance; energy market

1. Introduction

The United Nations Framework Convention on Climate Change (UNFCCC) Paris Agreement promotes the transition to low carbon economy by replacing conventional by renewable energies such as wind-, solar-, and hydro-power. In Europe, optimistic scenario by the European Climate Foundation foresees 100% renewable energy supply at the horizon 2050 [1]. Some countries such as Sweden, Spain and Austria are already well engaged for reaching this objective even before this deadline [2]. This issue is also relevant at regional scale level as highlighted in Northern Italy by reference (Ref.) [3].

Thanks to its huge resources, Norwegian electricity generation already comes for about 95.3% from hydropower [4]. Norway has an excess of energy at annual scale and presents on average a positive balance between importation and exportation [5]. On account for its high water storage capacity, Norwegian reservoirs are sometimes considered as the future Blue Battery of Europe. Gullberg [6], for instance, explains that thanks to its actual hydropower capacity, Norway might balance power in Europe. In the longer term, new transmission lines and pumped-storage hydropower in Norway would provide a backup capacity to the expected future high solar and wind power capacity in Europe [6,7].

The positive energy balance for Norway hides significant regional disparities. Mid-Norway is the most illustrative example (region 9 on Figure 1). Like the rest of the country, its electricity system is mainly based on hydropower with reservoirs that store high river flow during the snowmelt season in spring and summer, and then generate hydropower in winter (i.e., when electricity consumption and prices are much higher). Mid-Norway experiences an energy deficit almost every year [8]. Residual demand is satisfied with energy import from other parts of Norway and other countries of the Nordic Energy market (Norway, Sweden, Denmark and Finland). Due to high electricity prices, the energy deficit is moreover critical during the winter season since most buildings use electrical heaters. The winter 2002/2003, dubbed as "electricity crisis" by Norwegian media, is the most illustrative example [9]. The low hydropower resource resulting from the exceptionally dry 2002 fall, the high winter energy demand of the cold subsequent winter and the limited transmission lines with the neighbouring regions led electricity prices to double [10]. Even though such a situation is unusual, its frequency and intensity are both expected to increase in the future as a result of the increasing demand from the industry sector and electric cars [9].

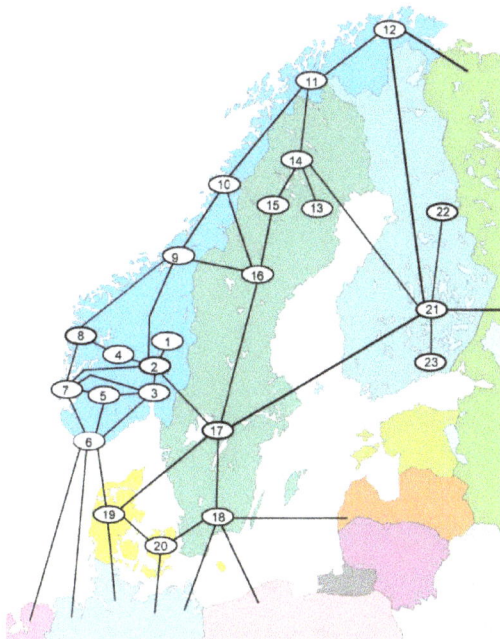

Figure 1. Simplified Nordic energy market grid as seen by EFI's Multi-area Power market Simulator (EMPS) model. Black lines represent transmission lines among the different regions. Mid-Norway region is the region number 9. It is connected to East-Norway, West-Norway, Helgeland and Inndalselven (Sweden) regions (respectively, regions 2, 8, 10 and 16). EFI: Norwegian Electric Power Research Institute.

To reduce the Mid-Norway energy balance deficit and thus the risk of energy shortcuts, local policy makers have been strongly motivated to increase wind power capacity. Wind resource is actually important in Mid-Norway and was estimated to be a relevant supplement to hydropower in Nordic areas (e.g., [11]). Increasing wind power capacity in Mid-Norway is also fully consistent with the objective of the Norwegian-Swedish Electricity Certificate Market aiming to increase the rate of renewable energy in the whole Nordic energy market where conventional energy sources are still used in Sweden and Finland [12]. In the present state, Mid-Norway regularly imports/exports energy from/to neighbouring regions. The high power generation variability obtained with more variable renewable energy obviously requires increasing transmission lines capacity [13]. Upgrading transmission lines is planned between Mid-Norway and West-Norway (http://www.statnett.no). As highlighted during a stakeholder meeting organized for the region within the COMPLEX EU research project (http://owsgip.itc.utwente.nl/projects/complex/), public acceptance for wind power and transmission development is however not straightforward. It is high if the benefits from the project mainly go to the regional industry and trade development. It is however rather low if the generated power has to be exported to neighbouring regions.

The first aim of the present work is to assess the effect of the development of additional wind farms and of the development of a new transmission line on the energy balance of Mid-Norway. Next, it is to explore the alternative question raised by local stakeholders about the finality of wind power development—deficit limitation or export growth?

The second main objective of this study is to assess the ability of the Mid-Norway system to cope with a modification of the energy balance due to climate change. Climate change could first impact the mean energy production via changes in wind- and hydro-power potential. In Nordic countries, change in wind power potential is expected to be very small with a lower than 5% decrease (a strong agreement was obtained between Global Circulation Models (GCMs) as highlighted by Ref. [14]. Changes in hydropower potential are conversely expected to be quite large as a result of both precipitation and temperature increase. The significant increase in precipitation expected for the region [15] should actually lead to an increase in river flows. Regional warming should additionally modify hydrological regimes with shorter winter droughts, earlier and smaller snowmelt flows [16]. On the other hand, climate change could also modify the electricity demand. Regional warming should especially lead to less heating needs in winter and to reduce the demand seasonality. The sensitivity of Mid-Norway system performance to changes in mean precipitation and temperature is thus definitively important to analyse.

Our analysis is carried out with the decision scaling approach developed by Ref. [17]. This approach is based on sensitivity analyses of system responses to a set of synthetic climate change scenarios. In the present study, we consider changes in mean precipitation and temperature. The objective is to build Climate Response Functions (hereafter noted as CRFs) putting in perspective: (i) either a given statistics of interest or an indicator of success of the considered system obtained via a set of synthetic scenarios implemented with a sensitivity analyses of its drivers (i.e., in our case, temperature and precipitation variables); and (ii) the expected changes of the drivers obtained from GCMs.

We use the EMPS (EFI's Multi-area Power market Simulator, EFI: Norwegian Electric Power Research Institute) power market model for simulating the Nordic energy market for the present and synthetic future climate scenarios and different wind power and transmission line capacities. The main indicator we account for is the one discussed by local stakeholders, namely the energy balance deficit in Mid-Norway. We focus on both the annual scale and the winter season.

The article is organized as follows. Section 2 gives the description of the Mid-Norway case study and of the considered future Mid-Norway electricity system scenario. Section 3 details the database and models used. Section 4 illustrates current situation in Mid-Norway while Sections 5 and 6 give results obtained for future electricity system and future climate. Section 7 concludes and gives some highlights for further research.

2. Case Study

All Nordic countries have liberalized their electricity markets, opening both electricity trading and electricity production to competition. For a given region, this means that regional electricity prices are determined by the energy balance and exchange capacity from/to neighbouring regions.

The Mid-Norway region covers the counties Møre og Romsdal, Sør-Trøndelag and most of Nord-Trøndelag. The region includes a set of fjords and mountains with altitudes ranging from 0 to 1700 m.a.s.l. The climate is relatively wet with annual precipitation ranging with the altitude from

500 mm/year in coastal areas to 3000 mm/year in inland mountains. Temperature also varies along this climate transect with an annual average ranging from +7 to −6 °C according to altitude.

Watercourses range from small coastal waterways to major mountainous rivers in the east, where catchment areas with large hydropower reservoirs are located. The hydrological regime also moves from an Atlantic regime in coastal areas (i.e., major flows in late autumn and winter) to an Alpine regime in the inland areas (i.e., low winter flows and high flows in late spring and summer due to snow melt and rainfall events). The period of snow accumulation lasts several months and the snow melting period usually starts in late March in the lowland regions and in June in high elevation areas.

Primary activities such as agriculture, fishery and forestry play a role in all the counties. Engineering industry, woodworking factories, fish farming, shipping trade and food industry are other important activities in the region. Energy-intensive industries and petroleum activity have large demand of electricity, which has increased over the last 20 years and probably will continue to do so.

The region produces on average 14 TWh per year and consumes about 21 TWh. Mid-Norway region continuously buys electricity from the Nordic market (http://www.Statnett.no). Its storage capacity is about 8% of Norway's total capacity which represents about one third of the annual consumption in the region (i.e., 6.7 TWh).

When this study was initiated in 2014, Mid-Norway wind power capacity equalled 1090 MW. Additional 4552 MW was already under construction or close to be, while concessions for another 1100 MW power capacity were asked (for details see: https://www.nve.no/). In this study, we consider two wind power capacity configurations. The first considers the installed wind power capacity in 2014 (hereafter, this scenario is denoted W1). The second one considers the additional planned and asked wind power capacity (meaning a total wind power capacity equal to 6742 MW, denoted as W2 scenario). Even though all asked concessions might not be accepted, W2 scenario gives a good guess about wind power capacity evolution.

Two transmission line scenarios are also considered. The first scenario, denoted as G1, considers the current line capacity: Mid-Norway is connected with East-Norway, West-Norway, Helgeland and Inndalselven (Sweden) regions (respectively, region 2, 8, 10 and 16 in Figure 1). The second scenario, denoted G2, takes into account the increased transmission line capacity which will be achieved between Mid-Norway and West-Norway within the next few years (see, for example, http://www.statnett.no/). Corresponding transmission line capacities are given in Table 1.

Table 1. Transmission line capacity scenarios between the Mid-Norway and its four neighbouring regions. Numbers in brackets refer to the market areas on Figure 1.

Mid-Norway [9]	East-Norway [2]	West-Norway [8]	Helgeland [10]	Inndalselven [16]
Scenario G1	600 MW	500 MW	900 MW	1950 MW
Scenario G2	600 MW	2000 MW	900 MW	1950 MW

3. Data and Models

The meteorological years used as reference cover 1961–1982. Along this period, the Nordic energy system has been evolving with, among other things, construction of many hydropower plants and water reservoirs. For our analyses, we consider fixed system configurations (we disregard any evolution of the system state during the considered time period). The reference configuration corresponds to the current one: for the whole Nordic market, the hydropower, wind-power and transmission line capacities are those available in 2014.

3.1. Mid-Norway Energy Balance Modelling

The EMPS model is a hydrothermal optimization and simulation model used by most players in the Nordic energy market for long and medium-term price forecasting, hydro scheduling, system and investment analysis. One of the main advantages of this model is that it includes a detailed representation of the hydro-system (i.e., power stations, reservoirs, diversions, etc.). The optimization aims at minimizing the overall system costs via a variant of the so-called water value method (see Ref. [18] for an early reference and Ref. [19] for a recent one). This method aims to balance the income related to the immediate use of stored water against the future income expected from its later use. A mathematical description of both optimization and simulation stages within EMPS modelling is given in Ref. [20].

In this study, the EMPS model is used to simulate consequences of increasing the capacity of: (i) wind power generation; and (ii) transmission lines in Mid-Norway region for the current and a variety of future climates. The model is set up for the whole Nordic energy market, to which Mid-Norway region belongs. The Nordic energy market is divided into 23 areas (Figure 1). Each area is characterized by transmission constraints and hydropower system properties. The different connections among areas reflect physical transmission lines. The Nordic energy market includes more than a thousand reservoirs and several hundred hydropower plants in more than 50 different river systems.

In the considered set up of the EMPS model, input data are: (i) weekly unregulated water discharge time series for a set of river basins in each market area; and (ii) weekly wind power time series for each market area. EMPS model simulations are typically done for an ensemble of historical weather years assumed to have equal probability. The weather years provide physically consistent weather scenarios (i.e., scenarios with consistent space/time correlation among surface weather variables) and in turn physically consistent scenarios for the various hydro-meteorological variables (river discharges, wind, solar radiation, temperature) that affect the energy production, the demand and then the market balance.

Hydro-meteorological time series scenarios required for the climate change impact analysis are obtained with a pattern scaling approach from the observed time series available for the reference period (e.g., [21–23]). This approach is expected to rather well preserve space/time correlations between variables. In the present case, the pattern scaling is carried out on a weekly basis. For river discharge, each reference river basin for which unregulated discharge time series are required in EMPS is considered separately. Fifty-two weekly scale factors C are first estimated from hydro-meteorological simulations forced with a set of future climate change scenarios indexed by i. They are then used to derive future discharge series from the observed ones as follows:

$$Q_{Future}(w,y,k,i) = Q_{Obs}(w,y,k)C(w,k,i) \tag{1}$$

with $Q_{Future}(w,y,k,i)$ the weekly water discharge for the w-th week of year y and the k-th reference river basin; Q_{Obs} is the observed weekly time series water discharge within the EMPS archive and $C(w,k,i)$, $k = 1–52$ are the scaling factors of weekly river discharges for the 52 calendar weeks.

For each market area and each future climate scenario i, scaling factors C are the ratios between future and present average regional runoff. Future regional runoff time series are obtained via hydrological simulation, for each grid cell of each market area, with a distributed version of the GSM-Socont hydrological model [24] (Glacier and SnowMelt—Soil CONTribution model). This model simulates the snowpack dynamic (snow accumulation and melt), water abstraction from evapotranspiration, slow and rapid components of river flow from infiltrated and effective rainfall respectively. For each climate change scenario i, future meteorological scenarios used for the hydrological simulation are obtained with the scaling approach. Historical time series of precipitation and temperatures are modified with a multiplicative (additive) perturbation prescribed from the relative (absolute) change in annual precipitation (temperature) (hereafter denoted as ΔP and ΔT, respectively). As described later, a number of climate change scenarios, in terms of change in mean annual precipitation and temperature, are considered.

Historical precipitation and temperature data comes from the European Climate Assessment & Dataset (ECAD, [25]) with a 0.25° space resolution. The hydrological model required the calibration of some parameters. We use a unique set of parameters for all market areas. It was calibrated by comparing the EMPS discharge archive with the discharges simulated from historical meteorological data.

In EMPS, the energy demand is indexed by the temperature, showing higher consumption during cold days. The energy demand is linked to a lesser extent to price dependent contracts. This link is represented by an additional demand which activates only if the price is low enough. A typical example of price dependent contract is a boiler-power contract where the customer has an oil heater connected in parallel with an electrical heater which is only used when the price is low. At annual scale, this price dependent demand represents about 8% of the temperature dependent demand. For future scenarios, the demand is estimated on a weekly basis from future temperature, obtained from the perturbation approach mentioned above, and from the energy price derived within EMPS for the current simulation time step.

We compute the weekly wind power generation time series with the wind speed database NORA10 [26] (NOrwegian Reanalysis Archive). NORA10 is currently the best near-surface (i.e., 10 m altitude) wind speed reanalysis over Scandinavian countries with a 10 km^2 space resolution. Wind speed data are available from 1957. We used the wind power generation model developed by Ref. [27] at daily time step. This model considers a nonlinear relationship between wind speed u (m·s^{-1}) and wind power generation, hereafter noted P_W (MWh). Below a given threshold (3 m·s^{-1} in this study), the wind speed is not sufficient to enable power generation. The power generation is then a third order polynomial function of wind speed and reaches the maximum wind turbine efficiency at a second threshold (13 m·s^{-1}). Above a third threshold (25 m·s^{-1}), the power generation has to be stopped in order to avoid any damages on the wind turbine. The 70 m altitude wind speed time series used for computing wind power time series were estimated from the 10 m altitude NORA10 wind following the scaling equation:

$$u_1 = u_2 \left(\frac{h_1}{h_2} \right)^\alpha \tag{2}$$

with u_1 and u_2 the wind speeds (m·s^{-1}) at the altitudes h_1 and h_2 (m), respectively. α is an air friction coefficient chosen equal to 1/7 (no dimension) [28]. Simulated daily wind power generation time series are then aggregated at weekly time scale.

3.2. Climate Response Functions

CRFs are expressed in a two dimensional climate change space defined from changes in temperature and precipitation. The climate change factors we considered range from -20% to $+50\%$ for precipitation (with 10% step) and from 0 to $+6\,^\circ$C for temperature (with 1 $^\circ$C step) in regards with the reference period 1961–1982. CRFs are built from the 8 × 7 hydro-climatic time series scenarios obtained for these climate change scenarios via the scaling approach presented in the previous section. The reference period corresponds to the scenario with no change in temperature (i.e., $+0\,^\circ$C) and no change in precipitation (i.e., $+0\%$).

Positioning on the CRFs the future projections of climate experiments available from the latest GCMs allows discussing the expected effects of climate change for different future prediction periods. In the present case, changes in future annual precipitation and temperature are estimated from the outputs of an ensemble of 23 GCM projections from CMIP5 experiments [29] (Coupled Model Intercomparison Project Phase 5). Using several GCM projections illustrates uncertainty on precipitation and temperature changes over the next decades and its amplitude in regard to the studied effects.

In recent studies, climate change factors for a given climate experiment are classically estimated from the change of the raw climate model outputs between a future and a reference period. A limitation to the robust estimation of change factors is the critical role of multi-decadal variations in the evolution of the climate system. These low-frequency variations, commonly termed as climate internal variability, can, temporarily, worsen, reduce or even reverse the long-term impact of climate change. Internal variability was found to be a major source of uncertainty in climate projections for the coming decades, especially for regional precipitation (e.g., [30,31]). A robust estimate of expected changes actually requires a noise-free estimate of the climate response from the modelling chains. In the present case, we estimate the climate response of each GCM using all data of the transient simulations available for the model (150 years from 1950 to 2099). For each GCM, a trend model is first fitted to the raw climate projections following Ref. [32] (piece-wise linear function of time for precipitation and a 3rd order polynomial trend for temperature). The expected change for any future period is obtained from the change between trend estimates obtained respectively for this period and for the reference time period. We consider three future 20-year time periods: 2040–2059, 2060–2079 and 2080–2099.

Figure 2 shows annual temperature and precipitation changes expected in Mid-Norway for each future period. Model uncertainty for precipitation changes is very large although a significant increase is consistently foreseen; only one GCM gives a slight decrease in precipitation. Changes in temperature are more univocal showing an increase along the century. Note however that the dispersion among models grows with the projection time horizon.

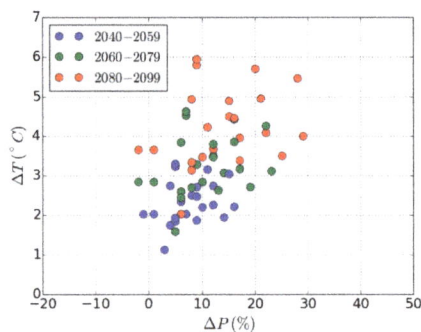

Figure 2. Scatterplot of average changes in precipitation and temperature for 23 GCM projections between control period (1961–1982) and three future time periods (i.e., 2040–2059, 2060–2079, and 2080–2099).

4. Mid-Norway Energy Balance in the Current System

This section presents the results obtained by EMPS for the control configuration, i.e., with W1G1 scenario for the control climate (2014 wind power and transmission capacity, observed meteorological forcing). We first note that weekly wind and hydropower generation are much more variable than the demand (see coefficients of variation in Table 2). This result agrees with the literature related to the variability of renewable energies (e.g., [33]). In Mid-Norway, unregulated wind power generation is positively correlated with electricity consumption; the winter generation is higher than the summer one (Figure 3b,d). Mid-Norway reservoirs are handled so that stored inflows are mainly released during the winter season, making winter hydropower generation higher (see energy storage scheme and hydropower generation time series on Figure 3a,c). On average, the total production (i.e., sum of hydro and wind power) is not sufficient to supply the load, neither at yearly scale nor for winter and summer seasons (Table 2). As a result of the reservoirs' management, the average energy balance deficit is higher in summer than in winter (Table 2). However, summer deficit is less critical than winter deficit since market prices are lower in summer (Figure 3e). For some years, electricity prices collapse, falling to 1 €cent/kWh during the spring and summer seasons. These situations correspond to periods where reservoirs are almost full and present a high risk of spill. When looking at the statistical distribution of the energy balance on Figure 3f, we note that only 10% of winter weeks present a positive energy balance while this number is lower than 5% during the summer season. Considering the whole year, less than 10% of all weeks have a positive energy balance.

Table 2. Average yearly, winter and summer Mid-Norway weekly energy balance components for W1G1, W2G1 and W2G2 scenarios and for the control period 1961–1982. Number within brackets give coefficient of variation (CV, defined as the ratio between the standard deviation and the mean).

W1G1 Scenario	Year	Winter (Week 43 → 10)	Summer (Week 21 → 35)
Hydro Power P_H	273 GWh (0.48)	372 GWh (0.25)	145 GWh (0.48)
Wind Power P_W	61 GWh (0.60)	84 GWh (0.45)	35 GWh (0.55)
Total Consumption	497 GWh (0.19)	590 GWh (0.06)	380 GWh (0.08)
Energy Balance	−163 GWh (0.60)	−134 GWh (0.74)	−200 GWh (0.36)
W2G1 Scenario	**Year**	**Winter (Week 43 → 10)**	**Summer (Week 21 → 35)**
Hydro Power P_H	272 GWh (0.47)	365 GWh (0.25)	152 GWh (0.48)
Wind Power P_W	160 GWh (0.63)	222 GWh (0.49)	90 GWh (0.52)
Total Consumption	499 GWh (0.19)	592 GWh (0.06)	381 GWh (0.08)
Energy Balance	−66 GWh (1.84)	-5.8 GWh (20.7)	−137 GWh (0.62)
W2G2 Scenario	**Year**	**Winter (Week 43 → 10)**	**Summer (Week 21 → 35)**
Hydro Power P_H	273 GWh (0.47)	366 GWh (0.25)	150 GWh (0.48)
Wind Power P_W	160 GWh (0.63)	222 GWh (0.49)	90 GWh (0.52)
Total Consumption	499 GWh (0.19)	592 GWh (0.06)	381 GWh (0.08)
Energy Balance	−66 GWh (1.88)	−4.5 GWh (27.4)	−141 GWh (0.59)

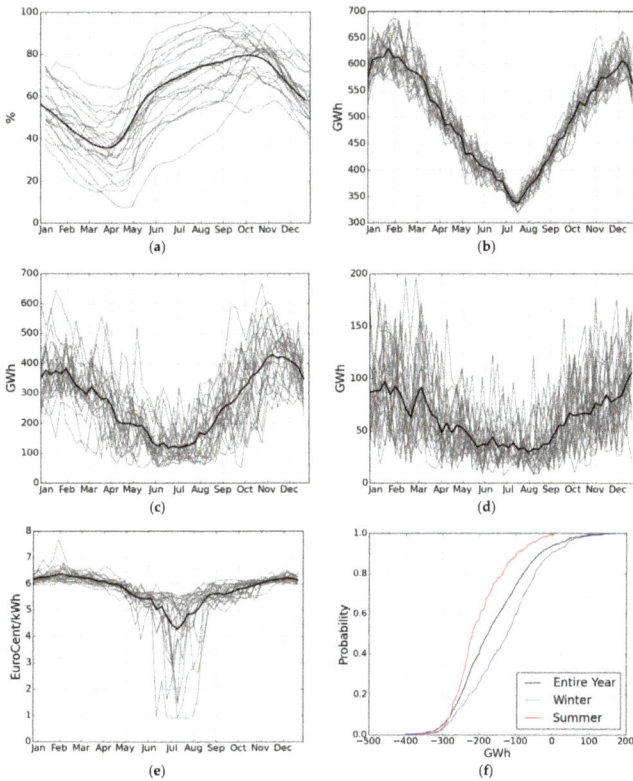

Figure 3. Weekly: (**a**) Aggregated energy storage expressed as ratio of the total storage capacity; (**b**) total electricity consumption (i.e., temperature and price dependent consumptions); (**c**) hydropower (from regulated + unregulated power plants); (**d**) wind power generation; and (**e**) electricity prices time series in Mid-Norway for the period 1961–1982. Note that only the fraction of electricity consumption temperature dependent is plotted. Grey curves represent week-to-week values for each year and the black curve represents the average annual cycle; (**f**) Cumulative distribution function of Mid-Norway weekly energy balance. Winter season is defined as from Week 43 to Week 10 of the following year and summer season from Week 21 to Week 35.

As a result of the deficit, Mid-Norway imports electricity from the regions of Inndalselven and Helgeland over, respectively, 90% and 70% of the weeks (Figure 4). Meanwhile, the region exports electricity to East-Norway 90% of the time. The line between Mid- and West-Norway is used for both importing and exporting electricity. Note that 20% of the weeks the transmission line is not used at all (Figure 4). Similar distributions are obtained during winter season. Note that these simulated results are consistent with the current deficit and exchange situation of the region as presented in the previous sections.

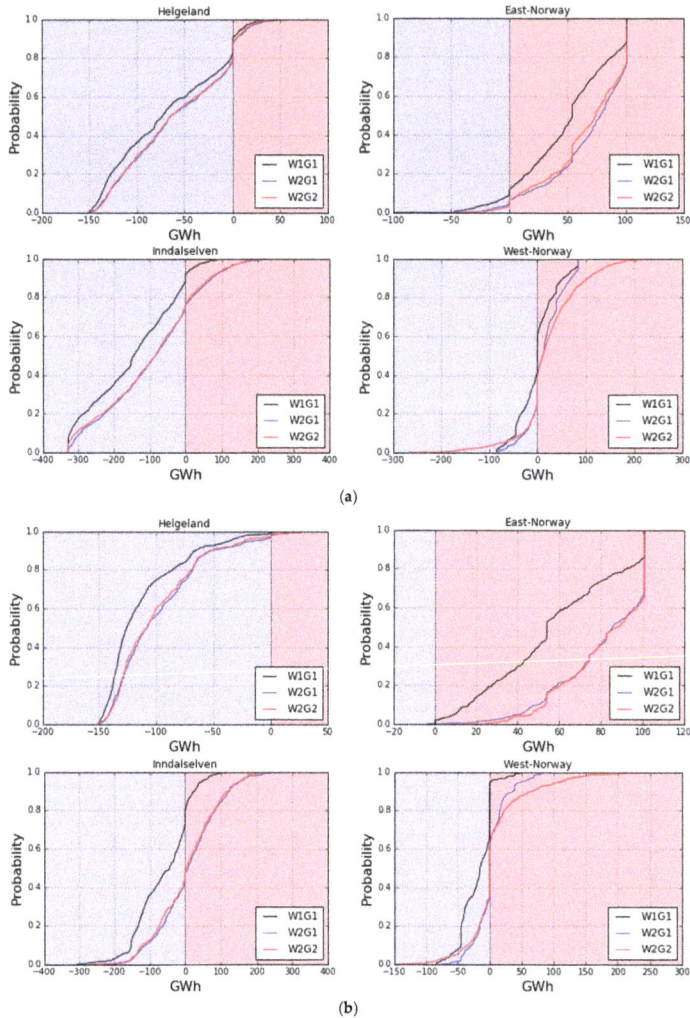

Figure 4. Cumulative distribution functions of weekly energy exchanges between Mid-Norway and the neighbouring regions during: (**a**) the whole year; and (**b**) the winter season only. Negative (blue background) and positive (red background) values, respectively, show importation to and exportation from Mid-Norway. Note than only the line with West-Norway is reinforced in G2 scenario. Note that some distributions clearly show when the full capacity is reached for the whole week (e.g., East-Norway).

5. Increasing Wind Power and Transmission Capacities

This section focuses on the evolution of the energy balance by considering, firstly, an increase of wind power capacity in Mid-Norway (i.e., W2G1 scenario), and secondly, an increase of both wind power capacity and transmission line capacity between Mid- and West-Norway (i.e., W2G2 scenario).

Additional wind power capacity almost triples the average generation from 61 to 160 GWh per week. The weekly generation increases from 84 to 222 GWh in winter season and from 35 to 90 GWh in summer (Table 2). Higher energy generation obviously reduces the energy balance deficit (Table 2). For winter, the deficit is close to 0 (i.e., −5.8 GWh/week). It remains rather important in summer (−137 GWh/week) as well as at annual scale (−66 GWh/week). Comparing W1G1 and W2G1 scenarios, we note that, on average, Mid-Norway exports every week 2 GWh more, which corresponds to roughly 2% of the additional wind generation. In winter, average export reaches 10 GWh (about 7% of the additional generation at this season). One can note that this exported electricity could have been used to further reduce Mid-Norway deficit.

Wind power generation being highly variable (see CV in Table 2), increasing wind power generation implies a higher temporal variability of the energy balance (the CV of the weekly energy balance increase by a factor of 3 during the whole year and by a factor 28 during the winter season; Table 2). Such variability requires systematically more important energy exchanges between Mid-Norway and all its neighbouring regions (Figure 4), even when the average deficit is close to 0 as it is the case during the winter, for instance. Transmission lines are effectively more often used for exporting energy and they are more often used at full capacity. For instance, Mid-Norway exports at full capacity to East-Norway during more than 25% of the weeks during the year (35% in winter). Another example is the number of winter weeks during which Mid-Norway exports electricity to West-Norway (60% of the weeks for W2G1 against roughly 40% for W1G1).

The increased transmission capacity of the line between Mid- and West-Norway (i.e., W2G2 scenario) has no significant effect on the mean annual deficit (Table 2). Note that the winter deficit slightly decreases to −4.5 GWh/week. Energy exchange distribution functions obtained with W2G2 scenario roughly overlap the ones obtained with W2G1 except for the reinforced line (Figure 4). Although the increased capacity is only used about 10% and 15% of the time, it allows exporting an important amount of energy. This mainly occurs during high wind power generation periods and/or when reservoirs are close to full in spring.

6. Evolution of Mid-Norway Energy Balance in a Changing Climate

This section focuses on climate change impact on Mid-Norway energy balance. As discussed in introduction, climate change is expected to impact both the average and the time variability of electricity generation and consumption. Considering changes in temperature and precipitation, two components of the energy balance are modified: the river flows and the electricity consumption. This section presents first the raw changes in these two components and secondly the impacts on Mid-Norway energy balance and exchanges.

6.1. Climate Change Impacts on River Flows and Electricity Consumption

The main driver of change in river flow is precipitation; higher precipitation giving higher river flows. As illustrated on Figure 5a, river flow increases linearly with precipitation change and higher temperatures increase evaporation and in turn reduce river flow. The effect of increasing temperature on mean annual discharge is rather weak. For instance, river flows slightly increase up to ΔT equal to +2 °C and then decrease when temperatures rise above this threshold. In any case, river flow modification is less than 5% whatever the change in temperature. However, increasing temperature significantly reduces river flow seasonality with higher discharges values in winter (due to a higher ratio of liquid precipitation resulting from higher temperatures) and lower values during the spring and summer seasons (due to less snowpack; not shown). Annual temperature and precipitation changes provided by 23 GCMs are also plotted on Figure 5a for three different future time periods. The CRF shows anticipated changes in water discharge as a function of temperature and precipitation estimates for each future period and each GCM. For instance, accounting for changes in temperature and precipitation obtained by most GCMs, future river flow would increase by 30% for the 2080–2099 time period. Only one GCM shows a small decrease in average water discharges related with a decrease in precipitation.

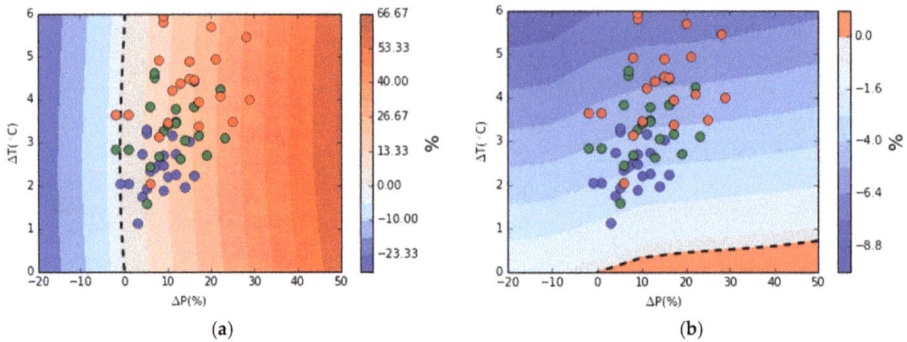

Figure 5. Climate Response Functions (CRFs) of the average annual changes (%) in: (**a**) river inflows; and (**b**) electricity consumption (obtained for W1G1 scenario) compared with control period. The dashed black curves show the "no change" edge. Dots show expected annual changes in temperature and precipitation change obtained from 23 GCM, as illustrated in Figure 2, for 2040–2059 (blue), 2060–2079 (green) and 2080–2099 (red).

Figure 5b shows the CRF obtained for the average annual electricity consumption. By construction, electricity consumption decreases almost linearly when temperature increases. Interestingly, increasing precipitation induces a slight increase of electricity consumption actually linked to price dependent contracts. More abundant water resource makes lower electricity prices (not shown) and stimulates consumption. Annual temperature changes obtained from the selected GCMs show a decrease in average electricity consumption, up to 8% for the time period 2080–2099.

6.2. Climate Change Impacts on Energy Balance and Exchanges in Mid-Norway

We only focus on differences between W1G1 and W2G2 scenarios considering that climate should change once wind power generation and transmission line capacity will be both developed.

Changes in water discharges and electricity consumption will modify the Mid-Norway energy balance deficit at both annual and winter season scales (Figure 6). Precipitation is the main factor of the deficit modification. Considering the current Mid-Norway electricity system (i.e., W1G1 scenario), the energy balance remains negative whatever the changes in precipitation and temperature. The energy balance might become positive during the winter season if changes in precipitation and temperature are quite drastic (from +40% to +50% in annual precipitation and from +4 to +6 °C in annual temperature).

When considering climate change with additional wind generation and stronger transmission lines (i.e., W2G2 scenario), the annual energy balance might become positive with less drastic changes than for the W1G1 scenario. For instance, the annual balance might become positive with 25% precipitation more and whatever the annual increase in temperature. Below 25% increase in precipitation, annual temperature must increase enough to reduce electricity consumption and to make the balance positive. For instance, an increase in annual temperature of +3.5 °C is required if precipitation increases by only 10%. During the winter season, the energy balance becomes positive but when precipitation decreases significantly or when a decrease in precipitation is conjugated with a low rise in temperature (which does not decrease significantly the electricity consumption). However, these two later configurations are not likely to appear according to GCMs projections as illustrated on Figure 6.

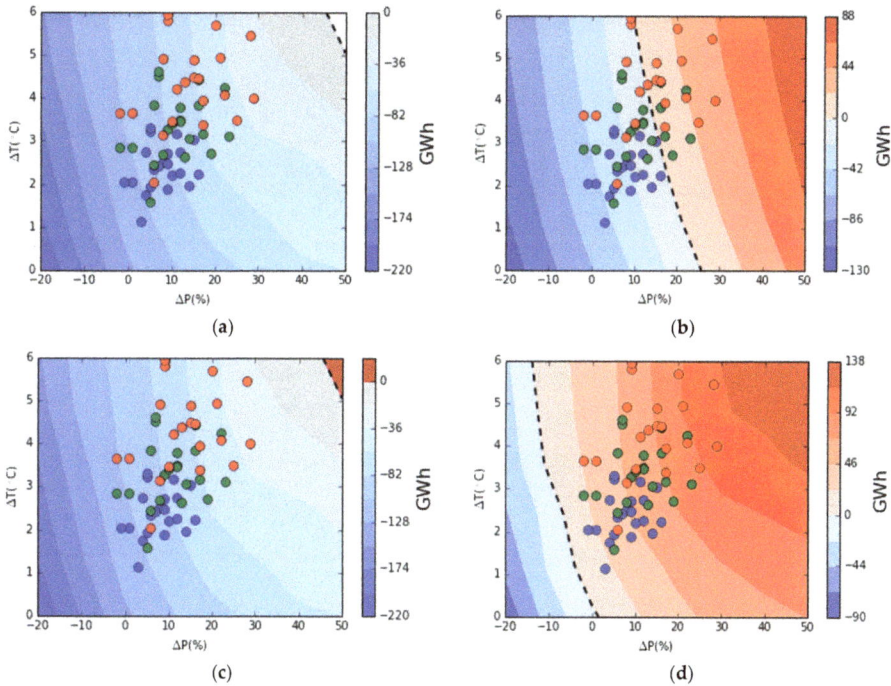

Figure 6. CRFs of the Mid-Norway weekly energy balance (GWh) obtained with the: (**a**) W1G1 scenario for the whole year; (**b**) W2G2 scenario for the whole year; (**c**) W1G1 scenario for the winter season; and (**d**) W2G2 scenario for the winter season. Nil energy balance curves are highlighted with dashed black lines. The coloured dots give temperature and precipitation changes from the 23 considered GCMs and three considered future time period (blue: 2040–2059; green: 2060–2079; and red: 2080–2099).

Changes in Mid-Norway energy balance deficit imply modifications of energy exchanges with the neighbouring regions. Energy imports from Helgeland should grow over the next decades (not shown). This results from both higher production in Helgeland (due to increasing precipitation) and lower consumption (due to higher temperatures; not shown). In association with higher in-situ generation, Mid-Norway region is able to export more electricity and then to strengthen its hub role in the Nordic energy market. As a consequence, electricity exports to East-Norway linearly increase with precipitation (and thus with hydropower generation) within Mid-Norway region (not shown). Note that East-Norway does not produce electricity. We note on Figure 7 that for both W1G1 and W2G2 scenarios, Mid-Norway keeps importing on average electricity from Inndalselven region at annual scale. However, thanks to the development of wind generation, the region might export more electricity to Inndalselven than it imports during the winter season (Figure 7). As discussed in the previous section, Mid-Norway imports and exports electricity from/to West-Norway, with a slightly negative balance, especially during the winter season. With the W2G2 scenario under future climate, exports from Mid-Norway to West-Norway are expected to increase significantly. We note that temperature changes impact average exportation from Mid-Norway to West-Norway more than the reinforcement of the line (Figure 7).

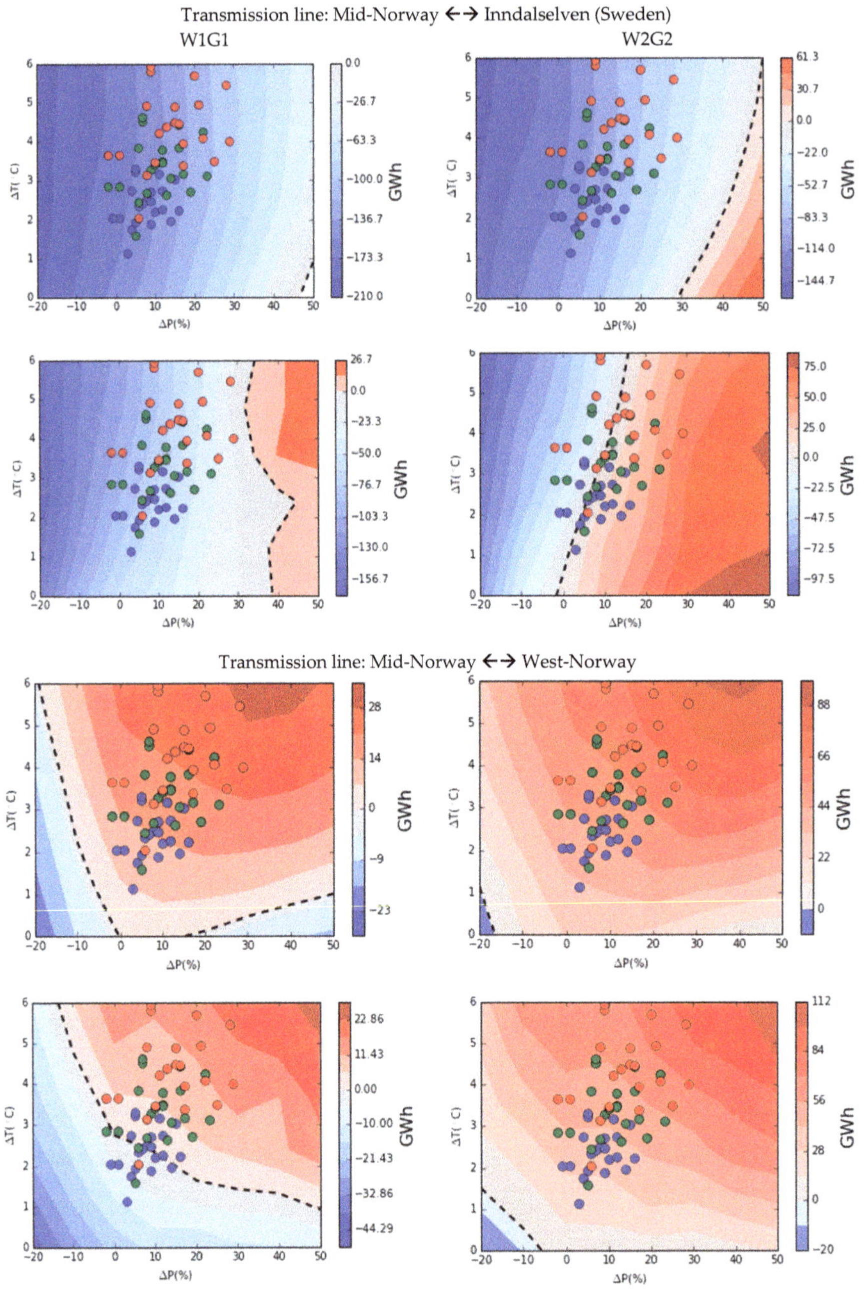

Figure 7. CRFs of the Mid-Norway energy exchanges with: Inndalselven (Sweden) (**top**); and West-Norway (**bottom**). CRFs left and right columns are obtained with W1G1 and W2G2 scenarios respectively. For each transmission line, top CRFs are for the whole year and the bottom for the winter season. Nil energy balance curves are highlighted with dashed black lines. For more details, see Figure 6 caption.

7. Discussion and Conclusions

Norwegian reservoirs are likely to be used as backup capacity for increasing wind and solar power in Europe. However, important space variability exists and some regions show an important energy balance deficit such as Mid-Norway.

Using the EMPS model to simulate the Nordic energy market, we show that increasing wind power capacity in Mid-Norway can reduce the energy balance deficit. The deficit becomes almost nil during high consumption/price period, i.e., in winter, although the deficit remains important at yearly scale (Table 2). Simulations also show that generation from new wind power plants in Mid-Norway is almost totally used for reducing the deficit. Only 2% of the additional wind generation is exported during the whole year (7% during winter season). Such a result should please Mid-Norway stakeholders about the finality of on-going wind power plant construction.

Increasing transmission line capacity between Mid- and West-Norway does not change drastically the export/import patterns from/to Mid-Norway. The increased capacity is actually used only few times during the year (less than 15% of the weeks for exporting and less than 10% for importing electricity; Figure 4). Although this increased capacity is not often used, it limits spillage when the reservoirs are full, in spring season especially.

Regarding climate change impact in Mid-Norway region, temperature is expected to rise in the next decades as well as precipitation (only one GCM out of 23 gives a slight decrease of annual precipitation; Figure 2). These changes have positive impact on Mid-Norway energy system components. More precipitation makes higher river flows and thus higher hydropower potential and higher temperatures lead to lower electricity consumption.

We assess the joint effect of increasing wind and transmission capacities with climate change with the Decision Scaling approach as developed by Ref. [17]. The Cumulative Distribution Functions (CDFs) of the weekly energy balance, calculated from changes in precipitation and temperature given by the GCMs, are illustrated in Figure 8. For the considered GCMs, the average energy balance deficit should decrease in time, highlighting that Mid-Norway climate will become increasingly favourable to the local balance between demand and generation. For instance, at annual scale, one third of the considered GCMs foresee an average positive balance during the 2060–2079 time period and two thirds during the 2080–2099 time period (Figure 8a).

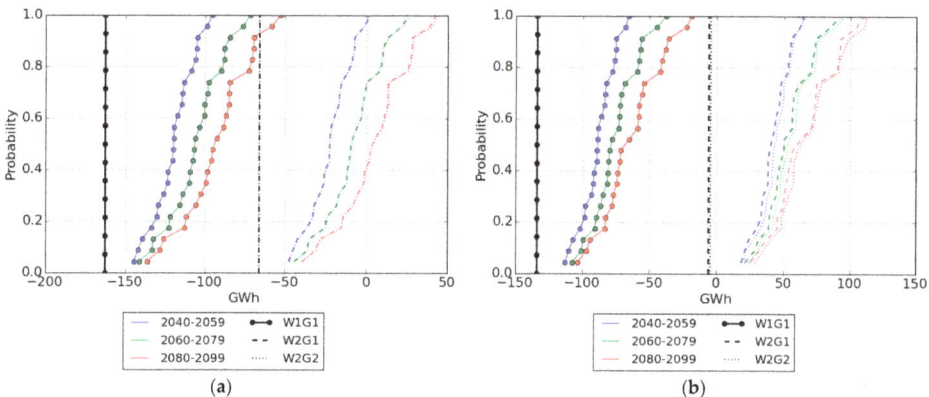

Figure 8. Cumulative distribution function (CDF) of the weekly energy balance for the whole year (GWh) (**a**) and for the winter season (**b**) calculated from annual changes in precipitation and temperature provided by GCMs (blue: 2040–2059; green: 2060–2079; red: 2080–2099). Vertical solid dotted, dashed and dotted black lines shows energy balance values obtained under present climate for W1G1, W2G1 and W2G2 scenarios, respectively. Note that for the whole year, CDFs of W2G1 and W2G2 scenarios are overlapping.

Returning to the question that motivated this study, we conclude that coupling effects from both climate change and increasing wind power and transmission lines capacities appear to lead to a win-win situation: Mid-Norway average energy balance deficit is reduced and would become positive in the next decades allowing the region to increase its exportation, especially during winter season when prices are high.

To our knowledge, this work was the first attempt to applying the Decision Scaling approach to electricity systems analysis. The variety of the results and the easiness of CRF reading can make Decision Scaling an interesting tool for any stakeholder willing to assess its system's vulnerability under climate change. Further research might consider applying the Decision Scaling approach in other climate conditions and or other market contexts (e.g., remote area with no transmission line, using other renewable energy sources such as solar power). This work is based on a number of assumptions, data and modelling choices which potentially lead to some degree of uncertainty in the presented results. Although comprehensive analysis of these uncertainties is out of the scope of this study, it is worth mentioning them.

First, the current consumption modelling within EMPS model does not account for cooling system usage during hot days. The reason is that, regarding of temperature range at Nordic latitudes, the usage of such systems is not common nowadays. However, expected temperatures for the next decades might lead to a growth in cooling system equipment and usage. This could slightly modify consumptions in summer and, eventually, the electricity prices at this period. These effects might deserve specific works although load modification should be weak at these latitudes. Accounting for the non-climatic factors that are also likely to influence the demand (e.g., demand-side management) would be obviously of interest for a more comprehensive view of possible changes in the future electricity balance.

Next, extended and deeper analyses should probably be based on other and/or additional weather scenarios. For instance, the scaling approach we used for generating time series of future weather might be reconsidered. Even though it presents the advantage to preserve the correlation in space and time among weather variables, it does not allow estimating changes in variability. This might be an important issue, especially for precipitation. In Nordic countries, a warmer climate is for instance likely to lead to much more convective precipitation events than today. Although this change in precipitation regime should be, somehow, smoothed by high reservoir capacity, its impacts on energy balance requires further investigation.

This study analyses only changes in generation due to mean changes in precipitation and temperature. Although the change in mean wind potential and in weekly wind variability should remain low over the next decades in this area [14], quantifying their impact on system performance would be valuable.

Accounting for the sub-weekly variability of wind power generation should also be considered. In the current EMPS set up, the sub-weekly variability of wind is disregarded. Wind power generation is estimated on a weekly basis and equally distributed along the week. High frequency variability of wind power generation could obviously limit wind integration into the grid resulting in an energy deficit in Mid-Norway larger than the one obtained in this study. Transmission lines from/to Mid-Norway would also play a major role in wind power integration, which is also impossible to check at weekly time scale.

In addition to climate change analyses, further analyses should also consider the low-frequency variability of weather variables, resulting from the internal variability of the climate system. The year to year to multi-decadal variability of weather variables, precipitation and river flow especially are expected to have a large influence on the potential of renewable and on system performance (e.g., [34]). This would be worth extra investigation. The weather generator developed by Ref. [35] could be considered for such an assessment in future works.

Since this study mainly focuses on aggregated indicators (i.e., computed over the whole period), adding forecast within the considered analysis framework should not significantly improve climate change impact assessment, as shown by Ref. [36]. However, future researches should also consider investigating on the effects of extreme events/periods. High wind power generation periods as illustrated on Figure 3d may have impact on the whole energy systems and especially on the energy exchange among regions. Considering the likely increases in extreme events, further analyses on their impact are required (McInnes et al. [37] give for instance an increase by more than 10% of extreme wind speed in Mid-Norway). Improving forecast of such events and integrating them in the analysis framework is also an important research perspective of this work.

Acknowledgments: This work is part of the FP7 project COMPLEX (knowledge based climate mitigation systems for a low carbon economy; Project FP7-ENV-2012 Number: 308601; http://owsgip.itc.utwente.nl/projects/complex/).

Author Contributions: Baptiste François designed the methodology, carried out the wind power and discharge modelling, analysed the market modelling results and wrote the article. Sara Martino ran the market optimization modelling and contributed to the results' analysis and to composing the final text. Lena S. Tøfte collected hydrological and wind speed data and contributed to the final text. Benoit Hingray carried out the climate change scenario modelling, and contributed to the results' analysis and composing the final text. Birger Mo contributed to setting-up the market modelling. Jean-Dominique Creutin coordinated the project and contributed to the results' analysis and composing the final text.

Conflicts of Interest: The authors declare no conflict of interest. The founding sponsors had no role in the design of the study, in the collection, analyses, or interpretation of data; in the writing of the manuscript, and in the decision to publish the results.

References

1. Roadmap 2050: A Practical Guide to a Prosperous, Low-Carbon Europe. Available online: http://www.roadmap2050.eu/ (accessed on 10 February 2017).
2. Šturc, M. *Renewable Energy: Analysis of the Latest Data on Energy from Renewable Sources*; Eurostat, European Union: Luxembourg, 2012; Available online: http://ec.europa.eu/eurostat/web/products-statistics-in-focus/-/KS-SF-12-044 (accessed on 10 February 2017).
3. François, B.; Borga, M.; Creutin, J.D.; Hingray, B.; Raynaud, D.; Sauterleute, J.F. Complementarity between solar and hydro power: Sensitivity study to climate characteristics in Northern-Italy. *Renew. Energy* **2016**, *86*, 543–553. [CrossRef]
4. International Energy Agency (IEA). *Technology Roadmap: Hydropower*; IEA: Paris, France, 2012.
5. Cherry, J.; Cullen, H.; Visbeck, M.; Small, A.; Uvo, C. Impacts of the North Atlantic Oscillation on Scandinavian hydropower production and energy markets. *Water Resour. Manag.* **2005**, *19*, 673–691. [CrossRef]
6. Gullberg, A.T. The political feasibility of Norway as the "green battery" of Europe. *Energy Policy* **2013**, *57*, 615–623. [CrossRef]
7. Piria, R.; Junge, J. *Norway's Key Role in the European Energy Transition*; Smart Energy for Europe Platform: Berlin, Germany, 2013.
8. Gleditsch, M. Balancing of Offshore Wind Power in Mid-Norway: Implementation of a Load Frequency Control Scheme for Handling Secondary Control Challenges Caused by Wind Power. Master's Thesis, Department of Electrical Engineering, Technical University of Denmark, Copenhagen, Denmark, 2009.
9. Karlström, H. When a Deregulated Electricity System Faces a Supply Deficit: A Never-Ending Story of Inaction? Available online: https://henrikkarlstrom.files.wordpress.com/2012/11/karlstrc3b8m-supply-deficits-in-a-deregulated-electricity-system.pdf (accessed on 10 February 2017).
10. Grønli, H.; Costa, P. The Norwegian Security of Supply Situation during the Winter 2002-03. Part I—Analysis. Available online: http://www.ceer.eu/portal/page/portal/EER_HOME/EER_PUBLICATIONS/CEER_PAPERS/Electricity/2003/WEBNORWAY20030702_PARTI.PDF (accessed on 10 February 2017).
11. Vogstad, K.-O. Utilising the complementary characteristics of wind power and hydropower through coordinated hydro production scheduling using the EMPS model. In Proceedings of the 2000 Nordic Wind Energy Conference, Trondheim, Norway, 13–14 March 2000.
12. *The Norwegian-Swedish Electricity Certificate Market: Annual Report 2013*; Norwegian Water Resources and Energy Directorate (NVE): Oslo, Norway; Swedish Energy Agency: Eskilstuna, Sweden, 2014; p. 44. Available online: https://www.energimyndigheten.se (accessed on 10 February 2017).

13. Weitemeyer, S.; Kleinhans, D.; Wienholt, L.; Vogt, T.; Agert, C. A European perspective: Potential of grid and storage for balancing renewable power systems. *Energy Technol.* **2015**, *4*, 114–122. [CrossRef]
14. Tobin, I.; Jerez, S.; Vautard, R.; Thais, F.; van Meijgaard, E.; Prein, A.; Déqué, M.; Kotlarski, S.; Maule, C.F.; Nikulin, G.; et al. Climate change impacts on the power generation potential of a European mid-century wind farms scenario. *Environ. Res. Lett.* **2016**, *11*. [CrossRef]
15. Jacob, D.; Petersen, J.; Eggert, B.; Alias, A.; Christensen, O.B.; Bouwer, L.M.; Braun, A.; Colette, A.; Déqué, M.; Georgievski, G.; et al. EURO-CORDEX: New high-resolution climate change projections for European impact research. *Reg. Environ. Chang.* **2014**, *14*, 563–578. [CrossRef]
16. Sælthun, N.R.; Aittoniemi, P.; Bergström, S.; Einarsson, K.; Jóhannesson, T.; Lindström, G.; Ohlsson, P.-E.; Thomsen, T.; Vehviläinen, B.; Aamodt, K. *Climate Change Impacts on Ryunoff and Hydropower in the Nordic Countries, TemaNord*; Nordic Council of Ministers: Copenhagen, Denmark, 1998; Volume 552, p. 170.
17. Brown, C.; Ghile, Y.; Laverty, M.; Li, K. Decision scaling: Linking bottom-up vulnerability analysis with climate projections in the water sector. *Water Resour Res.* **2012**, *48*. [CrossRef]
18. Hveding, V. Digital simulation techniques in power system planning. *Econ. Plan.* **1968**, *8*, 118–139. [CrossRef]
19. François, B.; Hingray, B.; Hendrickx, F.; Creutin, J.D. Seasonal patterns of water storage as signatures of the climatological equilibrium between resource and demand. *Hydrol. Earth Syst. Sci.* **2014**, *18*, 3787–3800. [CrossRef]
20. Wolfgang, O.; Haugstad, A.; Mo, B.; Gjelsvik, A.; Wangensteen, I.; Doorman, G. Hydro reservoir handling in Norway before and after deregulation. *Energy* **2009**, *34*, 1642–1651. [CrossRef]
21. Hingray, B.; Mouhous, N.; Mezghani, A.; Bogner, K.; Schaefli, B.; Musy, A. Accounting for global warming and scaling uncertainties in climate change impact studies: Application to a regulated lakes system. *Hydrol. Earth Syst. Sci.* **2007**, *11*, 1207–1226. [CrossRef]
22. Mo, B.; Doorman, G.; Bjørn, G. *Climate Change—Consequences for the Electricity System: Analysis of Nord Pool System*; Tech Report CE-Project; Hydrological Service—National Energy Authority: Reykjavik, Iceland, 2006; p. 157.
23. Bergström, S.; Jóhannesson, T.; Aðalgeirsdóttir, G.; Andreassen, L.; Beldring, S.; Hock, R.; Jónsdóttir, J.; Rogozova, S.; Veijalainen, N. Hydropower. In *Impacts of Climate Change on Renewable Energy Sources—Their Role in the Nordic Energy System*; Fenger, J., Ed.; Report Nord; Nordic Council of Ministers: Copenhagen, Denmark, 2007.
24. Schaefli, B.; Hingray, B.; Niggli, M.; Musy, A. A conceptual glacio-hydrological model for high mountainous catchments. *Hydrol. Earth Syst. Sci.* **2005**, *9*, 95–109. [CrossRef]
25. Haylock, M.R.; Hofstra, N.; Tank, A.M.G.K.; Klok, E.J.; Jones, P.D.; New, M. A European daily high-resolution gridded data set of surface temperature and precipitation for 1950–2006. *J. Geophys. Res. Atmos.* **2008**, *113*. [CrossRef]
26. Furevik, B.R.; Haakenstad, H. Near-surface marine wind profiles from rawinsonde and NORA10 hindcast. *J. Geophys. Res.* **2012**, *117*. [CrossRef]
27. François, B.; Hingray, B.; Raynaud, D.; Borga, M.; Creutin, J.D. Increasing climate-related-energy penetration by integrating run-of-the river hydropower to wind/solar mix. *Renew. Energy* **2016**, *87*, 686–696. [CrossRef]
28. Johnson, G.L. Wind characteristics. In *Wind Energy Systems*; Prentice-Hall: Englewood Cliffs, NJ, USA, 1965; pp. 32–99.
29. Taylor, K.E.; Stouffer, R.J.; Meehl, G.A. An Overview of CMIP5 and the Experiment Design. *Bull. Am. Meteorol. Soc.* **2012**, *93*, 485–498. [CrossRef]
30. Hawkins, E.; Sutton, R. The potential to narrow uncertainty in projections of regional precipitation change. *Clim. Dyn.* **2011**, *37*, 407–418. [CrossRef]
31. Lafaysse, M.; Hingray, B.; Gailhard, J.; Mezghani, A.; Terray, L. Internal variability and model uncertainty components in a multireplicate multimodel ensemble of hydrometeorological projections. *Water Resour. Res.* **2014**, *50*, 3317–3341. [CrossRef]
32. Hingray, B.; Saïd, M. Partitioning internal variability and model uncertainty components in a multimodel multireplicate ensemble of climate projections. *J. Clim.* **2014**, *27*, 6779–6798. [CrossRef]
33. Engeland, K.; Borga, M.; Creutin, J.D.; François, B.; Ramos, M.H.; Vidal, J.P. Space-time variability of climate and hydro-meteorology and intermittent renewable energy production—A review. *Renew. Sustain. Energy Rev.* **2017**, submitted for publication.

34. François, B. Influence of winter North-Atlantic Oscillation on Climate-Related-Energy penetration in Europe. *Renew. Energy*, **2016**, *99*, 602–613. [CrossRef]
35. Steinschneider, S.; Brown, C. A semiparametric multivariate, multisite weather generator with low-frequency variability for use in climate risk assessments: Weather Generator for Climate Risk. *Water Resour. Res.* **2013**, *49*, 7205–7220. [CrossRef]
36. François, B.; Hingray, B.; Creutin, J.D.; Hendrickx, F. Estimating Water System Performance under Climate Change: Influence of the Management Strategy Modeling. *Water Resour. Manag.* **2015**, *29*, 4903–4918. [CrossRef]
37. McInnes, K.L.; Erwin, T.A.; Bathols, J.M. Global Climate Model projected changes in 10 m wind speed and direction due to anthropogenic climate change. *Atmos. Sci. Lett.* **2011**, *12*, 325–333. [CrossRef]

energies

MDPI

Article

Combined Heat and Power Dispatch Considering Heat Storage of Both Buildings and Pipelines in District Heating System for Wind Power Integration

Ping Li, Haixia Wang, Quan Lv and Weidong Li *

Department of Electrical Engineering, Dalian University of Technology, Dalian 116024, China;
liping2014@mail.dlut.edu.cn (P.L.); whx@dlut.edu.cn (H.W.); lvquan@dlut.edu.cn (Q.L.)
* Correspondence: wdli@dlut.edu.cn; Tel.: +86-139-4269-8900

Academic Editor: Bahman Shabani
Received: 11 May 2017; Accepted: 28 June 2017; Published: 30 June 2017

Abstract: The strong coupling between electric power and heat supply highly restricts the electric power generation range of combined heat and power (CHP) units during heating seasons. This makes the system operational flexibility very low, which leads to heavy wind power curtailment, especially in the region with a high percentage of CHP units and abundant wind power energy such as northeastern China. The heat storage capacity of pipelines and buildings of the district heating system (DHS), which already exist in the urban infrastructures, can be exploited to realize the power and heat decoupling without any additional investment. We formulate a combined heat and power dispatch model considering both the pipelines' dynamic thermal performance (PDTP) and the buildings' thermal inertia (BTI), abbreviated as the CPB-CHPD model, emphasizing the coordinating operation between the electric power and district heating systems to break the strong coupling without impacting end users' heat supply quality. Simulation results demonstrate that the proposed CPB-CHPD model has much better synergic benefits than the model considering only PDTP or BTI on wind power integration and total operation cost savings.

Keywords: combined heat and power dispatch; pipelines' dynamic thermal performance (PDTP); buildings' thermal inertia (BTI); power and heat decoupling; wind power integration

1. Introduction

The installed capacity of wind turbines has been increasing recently in China [1], involving much uncertainty for the electric power system (EPS), which puts forward higher requirements for the system operational flexibility. The electric power generation range of the combined heat and power (CHP) units is severely restricted by the heat loads due to the power and heat coupling during heating seasons. The CHP units have to remain on certain constrained electric power output to meet the heat loads' demand, leaving little room for wind power integration during the wind power on-peak hours with low electric loads, but high heat loads. This makes the system operational flexibility become low, which results in heavy wind power curtailment, especially in the cold region with a high percentage of CHP units and abundant wind power energy, such as northeastern China. [1]. Therefore, breaking the power and heat coupling of CHP units to improve the system operational flexibility is crucial to reduce wind power curtailment.

It is an effective way to realize the power and heat decoupling by the optimal operation of multi-energy systems, which coordinate at least two different energy systems, such as electric power, heating, cooling or gas system, etc., with many advantages of lower operation cost, higher renewable energy integration and more reliable energy supply. Installing heat storage facilities [2–6] and introducing electricity-heat conversion devices [7–11] in CHP plants can coordinate the electric power and heat energy to utilize the flexibility of the heating system. Moradi et al. [12,13] introduced a novel approach for optimal management of multi-energy including heating, cooling and power in residential buildings to achieve high energy efficiency, low greenhouse gas emission and low generation cost. Ye et al. [14] proposed an integrated natural gas, heat and power dispatch model considering wind power and a power-to-gas unit to reduce wind power curtailment, fuel cost and CO_2 emissions.

The above approaches are greatly restricted by their high capital investment for introducing some other facilities, such as heat storage tanks, electric boilers, heat pumps, chillers, etc. Additionally, they abstract the pipeline network and heating buildings of the district heating system (DHS) into a single static heat load node model without considering their internal thermal characteristics. The urban DHS infrastructures already exist, including many insulated pipelines with a large capacity of internal heat water and a huge area of heating buildings with significant insulated envelope structures, which have plenty of heat storage capacity [15–17]. The DHS heat storage can be utilized to break the power and heat coupling without any additional investment in the scope of the combined heat and power system. Recently, several studies have focused on exploiting the DHS internal thermal characteristics including the pipelines' dynamic thermal performance (PDTP) and the buildings' thermal inertia (BTI) to improve the system operational flexibility.

Considering only PDTP, the thermal performance mainly refers to two factors including the water heat loss of pipelines and the water temperature time delays from heat sources to heat loads. The pipeline model was built considering heat loss [18–20] in the optimal supply and distribution of electric power and heat energy. Zhao et al. [21] studied the optimal operation of a CHP-type district heating system considering time delays in the distribution network. To further account for the effect of the pipelines' heat storage, the two factors were both considered [16], which established the pipeline model based on the node method [22] to describe the temperature dynamic profiles along the pipelines. Fu et al. [23] described the thermal performance by AMRA time series considering the DHS as a black box.

For considering only BTI, the potential of residential buildings as thermal energy storage in the DHS was studied through pilot tests [24–26]. Satyavada et al. [27,28] proposed an integrated control-oriented approach to describe the thermal characteristics of the heating, ventilating and air conditioning equipment in the buildings with modular models effectively. Yang et al. [29] utilized thermal energy storage and distributed electric heat pumps considering BTI to improve wind power integration. Wu et al. [30] proposed a novel day-ahead scheduling method and strategy by use of the indoor temperature adjustable region and BTI to reduce wind power curtailment. Pan et al. [31] proposed a modified feasible region method to give a new formulation of the DHS models similar to conventional power plants. Jin et al. [17] developed a building-based virtual energy storage system model to participate in the economic dispatch of the hybrid energy microgrid.

Coordinating the operation of both PDTP and BTI should be better than considering only one of them, since the district heating pipelines and buildings are connected together, which constitute the heat transmission, distribution and consumption sections of the DHS. The approaches involving both pipeline and building models are rarely studied. Li et al. [32] set up a simulation model of a single back pressure CHP plant-based district heating system with Ebsilon software, which analyzed the system performance by the simulation method. The economic operation of a district electricity and heating system was studied [33], which focused on the two systems' disturbance interaction effect on the system security. Further, they did not consider the coordinating effect of both PDTP and BTI in the optimal operation of EPS and DHS for wind power integration.

To bridge these gaps, this paper proposes a combined heat and power dispatch model considering both PDTP and BTI simultaneously (CPB-CHPD model) to reduce wind power curtailment and total operation cost, which meets the electric load and heat load demands, as well as satisfies the EPS and DHS constraints. This approach exploits the coordinating effect of both PDTP and BTI to break the strong linkage of power and heat supply of CHP units more effectively. The benefits of only PDTP, only BTI and both of them are separately evaluated in terms of improving wind power integration and reducing total operation cost to demonstrate the synergic benefits of both PDTP and BTI.

The main contributions of this paper are summarized as follows:

- A novel CPB-CHPD model is proposed with special emphasis on the coordinating operation of both PDTP and BTI aiming at breaking the power and heat coupling to significantly improve the system operational flexibility without any additional investment.
- A physical model of the DHS is proposed. The pipeline model is built considering heat loss, temperature time delays and network topology characteristics in terms of single and network level. The building model is formulated based on buildings' thermal equilibrium considering building characteristics' diversity and outdoor temperature variation.
- The synergic benefits of both PDTP and BTI on reducing wind power curtailment and total operation cost are evaluated, which are better than considering only one or neither of them.

This paper is organized as follows. In Section 2, the DHS is modeled regarding both PDTP and BTI. Then, the CPB-CHPD is formulated in Section 3. In Section 4, simulation cases are carried out to compare the four dispatch models (including considering both PDTP and BTI, or only one of them, or neither) to demonstrate the synergic effects of the CPB-CHPD model. Finally, the conclusions are given in Section 5.

2. System Model of the DHS

The typical DHS is composed of a heat source mainly referring to the high efficiency coal-fired CHP unit, a district heating pipeline network and many heat loads, which are usually space heating for the residential buildings especially in cold northeastern China.

2.1. Heat Sources

2.1.1. Electric and Heat Power Characteristics

The electric and heat power characteristics of both extraction condensing and back pressure turbine CHP units are shown in Figure 1. The operation points of the two kinds of CHP units are kept respectively inside the polygon region ABCD and on the line segment BC [34].

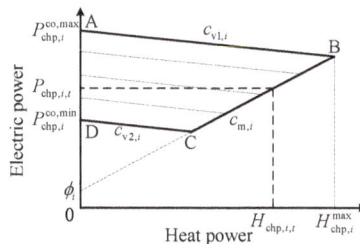

Figure 1. Electric and heat power characteristics of the CHP units.

The electric and heat power limits of the extraction condensing turbine CHP units are described in Equations (1) and (2).

$$\begin{cases} P_{chp,i,t} \geq \max \left\{ P_{chp,i}^{co,min} - c_{v2,i} H_{chp,i,t}, \ \phi_i + c_{m,i} H_{chp,i,t} \right\} \\ P_{chp,i,t} \leq P_{chp,i}^{co,max} - c_{v1,i} H_{chp,i,t} \end{cases} , \ \forall t \in N \tag{1}$$

$$0 \leq H_{chp,i,t} \leq H_{chp,i}^{max}, \ \forall t \in N \tag{2}$$

where the subscript i, the subscript t and the subscript chp denote the i-th CHP unit, the t-th dispatch period and the relevant variables of the CHP unit, respectively, $P_{chp,i,t}$ and $H_{chp,i,t}$ are the electric and heat power output (MW), $P_{chp,i}^{co,max}$ and $P_{chp,i}^{co,min}$ are the maximum and minimum electric power output in condensing operation condition (MW), $H_{chp,i}^{max}$ is the maximum heat power output (MW), $c_{v1,i}$ and $c_{v2,i}$ are the curve slope of electric power to heat power in the extraction operation condition, $c_{m,i}$ refers to back pressure operation condition, ϕ_i is the electric power value at the intersection between the extension of back pressure curve and the electric power axis (MW) and N is the index set of dispatch periods.

The back pressure turbine CHP units can be regarded as special operation conditions of the extraction condensing units when $c_{v1}=c_{v2}=0$ but $c_m \neq 0$. That is to say, the electric and heat power limits of the back pressure units can be described in Equation (3).

$$\begin{cases} P_{chp,i,t} = \phi_i + c_{m,i}H_{chp,i,t} \\ P_{chp,i}^{co,min} \leq P_{chp,i,t} \leq P_{chp,i}^{co,max} \end{cases}, \forall t \in N \tag{3}$$

2.1.2. Operation Cost

The operation cost of the CHP unit is expressed as a quadratic function of their electric and heat power output [29]:

$$\begin{cases} C_{i,t}^{chp} = \varphi \cdot f_{i,t}\left(P_{chp,i,t}, H_{chp,i,t}\right), \forall i \in S^{chp}, t \in N \\ f_{i,t}\left(P_{chp,i,t}, H_{chp,i,t}\right) = a_{chp,i}\left(P_{chp,i,t} + c_{v1,i}H_{chp,i,t}\right)^2 + b_{chp,i}\left(P_{chp,i,t} + c_{v1,i}H_{chp,i,t}\right) + c_{chp,i} \end{cases} \tag{4}$$

where $C_{i,t}^{chp}$ is the operation cost function, $f_{i,t}\left(P_{chp,i,t}, H_{chp,i,t}\right)$ is the coal consumption function, $a_{chp,i}$, $b_{chp,i}$ and $c_{chp,i}$ are the coal consumption coefficients (t/ $(MW^2 \cdot h)$, t/ $(MW \cdot h)$, t/h), φ is the price of the standard coal, 72.40 \$/t in this paper, and S^{chp} is the index set of CHP units.

2.2. District Heating Pipelines Network

Due to the lack of control devices at the end users in China, most of the DHSs are operated with constant flow and variable temperatures [31]. It is assumed that this operation mode is also utilized in this paper, where the mass flow rate is always constant, and the hydraulic conditions always keep stable. Then, we can focus on studying the thermodynamic model of the pipeline network with eliminating the nonlinear hydraulic model to simplify the solving [31]. This impact on the combined heat and power dispatch results is within acceptable limits [35]. In this paper, the PDTP is modeled at two levels, which are the single pipeline level and the pipeline network level.

2.2.1. Single Pipeline Level

The dynamic characteristics on a single pipeline level mainly are reflected in thermal conduction along the pipeline. Figure 2 shows the general structure of a single pipeline [36].

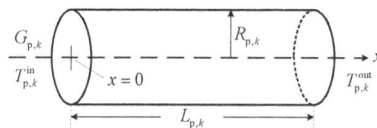

Figure 2. General structure of a single pipeline.

The thermal conduction in each pipeline, including heat loss and temperature time delays, can be modeled by a partial differential equation [16,36] as follows:

$$\frac{\partial T_{p,k,t}^x}{\partial t} + \frac{G_{p,k}}{\pi \rho_w R_{p,k}^2} \cdot \frac{\partial T_{p,k,t}^x}{\partial x} + \frac{2\mu_{p,k}}{c_w \rho_w R_{p,k}}\left(T_{p,k,t}^x - T_{p,k}^{soil}\right) = 0, 0 \leq x \leq L_{p,k}, \forall k \in S^{pipe}, t \in N \tag{5}$$

where the subscript k and the subscript p denote the k-th pipeline and the relevant variables of the pipeline, $T^x_{p,k,t}$ is the water temperature at a length of x from the inlet inside the pipeline (°C), $T^{soil}_{p,k}$ is the soil temperature outside the pipeline (°C), $G_{p,k}$ is the mass flow rate (kg/s), c_w is the specific heat capacity of the hot water (4.2×10^{-3} MJ/(kg $\cdot°$ C)), ρ_w is the density of the hot water (1.0×10^3 kg/m^3), $R_{p,k}$ and $L_{p,k}$ are the radius and length of the pipeline (m), $\mu_{p,k}$ is the thermal loss coefficient (W/(m$^2 \cdot°$ C)) and S^{pipe} is the index set of pipelines.

The solution of Equation (5) can be obtained [36,37] as follows:

$$T^{out}_{p,k,t+\Delta\tau_{p,k}} = T^{soil}_{p,k} + \left(T^{in}_{p,k,t} - T^{soil}_{p,k}\right) \exp\left(-\frac{2\mu_{p,k}}{c_w\rho_w R_{p,k}}\Delta\tau_{p,k}\right) \tag{6}$$

where $T^{in}_{p,k,t}$ and $T^{out}_{p,k,t}$ are the water temperature at the inlet and outlet of the pipeline (°C), and the delay time $\Delta\tau_{p,k}$ represents water flowing time from the inlet to the outlet of the pipeline, which can be comparable with one or several dispatch periods of the EPS. Since the mass flow rate is constant in this paper, $\Delta\tau_{p,k}$ is defined by:

$$\Delta\tau_{p,k} = \frac{\pi\rho_w L_{p,k} R^2_{p,k}}{G_{p,k}} \tag{7}$$

The subscript $\Delta\tau_{p,k}$ in Equation (6) is required to be an integer. In order to utilize Equations (6) and (7) in the discrete dispatch model, we rewrite Equation (6) as Equation (8).

$$T^{out}_{p,k,t+\lambda_{p,k}} = T^{soil}_{p,k} + \left(T^{in}_{p,k,t} - T^{soil}_{p,k}\right) \exp\left(-\frac{2\mu_{p,k}\lambda_{p,k}}{c_w\rho_w R_{p,k}}\Delta t\right) \tag{8}$$

where $\lambda_{p,k}$ is the multiples of the continuous delay time $\Delta\tau_{p,k}$ to the duration of the discrete dispatch period Δt, which is:

$$\lambda_{p,k} = \text{round}\left(\frac{\Delta\tau_{p,k}}{\Delta t}\right) \tag{9}$$

Though this approach will lose some accuracy, it can still describe the PDTP adequately with the advantage of reducing the solving complexity of the combined dispatch model. Additionally, the shorter Δt is, the more accurate the approach is. It requires some initial temperatures of a pipeline (i.e., temperatures before the first dispatch period, which can be known from measurement or prediction) to accomplish Equation (8) when $\lambda_{p,k} \geq 2$.

2.2.2. Pipeline Network Level

In the district heating pipeline network, heat energy is carried by the circulating hot water, which is transported from the heat sources to the heat loads.

Figure 3 shows the general structure of a node connecting with cross pipelines in the pipelines network [16]. In this paper, we define that the water inflowing side is the inlet of a pipeline, and the water outflowing side is the outlet correspondingly.

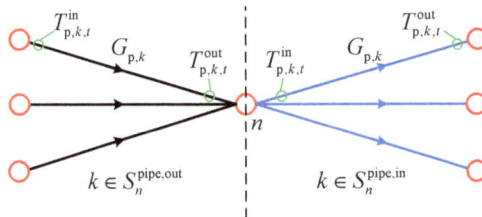

Figure 3. General structure of a node connecting with cross pipelines.

The pipeline network topology characteristics are described from four aspects as follows.

- Relationship between heat power and water temperatures:

The heat power of the hot water, flowing into the inlet and flowing out of the outlet, of pipeline k at period t is expressed respectively as follows:

$$\begin{cases} q_{p,k,t}^{in} = c_w G_{p,k} T_{p,k,t}^{in} \\ q_{p,k,t}^{out} = c_w G_{p,k} T_{p,k,t}^{out} \end{cases}, \forall k \in S^{pipe}, t \in N \tag{10}$$

where $q_{p,k,t}^{in}$ and $q_{p,k,t}^{out}$ are the heat power flowing into the inlet and flowing out of the outlet of the pipeline (MW).

- Supply and return water temperature limits:

The water temperatures in the water supply and return network should be kept within their limits:

$$T_{ps}^{min} \leq T_{p,k,t}^{in}, T_{p,k,t}^{out} \leq T_{ps}^{max}, \forall k \in S_{pipe}^{SN}, t \in N \tag{11}$$

$$T_{pr}^{min} \leq T_{p,k,t}^{in}, T_{p,k,t}^{out} \leq T_{pr}^{max}, \forall k \in S_{pipe}^{RN}, t \in N \tag{12}$$

where T_{ps}^{max} and T_{ps}^{min} are the upper and lower limits of water temperatures in the water supply network pipelines (°C), T_{pr}^{max} and T_{pr}^{min} are the upper and lower limits of water temperatures in the water return network pipelines (°C) and S_{pipe}^{SN} and S_{pipe}^{RN} are the index sets of pipelines in the water supply and return network.

- Mass flow rates' continuity and limits:

Similar to Kirchhoff's current law, for each node in the pipeline network, the total mass flow rates of all pipelines connecting to this node is zero:

$$\sum_{k \in S_n^{pipe,in}} G_{p,k} = \sum_{k \in S_n^{pipe,out}} G_{p,k} \tag{13}$$

where $S_n^{pipe,in}$ and $S_n^{pipe,out}$ are the index sets of pipelines whose inlet and outlet connect to pipeline network node n.

The mass flow rates at each period should not exceed their upper or lower limits:

$$G_{p,k}^{min} \leq G_{p,k} \leq G_{p,k}^{max}, \forall k \in S^{pipe} \tag{14}$$

where $G_{p,k}^{max}$ and $G_{p,k}^{min}$ are the upper and lower limits of the mass flow rate (kg/s).

- Node temperature characteristics:

According to the energy conservation law, the water temperatures of all pipelines flowing into the same node are mixed at this node, and the water temperatures of all pipelines flowing out of this node are equal to the mixed temperature at this node, as described in Equation (15).

$$\begin{cases} \sum_{k \in S_n^{pipe,out}} \left(c_w G_{p,k} T_{p,k,t}^{out} \right) = c_w T_{mix,n,t} \cdot \sum_{k \in S_n^{pipe,out}} G_{p,k} \\ T_{p,k,t}^{in} = T_{mix,n,t}, \forall k \in S_n^{pipe,in} \end{cases} \tag{15}$$

where $T_{mix,n,t}$ is the mixed temperature at node n in the water supply and return network (°C).

Equations (7)–(15) can adequately describe the PDTP including water heat loss, water temperature time delays and network topology characteristics in the combined heat and power dispatch model.

2.3. Buildings

Since there are many rooms with different structures from each other in a multi-story building and many different buildings in a heating region, it requires a huge calculation to model each room separately, which is almost impossible. In this paper, a lumped model is utilized to abstract a multi-story building or some adjacent buildings with similar characteristics as a large room for simplicity without affecting the thermal inertia performance. The lumped model can describe the BTI adequately for the combined heat and power dispatch model.

2.3.1. Relationship between Indoor Temperatures and Heat Power Supplied

The heat storage of a building is the difference of heat energy supplied and heat energy loss. Considering the winter heating scenario, the thermal equilibrium equation [17] of building j is shown as follows:

$$\Delta Q_{\mathrm{st},j,t} = \left(H_{\mathrm{hr},j,t} + H_{\mathrm{td},j} \right) - \left(H_{\mathrm{en},j,t} + H_{\mathrm{ca},j,t} \right), \ \forall j \in S^{\mathrm{bui}}, \ t \in N \tag{16}$$

- $\Delta Q_{\mathrm{st},j,t}$ denotes the change rate of the heat energy of the building, as expressed in Equation (17). When the indoor temperature increases, i.e., $\mathrm{d}T_{\mathrm{id},j,t}/\mathrm{d}t > 0$, the heat energy of the building increases, which means the building heat storage is charged. Oppositely, when the indoor temperature decreases, $\mathrm{d}T_{\mathrm{id},j,t}/\mathrm{d}t < 0$, the building heat storage is discharged.
- On the right side of Equation (16), the two items in the first parenthesis denote the building total heat energy supplied, where $H_{\mathrm{hr},j,t}$ and $H_{\mathrm{td},j}$ are the heat power supplied by district heating pipelines and by internal heat gains (such as the effect of indoor lighting, persons, appliances, etc.), respectively. Here, the heat power supplied by internal heat gains is assumed as 3.8 W/m^2.
- On the right side of Equation (16), the two items in the second parenthesis denote the building total heat energy loss, where $H_{\mathrm{en},j,t}$ is the sum of the heat power transfer through each side of the building envelope structures including doors, windows, walls, floors, roofs, etc., as expressed in Equation (18). Meanwhile, the solar radiation is appended to the heat power transfer by orientation correction, and the outdoor cold wind speed effect is also appended by its additional correction. $H_{\mathrm{ca},j,t}$ is the building heat power loss by cold air infiltration through the windows and doors gaps, as well as cold air intrusion from the opening windows and doors, as expressed in Equation (19); S^{bui} is the index set of buildings.

$$\Delta Q_{\mathrm{st},j,t} = I_{\mathrm{b},j} \frac{\mathrm{d}T_{\mathrm{id},j,t}}{\mathrm{d}t} \tag{17}$$

$$H_{\mathrm{en},j,t} = \left(1 + x_{\mathrm{h},j}\right) \sum_{\gamma \in S_j^{\mathrm{en}}} K_{\gamma,j} F_{\gamma,j} \delta_{\gamma,j} \left(T_{\mathrm{id},j,t} - T_{\mathrm{od},j,t} \right) \left(1 + x_{\mathrm{o},j} + x_{\mathrm{w},j}\right) \tag{18}$$

$$H_{\mathrm{ca},j,t} = 2.78 \times 10^{-4} \cdot c_{\mathrm{a}} \rho_{\mathrm{a}} V_{\mathrm{ca},j} \left(T_{\mathrm{id},j,t} - T_{\mathrm{od},j,t} \right) \tag{19}$$

where the subscript j and the subscript γ denote the j-th building and the γ-th side of the building envelope structures, $T_{\mathrm{id},j,t}$ and $T_{\mathrm{od},j,t}$ are the indoor and outdoor temperature ($^\circ$C), $I_{\mathrm{b},j}$ is the total heat capacity of the building (MJ/$^\circ$C), $K_{\gamma,j}$ is the heat transfer coefficient of the envelope structure (MW/(m$^2 \cdot^\circ$C)), $F_{\gamma,j}$ is the surface area of the envelope structure (m^2), $\delta_{\gamma,j}$ is the temperature difference correction coefficient of the internal envelope structure, $x_{\mathrm{h},j}$, $x_{\mathrm{o},j}$ and $x_{\mathrm{w},j}$ are the additional coefficient for height, orientation, and wind speed effect, c_{a} is the specific heat capacity of the outdoor cold air (1.0×10^{-3} MJ/(kg$\cdot^\circ$C)), ρ_{a} is the density of the outdoor cold air (1.29 kg/m^3), $V_{\mathrm{ca},j}$ is the total volume of the outdoor cold air flowing into the building per hour (m^3/h), 2.78×10^{-4} is for unit conversion (1 s $= 2.78 \times 10^{-4}$ h) and S_j^{en} is the index set of envelope structure sides of building j.

A concise equation can be obtained [31,38] from Equations (16)–(19) as follows:

$$\chi_{\mathrm{bt},j} t_{\mathrm{bs},j} \frac{\mathrm{d}T_{\mathrm{id},j,t}}{\mathrm{d}t} = \left(H_{\mathrm{hr},j,t} + H_{\mathrm{td},j} \right) - \chi_{\mathrm{bt},j} \left(T_{\mathrm{id},j,t} - T_{\mathrm{od},j,t} \right) \tag{20}$$

where $\chi_{bt,j}$ is the building total heat transfer coefficient between the indoor and outdoor air (MW/°C) and $t_{bs,j}$ is the building equivalent heat storage time coefficient (s), which indicate the building heat energy transfer and storage capacity respectively, as shown in Equations (21) and (22).

$$\chi_{bt,j} = \left(1 + x_{h,j}\right) \sum \delta_{\gamma,j} K_{\gamma,j} F_{\gamma,j} \left(1 + x_{o,j} + x_{w,j}\right) + 2.78 \times 10^{-4} \cdot c_a \rho_a V_{ca,j} \tag{21}$$

$$t_{bs,j} = \frac{I_{b,j}}{\chi_{bt,j}} \tag{22}$$

In the discrete combined dispatch model, we only focus on the changes of indoor temperatures at the beginning and end of each discrete dispatch period t, neglecting the changes of both the heat power supplied by pipelines to the building $H_{hr,j,t}$ and the outdoor temperature $T_{od,j,t}$ within the discrete dispatch period t. With a forward difference approximation on the time derivative, the differential Equation (20) can be converted to a finite difference equation [17], which can describe the coupling relationship of indoor temperatures, heat power supplied and discrete dispatch periods of the building in the discrete combined dispatch model, as expressed in Equation (23). The initial indoor temperatures before the first dispatch period can be known from measurement or prediction.

$$T_{id,j,t+1} = T_{od,j,t+1} + \frac{H_{hr,j,t} + H_{td,j}}{\chi_{bt,j}} + \left(T_{id,j,t} - T_{od,j,t+1} - \frac{H_{hr,j,t} + H_{td,j}}{\chi_{bt,j}}\right) \exp\left(-\frac{\Delta t}{t_{bs,j}}\right) \tag{23}$$

2.3.2. Indoor Temperatures Limits

In order to ensure the heat supply quality and the thermal comfort, indoor temperatures should be kept within their limits:

$$T_{id,j}^{min} \leq T_{id,j,t} \leq T_{id,j}^{max}, \ \forall j \in S^{bui}, \ t \in N \tag{24}$$

where $T_{id,j}^{max}$ and $T_{id,j}^{min}$ are the upper and lower limits of indoor temperature (°C).

Equations (21)–(24) can adequately describe the BTI considering building characteristics' diversity and outdoor temperature variation in the combined heat and power dispatch model.

2.4. Interfaces among Heat Sources, Network and Loads

2.4.1. Between Heat Sources and Pipelines Network

At the side of the heat sources, the return water is heated by the CHP unit heat exchanger and then pumped into the supply pipeline network. Heat energy is extracted from the heat sources and distributed to the pipeline network, as expressed in Equations (25) and (26).

$$H_{hs,i,t} = \eta_i \cdot H_{chp,i,t}, \ \forall i \in S^{chp}, \ t \in N \tag{25}$$

$$H_{hs,i,t} = q_{p,k1,t}^{in} - q_{p,k2,t}^{out}, \ \forall k1 \in S_n^{SN,pipe}, \ k2 \in S_n^{RN,pipe}, \ n = Node_i^{chp}, \ i \in S^{chp}, \ t \in N \tag{26}$$

where $H_{hs,i,t}$ is the heat power through the CHP unit heat exchanger to the pipeline network (MW), η_i is the efficiency of the CHP unit heat exchanger (0.97), $S_n^{SN,pipe}$ and $S_n^{RN,pipe}$ are the index sets of pipelines in the water supply and return network connecting to pipeline network node n and $Node_i^{chp}$ is the index of pipeline network node connecting to CHP unit i.

2.4.2. Between Pipeline Network and Heat Loads

At the side of the heat loads, i.e., buildings, the supply water releases heat energy to indoor air via heat radiators to maintain the indoor temperatures and then flows into the return pipelines.

Heat energy is extracted from the pipeline network and distributed to the heat loads, as expressed in Equation (27).

$$H_{hr,j,t} = q_{p,k1,t}^{out} - q_{p,k2,t}^{in}, \; \forall k1 \in S_n^{SN,pipe}, \; k2 \in S_n^{RN,pipe}, \; n = Node_j^{bui}, \; j \in S^{bui}, \; t \in N \quad (27)$$

where $Node_j^{bui}$ is the index of the pipeline network node connecting to building j.

3. Optimization Model of the CPB-CHPD

The CPB-CHPD model including wind farms is formulated in this section. The proposed CPB-CHPD model seeks the optimal dispatch by coordinating the electric power of every power generation unit and the heat power of every heat source aiming at the minimum total operation cost, which includes the penalty cost of wind power spillage, while meeting the electric loads and heat loads demands, as well as satisfying the EPS and DHS constraints.

3.1. Decision Variables

The decision variables in the CPB-CHPD model are composed of two parts, which are the electricity and heat decision variables. The electricity decision variables include the electric power output of CHP units ($P_{chp,i,t}$), condensing power (CON) units ($P_{con,i,t}$) and wind farms ($P_{wind,i,t}$). The heat decision variables include the heat power output of CHP units ($H_{chp,i,t}$), water temperatures at the inlet and outlet of pipelines ($T_{p,k,t}^{in}$ and $T_{p,k,t}^{out}$), mass flow rates of pipelines ($G_{p,k}$), heat power supplied by pipelines to buildings ($H_{hr,j,t}$) and indoor temperatures of buildings ($T_{id,j,t}$).

3.2. Objective Function

The objective function is the total operation cost consisting of the operation cost of thermal power units and the penalty cost of wind power spillage, as expressed in Equation (28).

$$\min \sum_{t \in N} \left(\sum_{i \in S^{chp}} C_{i,t}^{chp} + \sum_{i \in S^{con}} C_{i,t}^{con} + \sum_{i \in S^{wind}} C_{i,t}^{wind} \right) \quad (28)$$

- The operation cost of the CHP unit $C_{i,t}^{chp}$ is defined in Equation (4).
- The operation cost of the CON unit is expressed as a quadratic function of its electric power output [16]:

$$\begin{cases} C_{i,t}^{con} = \varphi \cdot f_{i,t} \left(P_{con,i,t} \right), \; \forall i \in S^{con}, \; t \in N \\ f_{i,t} \left(P_{con,i,t} \right) = a_{con,i} P_{con,i,t}^2 + b_{con,i} P_{con,i,t} + c_{con,i} \end{cases} \quad (29)$$

where the subscript i and the subscript con denote the i-th CON unit and the relevant variables of the CON unit, $C_{i,t}^{con}$ is the operation cost function, $f_{i,t}(P_{con,i,t})$ is the coal consumption function, $P_{con,i,t}$ is the electric power output (MW), $a_{con,i}$, $b_{con,i}$ and $c_{con,i}$ are the coal consumption coefficients (t/ $(MW^2 \cdot h)$, t/ $(MW \cdot h)$ and t/h) and S^{con} is the index set of CON units.

- The penalty cost of the wind farm is proportional to the wind power spillage:

$$C_{i,t}^{wind} = \sigma_i \cdot \left(P_{wind,i,t}^{max} - P_{wind,i,t} \right), \; \forall i \in S^{wind}, \; t \in N \quad (30)$$

where the subscript i and the subscript wind denote the i-th wind farm and the relevant variables of the wind farm, $C_{i,t}^{wind}$ is the penalty cost function, $P_{wind,i,t}$ is the wind power output (MW), $P_{wind,i,t}^{max}$ is the maximum available wind power (MW), σ_i is the penalty coefficient (79.64 \$/MWh) and S^{wind} is the index set of wind farms.

3.3. Constraints

The proposed CPB-CHPD model is subject to the EPS constraints and the DHS constraints.

3.3.1. EPS Constraints

The EPS constraints consist of the electric power balance constraints and the units operation constraints, etc.

1. Electric power balance constraints:

The system total electric power output and total electric loads are equal at each dispatch period:

$$\sum_{i \in S^{chp}} P_{chp,i,t} + \sum_{i \in S^{con}} P_{con,i,t} + \sum_{i \in S^{wind}} P_{wind,i,t} = \sum_{i \in S^{load}} P_{load,i,t} \tag{31}$$

where $P_{load,i,t}$ is the electric load demand (MW) and S^{load} is the index set of electric loads.

2. Units' operation constraints:

*   Generation range constraints:

    The electric and heat power limits constraints of extraction condensing and back pressure turbine CHP units are defined in Equations (1)–(3).

    The electric power output of the CON units must be kept within their limits:

$$P_{con,i}^{min} \leq P_{con,i,t} \leq P_{con,i}^{max}, \ \forall i \in S^{con}, \ t \in N \tag{32}$$

    where $P_{con,i}^{max}$ and $P_{con,i}^{min}$ are the maximum and minimum electric power (MW).

    The electric power output of the wind farms are limited by the maximum wind power:

$$0 \leq P_{wind,i,t} \leq P_{wind,i,t}^{max}, \ \forall i \in S^{wind}, \ t \in N \tag{33}$$

*   Ramping constraints:

    Within each dispatch period, the electric power output of thermal power units is limited by the ramping capability. Equations (34) and (35) are for the CHP and CON units, respectively.

$$\begin{cases} \left(P_{chp,i,t+1} + c_{v1,i} H_{chp,i,t+1} \right) - \left(P_{chp,i,t} + c_{v1,i} H_{chp,i,t} \right) \leq UR_{chp,i} \cdot \Delta t \\ \left(P_{chp,i,t} + c_{v1,i} H_{chp,i,t} \right) - \left(P_{chp,i,t+1} + c_{v1,i} H_{chp,i,t+1} \right) \leq DR_{chp,i} \cdot \Delta t \end{cases}, \ \forall i \in S^{chp} \tag{34}$$

$$\begin{cases} P_{con,i,t+1} - P_{con,i,t} \leq UR_{con,i} \cdot \Delta t \\ P_{con,i,t} - P_{con,i,t+1} \leq DR_{con,i} \cdot \Delta t \end{cases}, \ \forall i \in S^{con} \tag{35}$$

where UR_i and DR_i are the upward and downward ramping capability (MW/h).

In order to meet the requirements of different cases, some other EPS constraints may be needed, such as the wind power ramping constraints, the spinning reserve constraints, the system operation security constraints, the unit commitment constraints, etc.

3.3.2. DHS Consraints

The DHS constraints consist of the PDTP constraints, the BTI constraints and the interfaces constraints among heat sources, network and loads.

1. PDTP constraints:

*   Single pipeline constraints: Equations (7)–(9).

- Pipelines network constraints: Relationship between heat power and water temperatures: Equation (10). Supply and return water temperatures limits: Equations (11) and (12). Mass flow rates continuity and limits: Equations (13) and (14). Node temperature characteristics: Equation (15).

2. BTI constraints:

   - Relationship between indoor temperatures and heat power supplied: Equations (21)–(23).
   - Indoor temperatures limits: Equation (24).

3.3.3. Interfaces Constraints among Heat Sources, Network and Loads

1. Between heat sources and pipelines network: Equations (25) and (26).
2. Between pipelines network and heat loads: Equation (27).

4. Simulation Cases and Results Analysis

4.1. Simulation System Description

A simulation for the combined heat and power system shown in Figure 4 is carried out to demonstrate the effect of the proposed model, where the EPS consists of two CHP units, two CON units and one wind farm, and the DHS is composed of the two CHP units, twenty pipelines and six buildings. The two CHP units are coupling points between the EPS and DHS. The DHS has sixteen nodes, where Buildings 1–3 are heat supplied via Pipelines 1–10 by CHP 1, and Buildings 4–6 are heat supplied via Pipelines 11–20 by CHP 2.

The parameters of thermal power units, pipelines and buildings are listed in Tables 1–3, respectively. The upper and lower limits of water temperatures at every node in the district heating pipelines network are 130 °C and 50 °C. The upper and lower limits of the mass flow rates are 3700 kg/s and 800 kg/s. The standard indoor temperature for space heating is set as 18 °C, and the thermal comfort indoor temperature ranges of all buildings are set between 18 °C and 22 °C. These typical thermal power units are commonly used in northeastern China. The detailed parameters of pipelines and buildings, as well as the DHS operation data are from the typical design data of the standard design specification of heating, ventilation and air conditioning for civil buildings, which is established by the China Academy of Building Research.

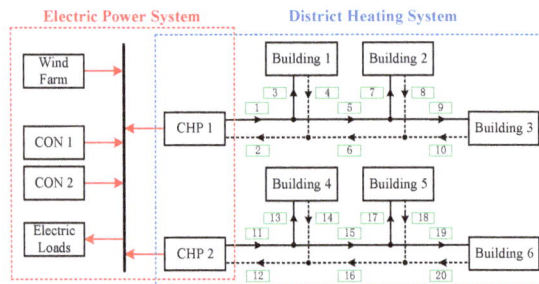

Figure 4. Configuration of the combined heat and power simulation system.

The typical day profiles of the total electric loads and forecast wind power, as well as the outdoor temperature are shown in Figure 5. The forecast wind power has almost the opposite peaks with the total electric loads, which is consistent with the characteristics of the EPS. With some modification, the outdoor temperature is from the historical data of the typical day during the medium heating season in northeastern China when the heat loads demand is high. These weather data were measured by the China Meteorological Administration during the past few years.

All simulation tests are considered for a 15-min operation scheduling over the course of 24 h.

Table 1. Parameters of thermal power units.

Type	CHP Units		CON Units	
Unit name	CHP1	CHP2	CON1	CON2
Capacity (MW)	300	200	500	200
$P_{chp,i}^{co,max}$ (MW)	323	212	/	/
$P_{chp,i}^{co,min}$ (MW)	150	100	200	80
$H_{chp,i}^{max}$ (MW)	357	241	/	/
$c_{v1,i}$	0.23	0.21	/	/
$c_{v2,i}$	0	0	/	/
$c_{m,i}$	0.45	0.44	/	/
Ramping rate (MW/h)	80	50	100	50

Table 2. Parameters of pipelines.

No.	$L_{p,k}$ (m)	$R_{p,k}$ (m)	$\mu_{p,k}$ (W/(m² ·° C))
1, 2, 11, 12	3250	0.8	32
3, 4, 5, 6, 13, 14, 15, 16	1500	0.6	32
7, 8, 9, 10, 17, 18, 19, 20	1050	0.5	32

Table 3. Parameters of buildings.

No.	$\chi_{bt,j}$ (MW/°C)	$t_{bs,j}$ (10^4 s)	Equivalent area (10^6 m²)
1	1.85	16.20	1.32
2	2.45	12.60	1.74
3	2.95	10.08	2.09
4	1.45	13.68	1.16
5	1.75	10.44	1.40
6	1.95	8.64	1.56

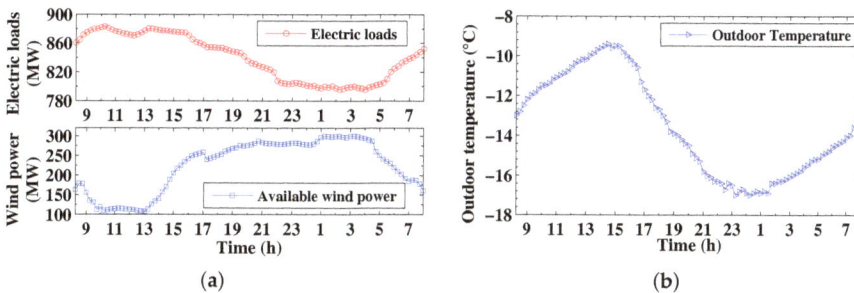

Figure 5. Profiles of the typical day during the heating season: (**a**) total electric loads and forecast wind power; (**b**) outdoor temperature.

4.2. Cases Settings

Four different dispatch models are given here including the CPB-CHPD, CP-CHPD, CB-CHPD and CED model, where the CPB-CHPD model refers to the model of combined heat and power dispatch considering both PDTP and BTI; the CP-CHPD model refers to the model of combined heat and power dispatch considering only PDTP; and the CB-CHPD refers to only BTI. The CED model refers to the conventional economic dispatch model, which just abstracts the whole of the pipelines and buildings as a simple static heat load node without considering their heat storage capacity. The objective functions of the other three models are the same as the CPB-CHPD model, but they are subject to different constraints listed as follows.

- The differences between the constraints of the CED and CPB-CHPD models are in two aspects. One is that Equations (7)–(15) and (26) should be replaced by Equation (36). The other is that Equations (21)–(24) and (27) should be replaced by Equation (37).

$$H_{hs,i,t} = \sum_{j \in S_{chp,i}^{bui}} H_{hr,j,t}, \ \forall i \in S^{chp}, \ t \in N \tag{36}$$

$$\begin{cases} H_{hr,j,t} = \chi_{b,j} \cdot \left(T_{id,j,t} - T_{od,j,t} \right) \\ T_{id,j,t} = T_{id,j}^{st} \end{cases} , \ \forall i \in S^{chp}, \ t \in N \tag{37}$$

where $T_{id,j}^{st}$ is the standard indoor temperature for space heating (°C) and $S_{chp,i}^{bui}$ is the index set of buildings connecting to CHP unit i via pipelines.
- The differences between the constraints of the CP-CHPD and CPB-CHPD models are in that Equations (21)–(24) and (27) should be replaced by Equation (37).
- The differences between the constraints of the CB-CHPD and CPB-CHPD models are in that Equations (7)–(15) and (26) should be replaced by Equation (36).

The simulation cases are set as follows: (1) Case 1 is utilized to describe the promotion effects on wind power integration and total operation cost savings of the CPB-CHPD model; (2) Case 2 is carried out to compare different results of the four dispatch models (CPB-CHPD, CP-CHPD, CB-CHPD and CED) based on Case 1 to demonstrate the synergic effects by coordinating PDTP and BTI.

4.3. Results Analysis

The electricity tariff is an important factor to the optimal results of the total operation cost. Since the electricity tariff is still regulated at present in China, we do not analyze its effect on the optimal results in this paper. The simulation results of the two cases are given below.

4.3.1. Case 1

This part is utilized to describe the promotion effects on wind power integration and total operation cost savings of the CPB-CHPD model. The optimization results are shown in Figures 6–9, respectively. Owing to the improved system operational flexibility by considering both PDTP and BTI, the total operation cost of the CPB-CHPD model is $521,741, reduced by nearly 11.86% based on the CED model whose operation cost is $591,929.

For the CED model, the heat power output of the CHP units must be always equal to the heat loads at each period, which are reflected in Figure 7. Comparing Figures 6 and 7 with Figure 8, during the wind power on-peak periods with low electric loads, but high heat load demand, the CHP units have to remain on certain constrained electric and heat power output to meet the high heat load demand because of the power and heat coupling and cannot be reduced any further; meanwhile other CON units have already been dispatched on their minimum technical generation, which results in that there is not enough space for wind power integration. Therefore, heavy wind power spillage occurs due to the inadequate downward spinning reserve.

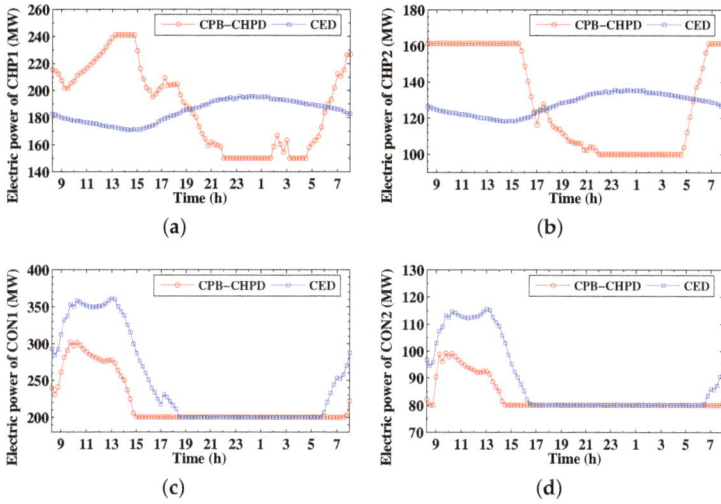

Figure 6. Electric power output of the thermal power units at each period in Case 1: (**a**) the first CHP unit CHP1; (**b**) the second CHP unit CHP2; (**c**) the first CON unit CON1; (**d**) the second CON unit CON2.

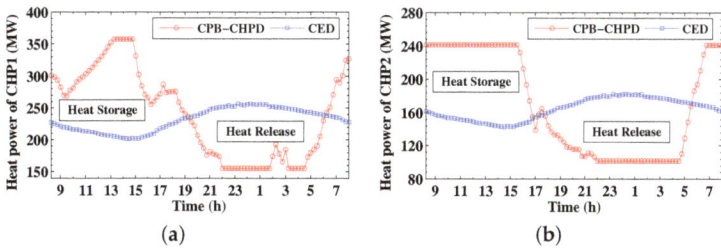

Figure 7. Heat power output of the CHP units at each period in Case 1: (**a**) the first CHP unit CHP1; (**b**) the second CHP unit CHP2.

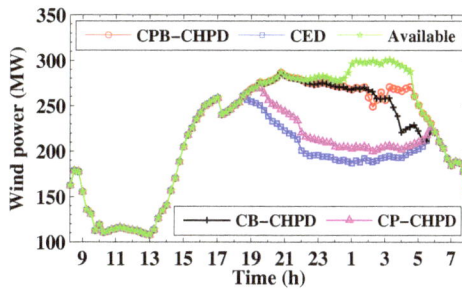

Figure 8. Electric power output of the wind farm of the four dispatch models at each period, including the CPB-CHPD, CB-CHPD, CP-CHPD models (combined heat and power dispatch models considering both PDTP and BTI, only BTI, only PDTP, respectively) and the CED model (conventional economic dispatch model considering neither PDTP, nor BTI).

For the CPB-CHPD model, the heat power output of the CHP units need not equal the heat loads at each period any more, which are required to satisfy the PDTP and BTI constraints instead. As shown in Figure 7, the heat power output of CHP1 and CHP2 is not restricted by the heat loads at each period. However, it is not indicated that they cannot meet the heating requirements of buildings. In contrast, all buildings' indoor temperatures at each period are kept within the thermal comfort range, as shown in Figure 9.

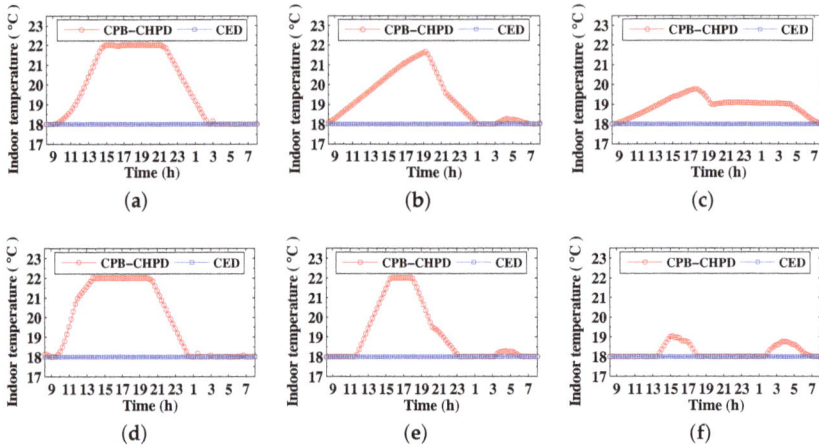

Figure 9. Profiles of indoor temperatures in Case 1 at each period: (**a**) Building1; (**b**) Building2; (**c**) Building3; (**d**) Building4; (**e**) Building5 ; (**f**) Building6.

The average indoor temperatures of all buildings are higher than the standard 18 °C, which indicates that there will be more heat energy stored in pipelines and buildings, resulting in more heat energy loss simultaneously. This requires the CHP units to produce more heat energy, which increases their operation cost. However, the total operation cost of the CPB-CHPD model decreases due to two reasons. One is that the operation cost of the CON units decreases greatly because their electric power output can be reduced significantly in the CPB-CHPD model, which can be seen in Figure 6c,d. The other is that a large amount of wind power energy can be integrated with saving much of the penalty cost of wind power spillage.

Pipelines and buildings both can be regarded as huge heat storage equipment. Their total heat storage/release capacity, as described in Figure 7, can be represented by the area that is enclosed by the red and blue curves when the red curve is higher/lower than the blue one.

The effect of heat storage in pipelines and buildings is illustrated in Figures 7 and 8 clearly. During the wind power off-peak periods with high electric loads, but low heat load demand, the CHP units can appropriately increase heat power output more than needed. The extra heat energy can be stored in pipelines and buildings, with indoor temperatures rising.

On the contrary, during the wind power on-peak periods with low electric loads, but high heat load demand, the CHP units can appropriately decrease electric and heat power output less than the constrained one of the CED model, which can provide an extra wind power integration space. As shown in Table 4, the CPB-CHPD model can utilize more wind power energy than the CED model by 689.71 MWh accounting for approximate 15.16%. Due to the lack of heat supply by the CHP units, indoor temperatures drop consequently, but not much, because they can be partly supplemented by heat release from pipelines and buildings. Since the operation status of the DHS changes very slowly, indoor temperatures will not change suddenly and dramatically, which can ensure the heat supply quality.

These simulation results demonstrate that, compared with the CED model, the CPB-CHPD model can break the power and heat coupling of the CHP units greatly, which can improve the system operational flexibility significantly.

4.3.2. Case 2

This part is carried out to compare different results of the four dispatch models (CPB-CHPD, CP-CHPD, CB-CHPD and CED) based on Case 1 to demonstrate the synergic effects by coordinating PDTP and BTI. The thermal comfort range of indoor temperatures are still 18–22 °C in the CB-CHPD model. Additionally, the indoor temperatures remain on the standard 18 °C in the CP-CHPD model, the same as the CED model.

Figure 8 shows the electric power output of the wind farm of the four dispatch models at each period. The optimization results of wind power integration and operation cost savings are given in Table 4. Based on these data, the histograms of the amount of abandoned wind power and operation cost savings are shown in Figure 10.

It is observed that, the capability of wind power integration increases in the order of the CED, CP-CHPD, CB-CHPD and CPB-CHPD models as shown in Figure 10a, and so do the operation cost savings, as shown in Figure 10b. In Table 4, based on the CED model, the other three models utilize more wind power energy by 159.54 MWh, 616.22 MWh and 689.71 MWh respectively, and they save more operation cost by $16,866, $61,455 and $70,172 respectively.

The amount of wind power energy integration and operation cost savings of the CB-CHPD model are more than those of the CP-CHPD model, which indicates that the CB-CHPD model makes the power and heat decoupling better than the CP-CHPD model. That is because the heat storage capacity of pipelines is much smaller than that of the buildings group. Further, the CPB-CHPD model can significantly exploit the synergic effects of PDTP and BTI to realize the power and heat decoupling more fully.

Table 4. Wind power integration and operation cost savings of the four dispatch models including the CPB-CHPD, CB-CHPD, CP-CHPD and CED models.

	Wind Power Integration (MWh)	Total Operation Costs ($)	Cost Savings Based on CED ($)	Saving Proportion Based on CED
CPB-CHPD	5239.68	521,741	70,172.55	11.86%
CB-CHPD	5166.19	530,458	61,455.19	10.38%
CP-CHPD	4709.51	575,062	16,866.53	2.85%
CED	4549.97	591,929	/	/

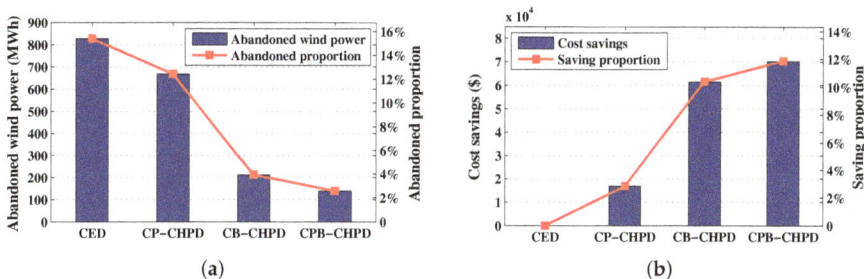

Figure 10. Optimal result comparison of the four dispatch models, including the CPB-CHPD, CB-CHPD, CP-CHPD and CED models: (**a**) abandoned wind power; (**b**) operation cost savings.

This case indicates that each of the other three models (the CP-CHPD, CB-CHPD and CPB-CHPD models) has a good performance on wind power integration and operation economic benefits, while the CPB-CHPD model involving PDTP and BTI together has much better synergic benefits.

5. Conclusions

The coordination of pipelines and buildings heat storage can be utilized to break the strong linkage of electric power and heat supply of the CHP units more effectively, which can improve the system operational flexibility, with enhancing wind power integration and reducing total operation cost significantly. Several simulation tests demonstrate that the proposed CPB-CHPD model has a good performance on power and heat decoupling. The detailed results are summarized as follows.

Coordinating the generation of every electric power and heat supply source in the combined heat and power system can introduce significant system operational flexibility just by considering PDTP and BTI without any additional investment. This method need not adjust the configuration of electric power and heat supply sources or impact end users' heat supply quality. The CPB-CHPD model can utilize more wind power energy than the CED model accounting for approximate 15.16%, and the total operation cost is reduced by nearly 11.86%.

The combined heat and power dispatch models considering PDTP or BTI can realize power and heat decoupling, where the effect of the latter BTI is more obvious than that of the former PDTP. However, the model considering both PDTP and BTI has much better synergic benefits. Based on the CED model, the CP-CHPD model can integrate more wind power energy and save more operation cost by approximate 3.51% and 2.85%, respectively, and the CB-CHPD model correspondingly 13.54% and 10.38%.

There is an interesting issue worth more study. In the large-scale electric power system, a simple equivalent model representing the heat storage capacity of pipelines and buildings in the district heating system may be concerned rather than the detailed and complex model to reduce the solving complexity of the combined heat and power dispatch model.

Acknowledgments: This work was supported by the National Natural Science Foundation of China (NSFC) (51607021).

Author Contributions: Ping Li and Weidong Li conceived of and designed the research problem. Ping Li wrote the whole manuscript. Weidong Li supervised the paper writing. Haixia Wang and Quan Lv participated in the results analysis.

Conflicts of Interest: The authors declare no conflict of interest.

Abbreviations

CHP	Combined heat and power
CON	Condensing power
EPS	Electric power system
DHS	District heating system
PDTP	Pipelines dynamic thermal performance
BTI	Buildings thermal inertia
CPB-CHPD	Combined heat and power dispatch considering both PDTP and BTI
CP-CHPD	Combined heat and power dispatch only considering PDTP
CB-CHPD	Combined heat and power dispatch only considering BTI
CED	Conventional economic dispatch considering neither PDTP, nor BTI

References

1. Chinese Renewable Energy Industries Association (CREIA). *China Wind Power Review and Outlook 2016*; CREIA: Beijing, China, 2016. (In Chinese)
2. Streckienė, G.; Martinaitis, V.; Andersen, A.N.; Katz, J. Feasibility of CHP-plants with thermal stores in the German spot market. *Appl. Energy* **2009**, *86*, 2308–2316. doi:10.1016/j.apenergy.2009.03.023.

3.   Celador, A.C.; Odriozola, M.; Sala, J.M. Implications of the modelling of stratified hot water storage tanks in the simulation of CHP plants. *Energy Convers. Manag.* **2011**, *52*, 3018–3026. doi:10.1016/j.enconman.2011.04.015.

4.   Rong, S.; Li, Z.; Li, W. Investigation of the promotion of wind power consumption using the thermal-electric decoupling techniques. *Energies* **2015**, *8*, 8613–8629. doi:10.3390/en8088613.

5.   Yuan, R.; Ye, J.; Lei, J.; Li, T. Integrated combined heat and power system dispatch considering electrical and thermal energy storage. *Energies* **2016**, *9*, 474. doi:10.3390/en9060474.

6.   Chen, H.; Yu, Y.; Jiang, X. Optimal scheduling of combined heat and power units with heat storage for the improvement of wind power integration. In Proceedings of the 2016 IEEE PES Asia-Pacific Power and Energy Engineering Conference (APPEEC), Xi'an, China, 25–28 October 2016; pp. 1508–1512. doi:10.1109/APPEEC.2016.7779742.

7.   Mathiesen, B.V.; Lund, H. Comparative analyses of seven technologies to facilitate the integration of fluctuating renewable energy sources. *IET Renew. Power Gener.* **2009**, *3*, 190–204. doi:10.1049/iet-rpg:20080049.

8.   Long, H.; Xu, R.; He, J. Incorporating the variability of wind power with electric heat pumps. *Energies* **2011**, *4*, 1748–1762. doi:10.3390/en4101748.

9.   Papaefthymiou, G.; Hasche, B.; Nabe, C. Potential of heat pumps for demand side management and wind power integration in the German electricity market. *IEEE Trans. Sustain. Energy* **2012**, *3*, 636–642. doi:10.1109/TSTE.2012.2202132.

10.   Chen, X.; Kang, C.; O'Malley, M.; Xia, Q.; Bai, J.; Liu, C.; Sun, R.; Wang, W.; Li, H. Increasing the flexibility of combined heat and power for wind power integration in China: Modeling and implications. *IEEE Trans. Power Syst.* **2015**, *30*, 1848–1857. doi:10.1109/TPWRS.2014.2356723.

11.   Zhang, N.; Lu, X.; McElroy, M.B.; Nielsen, C.P.; Chen, X.; Deng, Y.; Kang, C. Reducing curtailment of wind electricity in China by employing electric boilers for heat and pumped hydro for energy storage. *Appl. Energy* **2016**, *184*, 987–994. doi:10.1016/j.apenergy.2015.10.147.

12.   Moradi, H.; Abtahi, A.; Esfahanian, M. Optimal energy management of a smart residential combined heat, cooling and power. *Int. J. Tech. Phys. Probl. Eng.* **2016**, *8*, 9–16.

13.   Moradi, H.; Moghaddam, I.G.; Moghaddam, M.P.; Haghifam, M.R. Opportunities to improve energy efficiency and reduce greenhouse gas emissions for a cogeneration plant. In Proceedings of the 2010 IEEE International Energy Conference and Exhibition (EnergyCon), Manama, Bahrain, 18–22 December 2010; pp. 785–790. doi:10.1109/ENERGYCON.2010.5771787.

14.   Ye, J.; Yuan, R. Integrated natural gas, heat, and power dispatch considering wind power and power-to-gas. *Sustainability* **2017**, *9*, 602. doi:10.3390/su9040602.

15.   Andersson, S. Influence of the net structure and operating strategy on the heat load of a district-heating network. *Appl. Energy* **1993**, *46*, 171–179. doi:10.1016/0306-2619(93)90066-X.

16.   Li, Z.; Wu, W.; Shahidehpour, M.; Wang, J.; Zhang, B. Combined heat and power dispatch considering pipeline energy storage of district heating network. *IEEE Trans. Sustain. Energy* **2016**, *7*, 12–22. doi:10.1109/TSTE.2015.2467383.

17.   Jin, X.; Mu, Y.; Jia, H.; Wu, J.; Jiang, T.; Yu, X. Dynamic economic dispatch of a hybrid energy microgrid considering building based virtual energy storage system. *Appl. Energy* **2017**, *194*, 386–398. doi:10.1016/j.apenergy.2016.07.080.

18.   Awad, B.; Chaudry, M.; Wu, J.; Jenkins, N. Integrated optimal power flow for electric power and heat in a microgrid. In Proceedings of the 20th International Conference and Exhibition on Electricity Distribution (CIRED), Prague, Czech Republic, 8–11 June 2009; pp. 869:1–869:4. doi:10.1049/cp.2009.1037.

19.   Jiang, X.; Jing, Z.; Li, Y.; Wu, Q.; Tang, W. Modelling and operation optimization of an integrated energy based direct district water-heating system. *Energy* **2014**, *64*, 375–388. doi:10.1016/j.energy.2013.10.067.

20.   Li, J.; Fang, J.; Zeng, Q.; Chen, Z. Optimal operation of the integrated electrical and heating systems to accommodate the intermittent renewable sources. *Appl. Energy* **2016**, *167*, 244–254. doi:10.1016/j.apenergy.2015.10.054.

21.   Zhao, H.; Bohm, B.; Ravn, H.F. On optimum operation of a CHP type district heating system by mathematical modeling. *Euroheat Power* **1995**, *24*, 618–622.

22.   Zhao, H. Analysis, Modelling and Operational Optimization of District Heating Systems. Ph.D. Thesis, Technical University of Denmark, Copenhagen, Denmark, 1995.

23. Fu, L.; Jiang, Y. Optimal operation of a CHP plant for space heating as a peak load regulating plant. *Energy* **2000**, *25*, 283–298. doi:10.1016/S0360-5442(99)00064-X.

24. Wernstedt, F.; Davidsson, P.; Johansson, C. Demand side management in district heating systems. In Proceedings of the 6th International Joint Conference on Autonomous Agents and Multiagent Systems (AAMAS'07), Honolulu, HI, USA, 14–18 May 2007; pp. 1383–1389. doi:10.1145/1329125.1329454.

25. Kensby, J.; Trüschel, A.; Dalenbäck, J.O. Potential of residential buildings as thermal energy storage in district heating systems—Results from a pilot test. *Appl. Energy* **2015**, *137*, 773–781. doi:10.1016/j.apenergy.2014.07.026.

26. Brange, L.; Englund, J.; Lauenburg, P. Prosumers in district heating networks—A Swedish case study. *Appl. Energy* **2016**, *164*, 492–500. doi:10.1016/j.apenergy.2015.12.020.

27. Satyavada, H.; Baldi, S. An integrated control-oriented modelling for HVAC performance benchmarking. *J. Build. Eng.* **2016**, *6*, 262–273. doi:10.1016/j.jobe.2016.04.005.

28. Satyavada, H.; Babuška, R.; Baldi, S. Integrated dynamic modelling and multivariable control of HVAC components. In Proceedings of the 2016 European Control Conference (ECC), Aalborg, Denmark, 29 June–1 July 2016; pp. 1171–1176. doi:10.1109/ECC.2016.7810448.

29. Yang, Y.; Wu, K.; Long, H.; Gao, J.; Yan, X.; Kato, T.; Suzuoki, Y. Integrated electricity and heating demand-side management for wind power integration in China. *Energy* **2014**, *78*, 235–246. doi:10.1016/j.energy.2014.10.008.

30. Wu, C.; Jiang, P.; Gu, W.; Sun, Y. Day-ahead optimal dispatch with CHP and wind turbines based on room temperature control. In Proceedings of the 2016 IEEE International Conference on Power System Technology (POWERCON), Wollongong, Australia, 28 September–1 October 2016; pp. 1–6. doi:10.1109/POWERCON.2016.7753879.

31. Pan, Z.; Guo, Q.; Sun, H. Feasible region method based integrated heat and electricity dispatch considering building thermal inertia. *Appl. Energy* **2017**, *192*, 395–407. doi:10.1016/j.apenergy.2016.09.016.

32. Li, P.; Nord, N.; Ertesvåg, I.S.; Ge, Z.; Yang, Z.; Yang, Y. Integrated multiscale simulation of CHP based district heating system. *Energy Convers. Manag.* **2015**, *106*, 337–354. doi:10.1016/j.enconman.2015.08.077.

33. Pan, Z.; Guo, Q.; Sun, H. Interactions of district electricity and heating systems considering time-scale characteristics based on quasi-steady multi-energy flow. *Appl. Energy* **2016**, *167*, 230–243. doi:10.1016/j.apenergy.2015.10.095.

34. Andersen, T.V. Integration of 50% Wind Power in a CHP-Based Power System: A Model-Based Analysis of the Impacts of Increasing Wind Power and the Potentials of Flexible Power Generation. Ph.D. Thesis, Technical University of Denmark, Copenhagen, Denmark, 2009.

35. Liu, X.; Wu, J.; Jenkins, N.; Bagdanavicius, A. Combined analysis of electricity and heat networks. *Appl. Energy* **2015**, *162*, 1238–1250. doi:10.1016/j.apenergy.2015.01.102.

36. Sandou, G.; Font, S.; Tebbani, S.; Hiret, A.; Mondon, C. Predictive control of a complex district heating network. In Proceedings of the 44th IEEE Conference on Decision and Control, 2005 and 2005 European Control Conference (CDC-ECC'05), Seville, Spain, 12–15 December 2005; pp. 7372–7377. doi:10.1109/CDC.2005.1583351.

37. Arvastson, L. Stochastic Modelling and Operational Optimization in District Heating Systems. Ph.D. Thesis, Lund Institute of Technology, Lund, Sweden, 2001.

38. Lu, N. An evaluation of the HVAC load potential for providing load balancing service. *IEEE Trans. Smart Grid* **2012**, *3*, 1263–1270. doi:10.1109/TSG.2012.2183649.

energies

MDPI

Article

A Hybrid Genetic Wind Driven Heuristic Optimization Algorithm for Demand Side Management in Smart Grid

Nadeem Javaid [1,*], Sakeena Javaid [1], Wadood Abdul [2], Imran Ahmed [3], Ahmad Almogren [2], Atif Alamri [2] and Iftikhar Azim Niaz [1]

[1]   COMSATS Institute of Information Technology, Islamabad 44000, Pakistan; sakeenajavaid@gmail.com (S.J.); ianiaz@comsats.edu.pk (I.A.N.)
[2]   Research Chair of Pervasive and Mobile Computing, College of Computer and Information Sciences, King Saud University, Riyadh 11633, Saudi Arabia; aabdulwaheed@ksu.edu.sa (W.A.); ahalmogren@ksu.edu.sa (A.A.); atif@ksu.edu.sa (A.A.)
[3]   Institute of Management Sciences (IMS), Peshawar 25000, Pakistan; imran.ahmed@imsciences.edu.pk
*   Correspondence: nadeemjavaidqau@gmail.com; Tel.: +92-300-05792728

Academic Editor: K.T. Chau
Received: 8 November 2016; Accepted: 24 February 2017; Published: 7 March 2017

Abstract: In recent years, demand side management (DSM) techniques have been designed for residential, industrial and commercial sectors. These techniques are very effective in flattening the load profile of customers in grid area networks. In this paper, a heuristic algorithms-based energy management controller is designed for a residential area in a smart grid. In essence, five heuristic algorithms (the genetic algorithm (GA), the binary particle swarm optimization (BPSO) algorithm, the bacterial foraging optimization algorithm (BFOA), the wind-driven optimization (WDO) algorithm and our proposed hybrid genetic wind-driven (GWD) algorithm) are evaluated. These algorithms are used for scheduling residential loads between peak hours (PHs) and off-peak hours (OPHs) in a real-time pricing (RTP) environment while maximizing user comfort (UC) and minimizing both electricity cost and the peak to average ratio (PAR). Moreover, these algorithms are tested in two scenarios: (i) scheduling the load of a single home and (ii) scheduling the load of multiple homes. Simulation results show that our proposed hybrid GWD algorithm performs better than the other heuristic algorithms in terms of the selected performance metrics.

Keywords: Demand side management; priority scheduling; user comfort; heuristic optimization

1. Introduction

In order to make a robust and more reliable power grid, peak demand is taken into account rather than the average demand. As a consequence, natural resources are wasted, and the generation and distribution systems are under-utilized. Fast responding generators (e.g., coal and gas units), which are used to meet the peak demand, are not only expensive, but also have a high carbon emission rate. As a solution, different programs have been presented to shape the energy consumption profiles of users. Such programs aim to efficiently utilize the available generation so that new transmission and new generation infrastructures are minimally installed. These programs, known as demand side management (DSM) programs, aim either at scheduling consumption or reducing consumption [1].

A DSM program provides support towards power grid functionalities in various areas, such as electricity market control, infrastructure maintenance and management of decentralized energy resources [2]. In electricity markets, it informs the load controller about the latest load schedule and possible load reduction capabilities for each time step of the next day. Using this procedure, it schedules the load according to the objectives of interest associated with the power distribution systems [3,4]. The load shapes indicate the daily or seasonal electricity demands of industrial or residential consumers between peak hours (PHs) and off-peak hours (OPHs). These shapes can be modified by six techniques [5,6]: peak clipping, valley filling, load shifting, strategic conservation, strategic load growth and flexible load shape.

Peak clipping and valley filling are direct load control techniques. Peak clipping deals with the reduction of the peak loads, whereas valley filling considers the construction of loads for the off peak demands. Load shifting is the most effective and widely-used technique for load management in current power supply networks. It is concerned with shifting of the load from PHs to OPHs. Strategic conservation [5] applies demand reduction methods at the customer side for achieving optimized load shapes. If there is a larger load demand, then the daily responses are optimized by load growth techniques (distributed energy resources) [5–7].

The working of a generic DSM controller is shown in Figure 1. The figure shows that DSM aims for: (i) electricity cost minimization; (ii) energy consumption minimization; (iii) peak to average ratio (PAR) minimization; and (iv) user comfort (UC) maximization. In the literature, many DSM techniques are proposed [8–11] to achieve the aforementioned objectives. However, UC is not considered in most of these techniques, like [8,10,12–17]. In these works, [11,18,19] aim to reduce the electricity cost, and [20,21] focus on minimizing the aggregated power consumption using integer linear programming and mixed integer linear programming. Similarly, electricity bills and aggregated power consumption are reduced in [22] by using mixed integer non-linear programming. However, these techniques do not take into account the large number of different household appliances. Moreover, randomness in user load profiles makes the scheduling task more challenging.

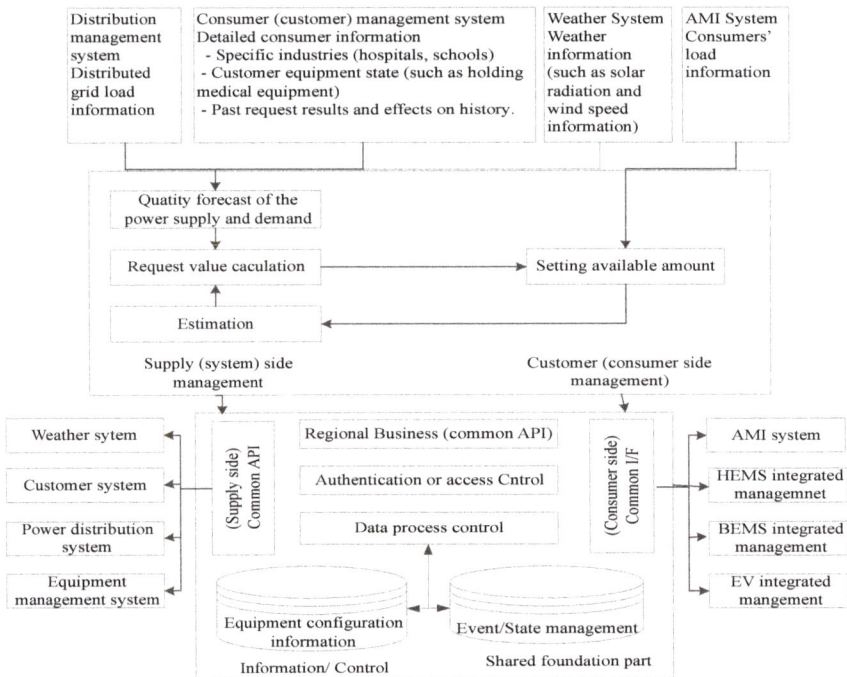

Figure 1. Working of demand side management (DSM). AMI: Advanced metering infrastructure, HEM: Home Energy Management.

In this paper, a heuristic algorithm-based DSM controller is designed for a residential area in a SG using the RTP scheme. In the designed DSM controller, five heuristic algorithms are implemented; GA, BPSO, wind-driven optimization (WDO), bacterial foraging optimization algorithm (BFOA) and our proposed hybrid genetic wind-driven (GWD) algorithm. These algorithms are chosen for implementation due to their flexibility for specified constraints and their low computational complexity [23]. More distinctively, prioritized load shifting is carried out between PHs and OPHs using a large number of appliances in the residential area. For effective scheduling and ease of implementation, the appliances are divided into two classes: (i) Class A (non-shiftable appliances) and (ii) Class B (shiftable appliances). Simulations are conducted in MATLAB such that all of the selected heuristic algorithms are compared in terms of electricity cost, energy consumption, PAR and UC. Results show that our proposed hybrid GWD performs better than the other compared techniques in terms of the selected performance metrics. It is worth mentioning that the nomenclature and list of abbreviations are given in Tables 1 and 3, respectively.

Table 1. Nomenclature.

Variables and Subscripts	Description
t	Time Interval
E_{ij}	Energy Consumption of an Appliance
$PR(t)$	Electricity Price at time t
A_i	Set of Appliances
S	Swarm Size
l_i	Length of Operation Time Counter
x_i	Position of Swarms
X	Appliance ON and OFF Status
g_{best}	Global Best Position of Particles
p_{best}	Local Best Position of Particles
P	Population Size
x_{new}	New Position of Particles
V_i	velocity of Particles
w	Weight of Particles
$EcostSavings$	Electricity Cost Savings
α	Cost Function Variable
β	Delay Function Variable
$delay$	Delay Function Counter
$EappUtil$	Appliance Utility
RT	RT Coefficient
g	Gravitational Constant
c	Constant in the Update Equation
$maxV$	Maximum Allowed Speed
H	Number of Homes
$pop1, pop2$	New Population
$Max.Cost$	Maximum Cost
$Gen.$	Generation
$tsize$	Total Size
$Maxgen$	Maximum Generations

Table 2. List of abbreviations.

Abbreviations	Definition
ANOVA	Analysis of variation
AC	Air conditioner
ACO	Ant colony optimization
ADA	Activity-dependent appliances
AMI	Advanced metering infrastructure
ANN	Artificial neural network
BPSO	Binary PSO
BFOA	Bacterial foraging optimization algorithm

289

Table 3. *Cont.*

Abbreviations	Definition
CAC	Central AC
CPP	Critical peak pricing
CN	Control node
CW	Clothes washer
DSM	Demand side management
DR	Demand response
DW	Dish washer
EMC	Energy management controller
EP	Energy price
F	Fan
FCFS	First come first serve
FF	Furnace fan
GA	Genetic algorithm
HG	Home gateway
HP	Heat pump
IHD	In-home display
IBR	Inclined block rate
LOT	Length of operation time
MC	Master controller
ODA	Occupancy-dependent appliances
OIA	Occupancy independent appliances
OPH	Off peak hour
PSO	Particle swarm optimization
PAR	Peak to average ratio
PH	Peak hour
PB	Priority bit
RAC	Room AC
RF	Refrigerator
RTP	Real-time pricing
SM	Smart meter
SH	Space heater
TOU	Time of use
UC	User comfort
WDO	Wind-driven optimization
WH	Water heater
WSN	Wireless sensor network

The rest of the paper is organized as follows. Section 3 briefly describes the related work. Section 4 formulates the problem. The system model is given in Section 5. Section 6 deals with the results and discussions. The paper is concluded in Section 7.

2. Related Work

In [10], the authors propose a technique for controlling the residential energy loads while maximizing UC and minimizing the electricity bill. A survey of home energy management for the residential customers is presented in [24], where the authors focus on different techniques relating to shiftable, non-shiftable load and peak shaving. They use various pricing schemes, like RTP, TOU, CPP, IBR, etc. In [25], a fully-automated EMSfor residential and commercial buildings is presented. They use the Q-learning algorithm for optimal DR mechanisms. Cristopher et al. [26] design a new framework. They use SMs to decide the appliance schedules based on their load or power consumption. After scheduling, all of the data are transferred to the aggregator module, where the power consumption of all of the appliances is determined. The concept of load clustering is introduced in this approach, which comprises three clusters for scheduling purposes, as the first cluster is from 1 a.m. to 7 a.m., the second from 8 a.m. to 3 p.m. and the third from 3 p.m. to midnight. Two battery scheduling scenarios are used as: (i) the FCFS scheduling policy and (ii) appliance first scheduling policy. In FCFS, requests to consume electricity from clients are assigned priorities based on their arrival, whereas in the appliance first scenario, all electrical devices' requests are given priority over battery charging.

Another methodology is proposed for minimizing the energy price under the dynamic pricing scheme to avoid PHs in [27]. Its architecture comprises SM, CN, WSN and IHD. AMI controls bidirectional data flow between the utility and SM. The SM operates between MC and AMI. The MC organizes and controls the schedules of both controllable and uncontrollable electrical appliances, such that the optimal schedule is transmitted to each CN via the WSN. IHD invigilates the whole process. In [8], GA is used to solve the scheduling problem under the RTP tariff in residential, commercial and industrial sectors. The authors present a novel approach known as the realistic scheduling mechanisms in [28] for minimizing the customer inconvenience using the TOU pricing scheme. They organize three categories of appliances (ADA, ODA, OIA) and the algorithms relevant to their working times. They also use the BPSO algorithm for the scheduling of these appliances. In [9], the researchers elaborate an efficient energy scheduling model and an algorithm based on artificial intelligence for residential area energy management in order to minimize the electricity cost. BPSO and GA are used for scheduling the optimal time of appliances and also for obtaining the best fitness values of the objective function.

For solving the numerically-constrained optimization problems, a review of BFOA is presented in [29]. The authors discuss the taxonomy of constraint handling techniques, the main steps and adaptations to different schemes, including search space, step size, tumble-swim operator and the elimination-reproduction process. In [30], a case study describes the electric demand model in rural households of Narino. Distributed privacy-friendly DSM is presented in [31], which preserves users' privacy by integrating data aggregation and perturbation. The authors describe that the users schedule their requests of appliances according to the aggregated energy consumption measurements as an additive white Gaussian process.

The authors in [32] focus on cost and emission minimization approaches in data centers and corresponding cloud network infrastructures. They use renewable energy generation capability to enhance the reliability and energy efficiency in SG. They also improve the latency using the ICTs. The decentralized system framework presents DR mechanisms for the residential users to minimize electricity bills, maximize the UC and privacy in [33]. In this framework, customers' SMs integrate home load management modules for exchanging the load profiles' information. Agents exchange information until they find an accurate load profile where the system does not get more improvement in the solution.

In [34], an energy consumption management approach considers household users in which each house consists of two types of requests or demands: (i) essential and (ii) flexible, where flexible demands are further delay sensitive and delay tolerant. To optimize energy for both delay-sensitive and delay-tolerant demands, a new centralized algorithm is presented for scheduling. This approach also aims to minimize the total cost and delay of the flexible demands for obtaining optimal energy decisions. The authors design a cost-efficient demand side day-ahead bidding process and RTP mechanisms by using fractional programming methods in [35].

In [36], the authors present a survey of DSM optimization methods for the residential customers. They classify the DSM techniques into three dimensions as: (i) DSM for individual users and cooperative consumers; (ii) DSM as a deterministic model versus the stochastic method; and (iii) day-ahead DSM versus real-time DSM. The dynamic load priority method presents priorities to modify load priorities during the occurrence of demand response events in [37]. A DR technique formulates the two-stage stochastic problem for energy resource scheduling; inciting the challenges of the renewable sources, electric vehicle and market price uncertainty. It reduces the overall operational cost of the energy aggregator by using stochastic programming [38]. In [39], global load balancing schemes describe the data center power management for minimizing the total electricity cost. They explain different components of the data centers as information technology equipment, the power delivery system and the cooling system in relationship with the SG's features (power delivery, sustainability, peak shaving, etc.). A multi-objective optimization solution is designed using the market operator and the distributed network operator for a microgrid in [40]. The generation of the price signal from the market operator and the power distribution system is specified using the Pareto-optimal solution. In [41], a novel pricing strategy is proposed to investigate the robustness against renewable energy source power inputs. This scheme also focuses on the marginal befits and marginal cost of the power market using all existing information related to electricity demand, supply and energy imbalance.

In short, the existing optimization techniques in [8,10,12–14] are unable to handle the complexity of cost minimization and UC maximization problems due to their non-flexible nature. In fact, the solution of these non-linear problems lead to high computational complexity. Therefore, we use heuristic algorithms (GA, BPSO, WDO and BFOA) to solve these two problems. These algorithms support the multi-objective optimization problems and have flexible constraints and parameters, which are easy to handle. These algorithms are similar to population-based search methods [42], which move from one population to another population in a number of iterations with improvement using a combination of deterministic and probabilistic rules. The comparison of the aforementioned techniques along with their achievements and drawbacks is listed in detail in Table 4.

Table 4. Recent trends: state of the art work.

Techniques	Targeted Area	Objective	Drawbacks
GA-Based DSM Scheme for SG [8]	Residential, Commercial and Industrial Area	Cost Minimization	Inconsideration of PAR and UC
Optimal Energy Consumption Scheduling Algorithm [9]	HEMS	Cost Minimization	Compromising the UC and RES
Residential Load Management in Smart Homes [10]	Residential Energy Load	Cost and PAR Reduction, UC Maximization	Explicit Pressure Values Degrade Performance
Home Energy Management for Residential Customers [24]	HEMS	Concentrates on UC, Energy Conservation and PAR	Commitments are Required for Effective Maintenance
Optimal DR Mechanisms [25]	Commercial and Residential Buildings	Considerations on DR Mechanisms	Do not Focus on Randomizing Automatic EMS
Smart Charging and Appliance Scheduling Approaches [13]	Appliance Scheduling and Storage	Cost Maximization and Maximum Storage Utilization	Inconsideration of Superclustering
Optimal Residential Appliance Scheduling via HEMDAS [27]	HEM	Cost Minimization and UC Maximization	Inconsideration of the Initial Installation Cost
Realistic scheduling mechanisms [18]	EMS	UC Maximization	Inconsideration of EC and PAR
BFOA in Constrained Numerical Optimization [11]	Residential Area	PAR Reduction and Cost Minimization	Inconsideration of Larger Population Size
Electricity Demand Modeling [30]	Rural Households	Energy Consumption Minimization	Inconsideration of Control Variables for Electric Demand
Enabling Privacy in a Distributed Game-Theoretical Scheduling Systems [31]	Game-theoretic DSM	Focused on Privacy, Electricity Bills Minimization and PAR Reduction	Inconsideration of Total Bill Reduction
Information and Communication Infrastructures [32]	ICTs	Energy Efficiency	Inconsideration of UC
Optimal Residential Load Management [33]	Residential Customers	Energy Efficiency	Inconsideration of Cost
Queuing-based Energy Consumption Management [34]	Residential SG Networks	Cost Minimization and Delay Reduction	Inconsideration of Parameters Tuning
Residential Load Scheduling in SG [35]	DSM	Concentrates on Energy	Inconsideration of Cost Minimization
SG and Smart Home Security [30]	DR	Energy Efficiency	Tradeoff between Demand Limit and UC

3. Problem Formulation

In this work, the major objectives are: (i) to reduce consumers' electricity cost by optimizing the energy consumption of end users; (ii) to maximize the UC of consumers. Here, the problem is formulated as an optimization problem with fixed, shiftable and elastic loads.

3.1. Cost Minimization

Cost minimization refers to the minimum charges for the consumed loads provided by the utilities to the customers. The elastic and shiftable loads are considered for the cost minimization problem, which is formulated as follows:

$$\text{Minimize} \sum_{i=1}^{N} \sum_{t=1}^{T} (X_{i,t} \times PR_{i,t}) \tag{1}$$

such that:

$$X_{i,t} = \begin{cases} 0, & if\ t \in H_1 \\ 1, & if\ t \in H_2 \end{cases} \tag{1a}$$

$$1 \leq t \leq T \tag{1b}$$

$$1 \leq i \leq N \tag{1c}$$

where $X_{i,t}$ represents the states of the appliances as ON or OFF (1 = ON and 0 = OFF) and $PR_{i,t}$ shows the price of the electricity consumed during any time interval t, which is the index for time upper bounded by T ($T = 24$) hours in a day. $H = \{1, 2, ..., T\}$, where H shows the time for the 24 h of a day, including PHs and OPHs. Here, $H_1 = \{7, 8, 9, 10\}$ indicates the PHs and $H_2 = \{H/H_1\}$ describes the OPHs. i denotes the appliances' index number, which is taken as $N = 12$.

3.2. UC Maximization

UC is modeled in terms of the minimum delay of appliances and optimal amounts for the electricity bills. Therefore, consumers always expect utilities with minimum delay and cost. Moreover, it also helps in minimizing the customers' frustrations when the energy consumption is high during the OPHs. In this scenario, the appliances are assigned a specific priority, and high priority appliances are scheduled at the first and foremost available time intervals during the OPHs. The operations of the low priority appliances can be canceled or delayed during the PHs. In this way, appliances' waiting time is minimized, and UC is achieved maximally. This is the multi-objective problem; several authors handle it using different approaches, as mentioned in the literature [12–17]. Here, it is handled by the metaheuristics for scheduling the residential area loads in order to reduce the electricity cost and maximize the UC. Energy cost is weighted at the minimum electricity bill, and UC weights are considered between [0, 1]. It is calculated by using the equations given below,

$$Maximize(EappUtil + EcostSavings) \tag{2}$$

such that:

$$EappUtil = (\alpha - (delay/24)) \tag{2a}$$

$$0.3 \leq \alpha \leq 0.7 \tag{2a.1}$$

$$1 \leq delay \leq 4 \tag{2a.2}$$

$$EcostSavings = \beta \times (cost/100) \times (Sch\_cost/Max.\ cost) \tag{2b}$$

$$0.3 \leq \beta \leq 0.7 \tag{2b.1}$$

$$\alpha + \beta = 1 \tag{2b.2}$$

α and β are the delay variables. Moreover, *delay* is the delay function, and it is restricted to four hours in our scenario. It is worth mentioning that these 4 h are chosen from PHs for elucidating the maximum delay of the appliances. If the delay is greater than 4 h, then the utility pays a penalty by either paying back to customers or providing them with reductions in the electricity bills. According to Constraint (2b.2), the sum of α and β is equal to one because UC ranges between zero and one. *Cost* is the cost function, and its values are between 20% and 70%. Below 20%, its values are assumed to be negligible, and cost is inconsiderable; and above 70% cost prices are used for the microgrids. *Sch_cost* is the cost of the appliances during the full day, and *Max.cost* is the cost of peak hours of the day; *Sch_cost* is obtained from the status bits of the appliance x power rating; *Max.cost* is also obtained from the hourly information updates. The values of α, β, *delay* and *Sch_cost* are taken from [28].

3.3. Multi-Objective Function

From the objective functions in Equations (1) and (2), it is clear that the optimization problem is multi-objective. We formulate the combined objective function as follows:

$$Minimize(c_1 \sum_{i=1}^{N} \sum_{t=1}^{T} (X_{i,t} \times PR_{i,t}) + c_2 \frac{1}{EappUtil + EcostSavings}) \tag{3}$$

where $c_1 = c_2 = 0.5$. Here, it is worth mentioning that the combined objective function in Equation (3) is subject to the respective constraints of objective functions in Equations (1) and (2).

4. Proposed Solution

The proposed DSM techniques deal with the load management in a residential area for single and multiple homes. Its architecture consists of the number of homes, SMs, AMI and the utility companies. Let multiple homes be connected with a utility and SMs be installed in all of the homes as shown in Figure 2. The AMI is used for bidirectional communication between SM and the utility. All homes have three types of appliances: (i) fixed; (ii) elastic; and (iii) shiftable. These appliances are also categorized into Class A and Class B based on their fixed or interruptible load profiles. Fixed load appliances are included in Class A, whereas elastic and shiftable are included in Class B. In other words, Class B contains interruptible appliances, which take part in the scheduling process.

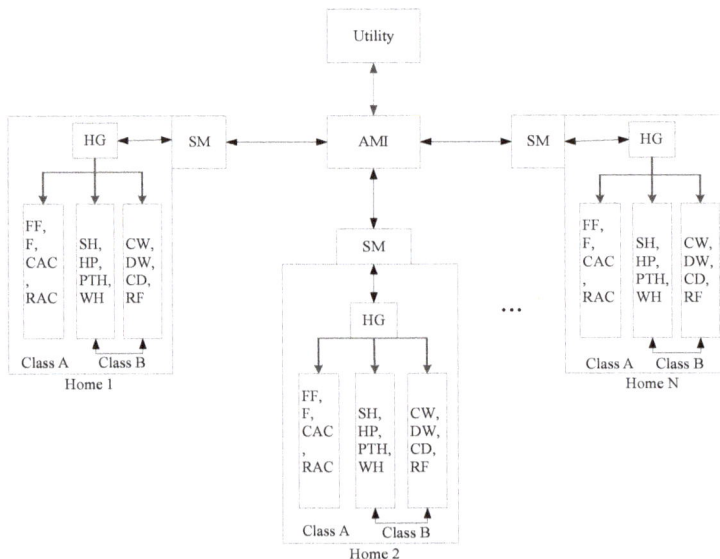

Figure 2. Proposed system design.

The RTP tariff model is used for tracking the pattern of the total hourly costs of the consumed energy. Figure 3 shows that the appliances are scheduled by the appliances' handler (EMC) during the specified time intervals using the given frame format. EMC schedules and checks appliances' PB using the frame format. Each frame format consists of an eight-bit pattern, such that each appliance uses a specified bit pattern relating to its class ID, appliance ID, scheduling bit, interruptible or non-interruptible bit and priority bit. Based on the operational status of an appliance, its hourly cost schedule is tracked. In each class, every attribute uses a single bit, except class ID and appliance schedule, which use three- and two-bit patterns, respectively. This scenario is specific to these sets of the appliances using the given frame format for the proposed system's test cases; however, it can be further extended to a larger set of appliances, and frame length can also be extended accordingly. Evolutionary algorithms are efficient in terms of computational complexity, however, at the cost of reduced accuracy. We prefer frame tracking over other evolutionary algorithms because it provides simple and efficient procedure in terms of relative accuracy and relative computational complexity. In the following subsections, the algorithms of GA, BPSO, WDO, BFOA and our proposed GWD algorithm are discussed in detail.

	Class ID	Appliance ID	Appliance schedule	Interruptible or not	Priority bit
Number of bits assigned for each attribute	2	1	3	1	1
Bit pattern Class A	00	0	000, ..., 111	1	0
Bit pattern Class B	11	1	000, ..., 111	1	1

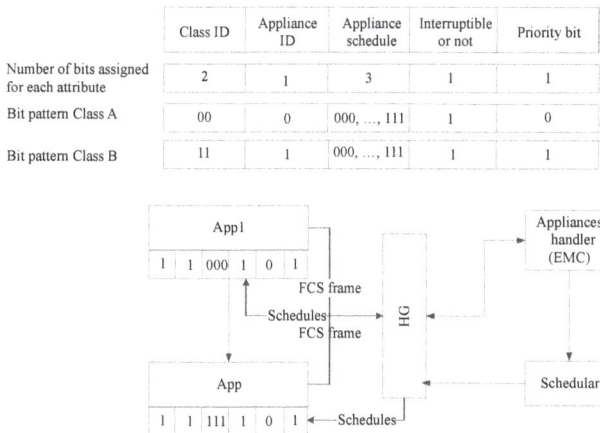

Figure 3. RTP price tracking system.

4.1. GA, BPSO, WDO and BFOA Algorithms

In this section, we modify the existing versions of GA, WDO, BPSO and BFAO to optimally schedule shiftable appliances. Firstly, the load is shifted to the OPHs subject to electricity cost minimization. In order to reduce peaks during the OPHs, each appliance is assigned a specific PB, which indicates the status (either ON or OFF) of the selected appliance. If an appliance is demanded to run in a specific time slot, its PB = 1; otherwise, its PB = 0. This status bit information is communicated via an RTP frame format.

The authors in [13] have proposed a GA-based home energy management controller for a single home in a residential area using RTP tariffs. In this manuscript, a modified GA (an improved form of [13]) is presented, which is shown in Algorithm 1. Objective functions (refer to Equations (1)–(3)) and their constraints are used by all of the selected optimization algorithms to find feasible solutions. Users input initial parameters for all appliances. GA creates a random population initially, which consists of a number of chromosomes represented by binary strings as the ON/OFF status of each appliance. Each chromosome is evaluated using Equations (1)–(3). RTP is used as the electricity pricing scheme. Key modifications that are implemented in GA (Algorithm 1 [13]) to achieve the objectives in the proposed scheme and its expected outcomes are given in Table 5.

Table 5. Modifications in GA.

Modifications	Expected Outcomes
Scheduling using PBs (refer to Equations (1)–(3)) with constraints	Curtails load Reduced PAR Enhanced UC
Use of RTP steps (10, 11, ..., 19)	Tracks the real-time behavior of system Minimizes the cost

---

Algorithm 1: GA algorithm.

Input: set of appliances A_i or P;
Initialization: PHs, OPHs, $t = 0$, H, PB = 0, 1;
for $t = 1$ to T do
  for $h = 1$ to H do
    Generate feasible P randomly;
    for $h = 1$ to P do
      Calculate fitness function using Equation (3) ;
      Select the best solutions in P, pop and save them in new pop1 ;
      Check status of A_i using PHs and OPHs while LOT, $X_i = 1$ and $l_i = l_i - 1$;
      if $t == PHs$ then
        wait until OPHs;
        if *EnergyConsumption* $== high$ then
          Check PB of appliances;
        else
          Check the remaining t of all A_i, LOT until 0 ;
        end
      end
    end
  end
  Generate new population;
  Perform crossover operation by randomly selecting two chromosomes from P;
  Save it in *pop2*;
  Perform mutation operation;
  Select a solution from *pop2*;
  Mutate each bit of solution and generate a new solution;
  if *solution is infeasible* then
    Update solution with a feasible solution by repairing solution;
    Update solution with solution in *pop2*;
  end
  Update pop best solution;
  Update $t = t + 1$ till 24 h;
  Terminate when $t = 24$ h;
  end
end

In [15], another energy management model is presented in which BPSO is used to meet the DSM challenges. The goal of this study is to minimize the electricity cost for residential area by scheduling shiftable loads. The authors use the TOU pricing model to calculate electricity bills of customers by investigating DR; however, they have ignored UC. Furthermore, in our proposed work, the objective function is formulated for cost minimization and UC maximization. BPSO is used to solve the designed optimization problem. RTP scheme is used for tracking the real-time behavior of the system. Thus, this proposed work gives a more significant solution for electricity bill minimization, PAR minimization and UC maximization. All steps of the proposed work are shown in Algorithm 2. Compared to [15], BPSO is modified according to the customers' requirements. Each particle in the generation is represented by a binary string denoted as states of an appliance. The proposed model is applicable for single and multiple homes in residential areas. In Table 6, some suitable modifications and expected results in response to those modifications for the BPSO algorithm are given.

Table 6. Refinements in BPSO.

Refinements	Expected Consequences
Addition of PBs for scheduling (refer to Equations (1)–(3)) with the required constraints	Reduce energy consumption Minimizes the PAR Boosts up UC
Use of RTP steps (21, 22, ..., 25)	Monitors the real-time behavior of the system Minimizes the cost

---

Algorithm 2: BPSO algorithm.

Input: number of particles, maximum iterations, electricity price;
Initialization: S, $t = 0$, H, PHs, OPHs, PB = 0, 1;
Specify LOT of appliances and power ratings;
Randomly generate population of particles;
for $t = 1$ *to* T **do**
  for $h = 1$ *to* H **do**
    Evaluate the value of electricity cost of A_i;
    Evaluate LOT;
    set p_{best};
    for $i = 1$ *to* M **do**
      if $f(x_i) > f(p_{best,i})$ **then**
        $f(p_{best,i}) = f(x_i)$;
        if $f(p_{best,i}) > f_{gbest,i}$ **then**
          $f(g_{best,i}) = f(p_{bset,i})$;
        else
          $f(g_{best,i}) = f(g_{best,i})$;
        end
      end
    end
  end
  if $t == PHs$ **then**
    Wait till OPHs;
    if $EnergyConsumption == high$ **then**
      Check PBs of appliances;
    end
    Evaluate fitness function using Equation (3);
    Decrement one from the total LOT of appliances;
  end
  for $j = 1 to P$ **do**
    Update w of the particles using piecewise linear function [15] ;
    Update V_j using sigmoid function ;
    Update position vector x_j using piecewise linear function [15];
    Increment time counter $t = t + 1$ until $t = 24$;
  end
  end
end

A WDO-based scheduling technique is presented in [10] for comfort maximization of residential users. By considering appliance classes, user preferences and weather status, they model the UC and electricity cost. The WDO algorithm is used for minimizing electricity cost and maximizing UC. This work also analyses peak cost reduction in electricity bills by considering the TOU tariff. In this proposed work, household appliances are categorized on the basis of LOT and appliance power consumption. In order to make the scheduling process more efficient, delay and PB criteria (which are not considered in [10]) are incorporated here for reducing electricity bills. In this study, WDO is enhanced in which LOT and the energy consumption of each appliance are calculated by evaluating the objective function (refer to Equations (1)–(3)) using constraints. Table 7 shows the enhancements made as per our proposed work and the expected results based on the enhancements. All steps of the implemented WDO algorithm are shown in Algorithm 3.

Table 7. Adaptations in WDO.

Adaptations	Expected Results
Incorporation of the PBs (refer to Equations (1)–(3)) by considering constraints	Minimizes energy consumption Reduces the PAR Improves UC
Use of RTP steps (10, 11, ..., 19)	Tracks the real-time behavior of the system Minimizes the cost

---

Algorithm 3: WDO algorithm.

Initialization: P, Maxgen, RT, g, c, max. V, particles' pressure, $t = 0$, PHs, OPHs, H and PB = 0, 1;
Generate initial random population;
for $t = 1$ *to* T **do**
    for $h = 1$ *to* H **do**
        for $i=1$ *to* P **do**
            Assign random positions and velocities to air particles;
            Evaluate fitness of each air parcel Equation (3);
            Identify the best solution among all air parcels;
            while *number of iterations reached to specified limits* **do**
                if $t == PHs$ **then**
                    swap (OPH, PH);
                    if *EnergyConsumption* $==$ *high* **then**
                        Check appliance PB;
                    else
                        Check velocity and speed values;
                        Update velocities and positions;
                    end
                end
            end
            Generate new population;
            Check the limits (t);
            Identify the best solution among all air parcels;
            Increment the generation count G = G + 1;
            Increment timeslots $t = t + 1$;
        end
        end
    end
end

In [17], the authors propose a BFOA technique for grid resource scheduling. This technique is based on the hyper-heuristic resource scheduling algorithm, which has been designed to effectively schedule jobs on available resources in a grid environment. The authors evaluate the performance of the proposed BFOA algorithm by comparing it with the existing heuristic scheduling algorithms (GA and simulated annealing) using the makespan and cost performance metrics. Experimental results show that the proposed algorithm outperforms the existing algorithms in terms of cost minimization. In comparison to [17], the proposed work introduces a new methodology of appliance scheduling for minimizing electricity cost, energy consumption and PAR, which benefits both customers and the utility. In this study, objective functions (refer to Equations (1)–(3)) and their constraints are modified according to the designed scenario. Table 8 contains the refinements made and their respective expected results. All steps of the proposed work are given in Algorithm 4.

Algorithm 4: BFOA algorithm.

Input: randomly initialize the swarm of bacteria $\theta^i(j,k,l)$;
Initialization: PHs, OPHs and $t = 0$, H, PB = 0, 1;
Generate initial population randomly;
for $t = 1$ to T **do**
    for $h = 1$ to H **do**
        for $i=1$ to P **do**
            Compute for $f(\theta^i(j,k,l))$;
            for $l=1$ to N_{ed} **do**
                for $k=1$ to N_{re} **do**
                    for $j=1$ to N_{sb} **do**
                        for $Gen._l = 1$ to $Gen._{tsize}$ **do**
                            if $t == PHs$ **then**
                                swap (OPH, PH);
                            else if *EnergyConsumption* $==$ *high* **then**
                                check appliance PB;
                            end
                            else
                                Evaluate objective functions using Equation (3);
                            end
                        end
                      Calculate $f(\theta^i(j,k,l))$;
                      Perform chemotactic procedure;
                      Check tumble-swim operations;
                      Each bacteria controlled by $\theta^i(j,k,l)$ in N_{sb} steps;
                  end
                end
            end
        end
        Check reproduction process by swapping;
        Remove weak bacteria;
    end
    Perform the elimination-dispersal by elimination;
    Each bacteria is based on $\theta^i(j,k,l)$ with $P_{ed}0 \leq P_{ed} \leq 1$;
    end
end

Table 8. Refinements in BFOA.

Refinements	Expected Achievements
Scheduling using PBs (refer to Equations (1)–(3)) along with their constraints	Reduce energy consumption Minimizes the PAR Increases UC
Use of RTP steps (12, 13, ..., 20)	Monitors the real-time behavior of the system Reduces the cost

4.2. Developing a Hybrid GWD Optimization Algorithm

In this algorithm, all of the stages of WDO are performed in a similar way as explained in Section 4.1; however, the velocity updating steps for the global air pressure is replaced with GA's crossover and mutation operations. In some cases, pressure values are very large, such that the updating velocities become too large, which degrade WDO's performance. Thus, we replace these with GA's crossover and mutation values. The scheduling procedure is followed as the same described in GA, BPSO, BFAO and WDO. It is evaluated with the help of the same objective functions (refer to Equations (1)–(3)). Detailed steps of this algorithm are shown in Algorithm 5. Modifications of the hybrid GWD and their respective expected outcomes are given in Table 9 [8,10].

Algorithm 5: GWD algorithm.

Initialization: P, Maxgen, RT, g, c, max. V, particles' pressure, $t = 0$, PHs, OPHs, H, crossover
rate = 0.9, mutation rate = 0.1, PB = 0, 1;
Generate initial random population;
for $t = 1$ *to* T **do**
  for $h = 1$ *to* H **do**
    for $h = 1$ *to* P **do**
      Assign random positions and velocities to air particles;
      Evaluate fitness of each air parcel using Equation (3);
      Identify the best solution among all air parcels;
      while *Stopping criterion is not satisfied* **do**
        if $t == PHs$ **then**
          swap(OPH, PH);
        else if *EnergyConsumption == high* **then**
          Check appliance PB;
        else
          Check velocity and speed values of particles;
          Apply crossover and mutation operation;
          Update velocities and positions;
        end
      end
    end
    Generate new population;
    Check the limits (t) until $t = 0$;
    Evaluate fitness of each air parcel;
    Identify the best solution among all air parcels;
  end
  end
end
end

Table 9. Modifications in GWD.

Modifications	Anticipated Outcomes
Enhancements	Expected Results
Using PBs for scheduling	Reduce energy consumption
(refer to Equations (1)–(3))	Minimizes the PAR
	Increases UC
Use of RTP	Tracks the real-time behavior of the system
steps (10, 11, ..., 20)	Minimizes the cost

The metaheuristic algorithms do not guarantee exact reachability of the global optimum solution. The obtained solution is dependent on the set of random variables generated at the start of the metaheuristic optimization process. In our scenario, PSO, BFOA and WDO suffer from the global optima, and GA is a relatively better suited algorithm for the global optimal solution. In order to filter out the effects of random initializations, simulation runs of these algorithms are increased in number. However, this filtration is achieved at the cost of increased computational time. We have presented the statistical analysis of all of the algorithms with respect to cost and user comfort using the ANOVA in the Results Section after taking the average of the 10 runs.

5. Results and Discussion

In order to evaluate the proposed work, simulations are conducted in MATLAB using the RTP scheme. The 24-h time period is divided into PHs and OPHs for tracking the real-time behavior of the system. Four hours are taken as PHs (from 7 p.m.–10 p.m.) such that the PHs vary from season to season [43]. From December–February, PHs are from 5 p.m.–9 p.m.; from March–May, PHs are 6 p.m.–10 p.m.; from June–August, PHs are 7 p.m.–10 p.m.; and from September-November, these vary accordingly. Four hours are used in this case (from 7 p.m.–10 p.m.) of one season, and the remaining all are included in OPHs.

There are two simulation scenarios that are discussed here: (i) single home and (ii) fifty homes. Each home has 12 appliances, and appliances are categorized into two classes: (i) Class A with fixed load appliances and (ii) Class B with shiftable and elastic load appliances, as shown in Table 10. Figure 4 shows the RTP rates during each hour of the full day. The parameters of GA, BPSO, WDO, BFAO and GWD are given in Tables 11–15, respectively. To evaluate the performance of these algorithms, the following performance metrics are used.

- Cost: Amount of electricity bills for the total number of units consumed per unit time in cents.
- Energy Consumption: It is calculated as the total energy utilized per unit time in kilowatts per hour.
- PAR: It is defined as the total peak load divided by average load during the whole day.
- UC: It is calculated in terms of minimum cost and minimum appliance delay.

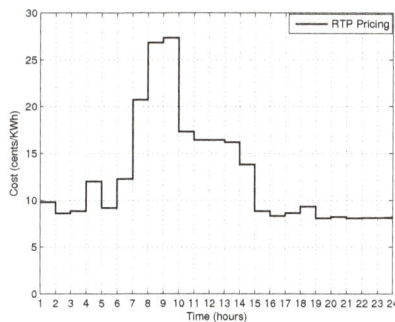

Figure 4. RTP price signal.

Table 10. Parameters and power ratings.

Class Name	Appliance Name	Power Rating	LOT	Deferrable Load
Class B	Space Heater	1	9	1
Class B	Heat Pump	0.11	4	1
Class B	Portable Heater	1.00	5	1
Class B	Water Heater	4.50	8	1
Class B	Clothes Washer	0.51	9	1
Class B	Clothes Dryer	5.00	5	1
Class B	Dishwasher	1.20	11	1
Class B	First-Refrigerator	0.50	24	1
Class A	Fan	0.5	11	0
Class A	Furnace Fan	0.38	8	0
Class A	Central AC	2.80	12	0
Class A	Room AC	0.90	5	0

Table 11. GA parameters and values.

Parameter	Value
Population Size	200
Selection	Tournament Selection
Elite Count	2
Crossover	0.9
Mutation	0.1
Stopping Criteria	Max. Generation
Max. Generation	1000

Table 12. BPSO parameters and values.

Parameter	Value
Swarm Size	20
Max. Velocity	4 ms
Min. Velocity	4 ms
Local Pull	2 N
Global Pull	2 N
Initial Momentum Weight	1.0 Ns
Final Momentum Weight	0.4 Ns
Stopping Criteria	Max. iteration
Max. Iteration	600

Table 13. WDO parameters and values.

Parameter	Value
Swarm Size	10
Max. V	4 m/s
RT-Coefficient	3
g	0.2
c	0.4
Dimensions	$[-1, +1]$
Stopping Criteria	Max. Iteration
Max. Iterations	500

Table 14. BFAO parameters and values.

Parameter	Value
Population Size	10
Maximum Number of Steps	30
Number of Chemotactic Steps	5
Number of Elimination Steps	5
Number of Reproduction Steps	25
Probability	0.5
Step Size	0.1
Stopping Criteria	Max. Generations
Max. Generations	100

Table 15. GWD parameters and values.

Parameter	Value
Particle Size	20
Number of Iterations	500
Max. V	0.4
Dimensions	[−1, +1]
RT-Coefficient	3.0
g	0.2
c	0.4
α	0.4
Crossover Rate	0.9
Mutation Rate	0.1

5.1. Single Home

The energy consumption of our proposed scheme hybrid GWD with respect to GA and WDO in unscheduled and scheduled cases is shown in Figure 5. This figure shows that the maximum energy consumption values are 16.2 kWh, 11.8 kWh, 8.2 kWh and 4.1 kWh for the unscheduled case, scheduled GA, WDO and the hybrid GWD approach, respectively. The energy consumption of all algorithms is below their unscheduled cases. The energy consumption in GA, WDO and GWD is 56.89%, 67.18% and 65.87%; which is obtained by dividing the scheduled cost and unscheduled cost with percentage. It is important to note that the hybrid GWD algorithm is better than the simple WDO and GA in terms of energy consumption. GWD uses crossover and mutation operations from the GA, which helps with the faster convergence for achieving optimized results, and WDO uses explicit pressure values; however, when velocities are high, pressure values become extremely large, which leads to performance degradation.

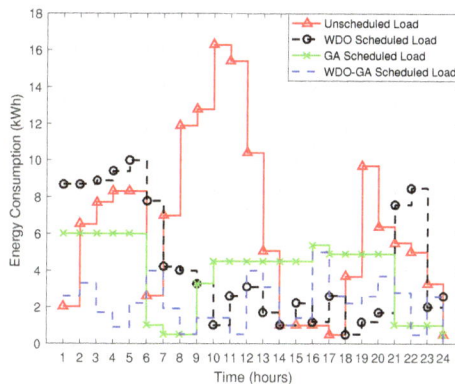

Figure 5. Energy consumption.

The maximum amount of the electricity bill in the unscheduled case is 318.88 cents, as shown in Figure 6. It is reduced to 78 cents in the case of GA, while it is reduced from 318 cents to 245 cents in WDO and up to 75 cents in GWD. The electricity cost in GA, WDO and GWD is 60%, 62% and 30%, respectively. During PHs, sufficient electricity cost reduction is achieved for all designed algorithms (GA, WDO and GWD). GWD performs better than the other algorithms in terms of the electricity cost reduction due to the amalgamation of crossover and mutation. The WDO's cost is high due to its high pressure values.

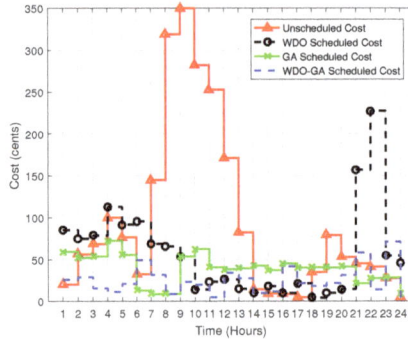

Figure 6. Total cost.

The PAR performance of all algorithms (GA, WDO and GWD) is shown in Figure 7. This figure shows that PAR is significantly reduced in hybrid GWD as compared to the GA, WDO and unscheduled case. Results prove that our proposed algorithm effectively tackles the peak reduction problem. The PAR graph for GA, WDO and hybrid GWD displays that the power consumption of appliances is optimally distributed without creating peaks during the OPHs and PHs of the day. The PAR in GA, WDO and GWD is 60%, 75% and 40%. WDO has higher PAR than GA because it has higher pressure values of the particles, and GA is more effective in PAR reduction due to its ability to generate new populations of more feasible solutions using crossover and mutation. From these results, it is shown that the hybrid GWD approach outperforms all other schemes, because it uses the best features of both. Peak formation is a major drawback in the traditional electric power system, as it causes customers to pay high electricity bills, and the utility also suffers from high demand, which leads to blackouts or load shedding. The performance of these algorithms in this scenario is improved due to load shifting using appliances' PBs, which causes utilities to fulfil the demands of customers and gives customers a chance to reduce their electricity bills.

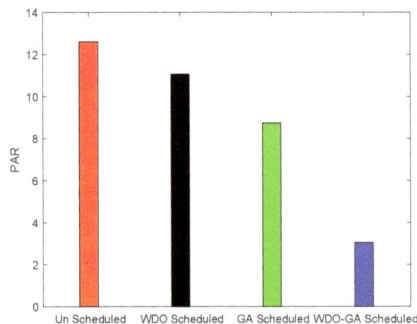

Figure 7. Scheduled and unscheduled PAR.

In our proposed hybrid scheme, we have achieved the desired UC as shown in Figure 8. It shows that UC is significantly reduced for GWD, GA and WDO as compared to the unscheduled case. By applying priority scheduling on the objective functions (refer to Equations (1)–(3)), this work enhanced the performance in terms of UC. UC of the unscheduled case is 98%, while in schedule WDO, GA and GWD, it is 60%. The maximum delay considered here is 4 h; otherwise, the utility has to pay a penalty for the users. There is a tradeoff in UC of all scheduled algorithms because only one scenario is considered here. However, the performance of this work is much better by considering the priority bits and minimum delay during scheduling.

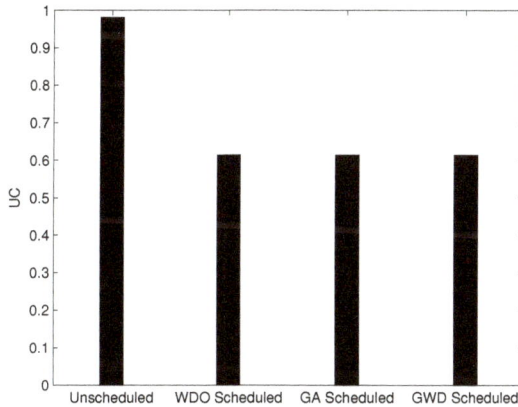

Figure 8. Scheduled and unscheduled UC.

All above simulations are performed for a single home; however, for testing the effects of the proposed scheme in multiple homes, multiple homes are taken in the next section. All of the modified algorithms (GA, BPSO, WDO and BFOA) are tested for 50 homes to investigate these in terms energy consumption minimization and electricity cost reduction. From Figure 13, it is clear that the proposed work achieves significant results. As these algorithms are designed to satisfy the constraints of the objective function in 24 h, so that residential users get facilitated by reducing their electricity bills and that utilities get the benefit by keeping demand under the power capacity of the grid.

5.2. Fifty Homes

The energy consumption of GA, BPSO, WDO and BFOA is 15.00 kWh, 7.90 kWh, 11 kWh and 14.5 kWh, respectively, which is less than the unscheduled case as 16.5 kWh, approximately; as shown in Figures 9–12. The energy consumption in GA, BPSO, WDO and BFOA is 79%, 47%, 45% and 88%. GA is efficient among all of the others, though it considers a larger population size. It uses a natural selection operator, which reduces the convergence time towards the efficient solution during scheduling. BFOA is faster than BPSO and consumes less energy because BFOA is faster for a small population size. On the other hand, BPSO is suitable for a larger population size, and it also escapes from the local minima. WDO consumes more energy as compared to BPSO, BFOA and GA, because it has explicit pressure values of particles, causing performance degradation.

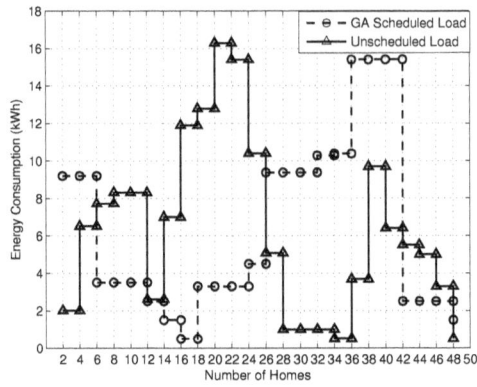

Figure 9. GA energy consumption.

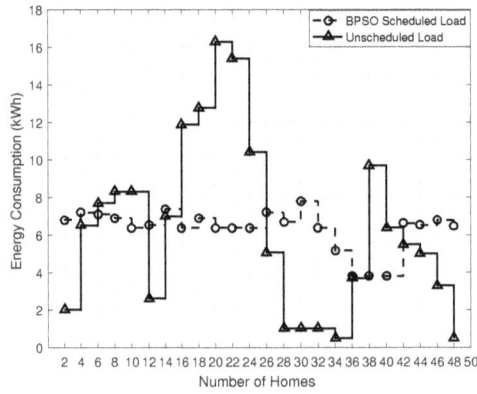

Figure 10. BPSO energy consumption.

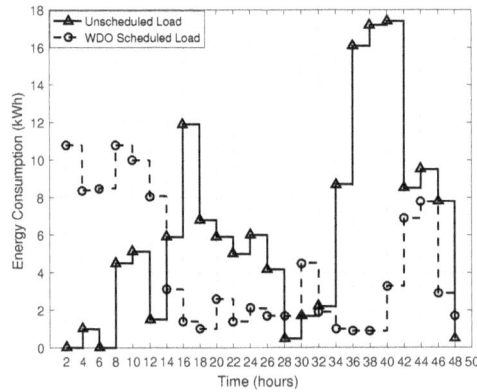

Figure 11. WDO energy consumption.

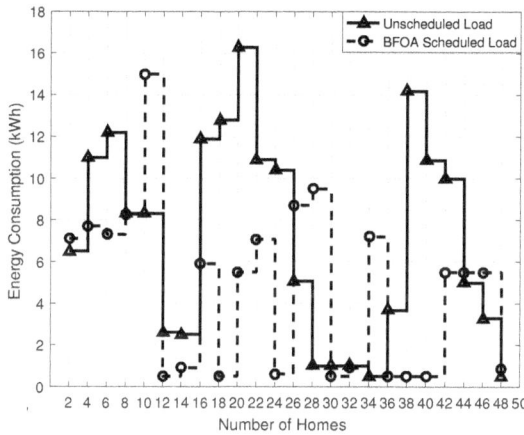

Figure 12. BFOA Energy Consumption.

The electricity cost of the simulated algorithms is shown in Figures 13–16, which is obtained during the scheduling process. In each case, the scheduled costs of all four algorithms, GA, BPSO, WDO and BFOA, are 125.20, 175, 215 and 160 cents, respectively, which are lower than the unscheduled cost of 350. Furthermore, by using the PBs during appliance scheduling, the overall cost is reduced as compared to the unscheduled cases. After scheduling, the obtained electricity cost by using GA, BPSO, WDO and BFOA is 35%, 50%, 61% and 45%, respectively; whereas, in the unscheduled case, it is 100%. In this case, GA is the most effective algorithm even considering a larger population size than the other algorithms. GA uses the crossover and mutation operation, which is efficient in convergence and at finding the global optimal solution. BPSO uses linear and piecewise functions instead of natural selection operators, and it is mostly used for a large population size to avoid local minima. BFOA is suitable for a small population size, and it is more efficient than BPSO and GA in terms of convergence and energy efficiency. WDO suffers from pressure values, so it gives a higher cost than the others.

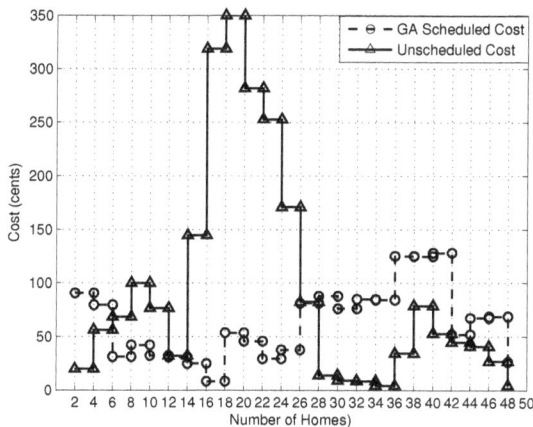

Figure 13. GA total cost.

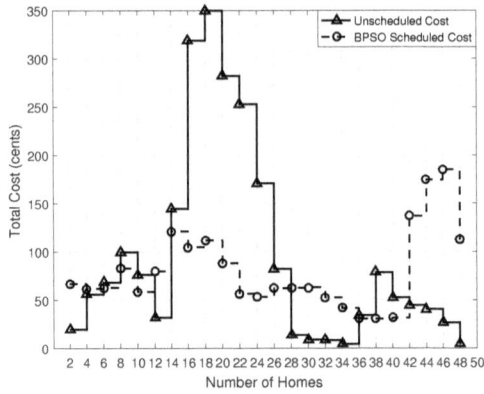

Figure 14. BPSO total cost.

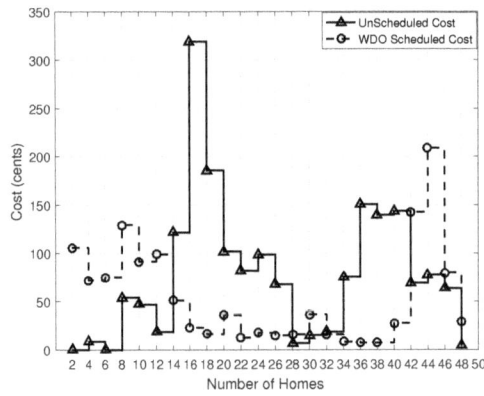

Figure 15. WDO Total Cost.

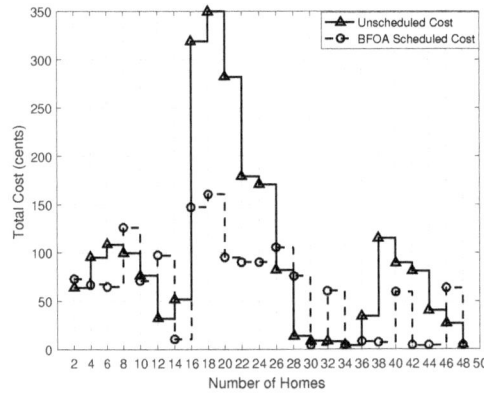

Figure 16. BFOA total cost.

Overall, the scheduled peak formation rate is better than the unscheduled cases, and the desired results of the load shifting are achieved by the scheduling. The PAR obtained in GA, BPSO, WDO, BFOA and the unscheduled case is 26%, 25%, 12%, 2% and 46%, respectively. All of the high profile appliances are scheduled to low price rate hours. If the consumed energy in OPHs is high (creating peaks), then appliances are scheduled according to their PBs for reducing load and avoiding peak formation even during the low pricing rate hours. PAR in WDO, BPSO and BFOA is better than GA because GA is tested for a large set of populations, whereas all of the others are tested for a small population size, as shown in Figure 17.

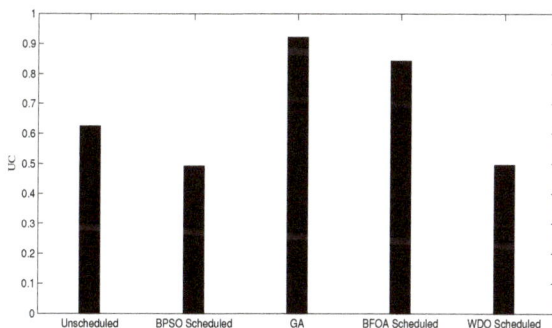

Figure 17. UC of GA, BPSO, WDO and BFOA.

UC achieved by GA and BFAO is significantly greater than BPSO, WDO and the unscheduled case as shown in Figure 18. The UC achieved in GA is nearly 0.9; BPSO is 0.5; WDO is 0.55; BFAO is 0.85; and it is 90%, 50%, 50% and 85%. Because during scheduling, all high power utilization appliances are shifted to OPHs, which facilitates the customers to pay less on the bill, so UC is maximized in BFOA and GA as compared to WDO and BPSO, which are the desired results obtained by the designed objective functions, and it is also beneficial for both customers and utilities.

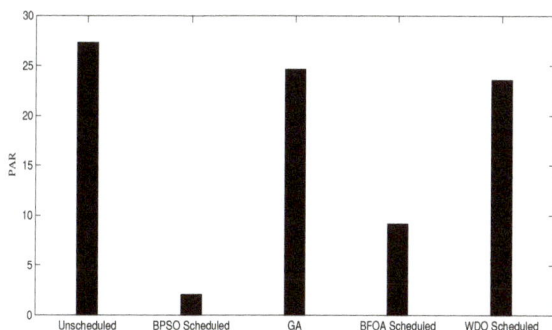

Figure 18. PAR of GA, BPSO, WDO and BFOA.

In order to quantify the computational burden of the algorithms, we have chosen algorithm execution time (in s) as a performance metric. Figure 19 shows the execution time of the five simulated algorithms: GA, BPSO, WDO, BFOA and GWD. From the figure, it is evident that BPSO has the maximum computational burden (execution time = 88 s), and BFOA has the minimum computational burden (execution time = 8 s); a difference of 80 s. Similarly, GA, WDO and GWD take 13 s, 43 s and 32 s (to execute), respectively. The previous figures in the simulation Results Section show that GWD is relatively better than the compared algorithms in terms of the selected performance metrics, and Figure 19 shows the execution time of GWD as relatively moderate (better than WDO and worse than GA). To sum up, the GWD pays the cost of moderate execution time to achieve a considerable increase in UC and a decrease in both PAR and price.

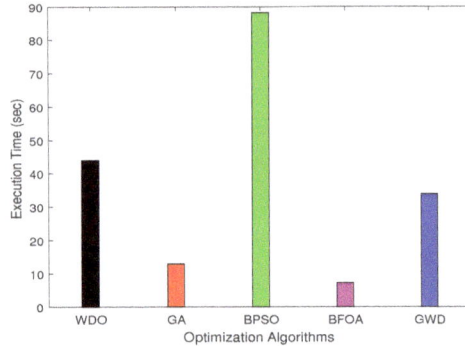

Figure 19. Execution time of GA, BPSO, WDO, BFOA and GWD.

5.3. Performance Trade-Offs in the Proposed Technique

After conducting the simulations, we have found some trade-offs and achievements. This approach is evaluated with the help of the following parameters: cost minimization, energy consumption minimization, UC maximization and PAR reduction. The achievements and trade-offs are mentioned in Table 16.

Table 16. Tradeoffs in the proposed algorithms.

Technique	Tariff Model	Achievement	Tradeoff
GA	RTP	Minimizes the cost up to 56% and reduces the PAR to 26% in individual testing and hybrid case cost is minimized up to 30% and PAR is reduced up to 49%	UC is compromised in scheduled case up to 60% in hybrid case while it is improved in individual testing to 90%
WDO	RTP	Reduces cost up to 67.18% and reduces the PAR to 26% in individual testing and hybrid case cost is minimized up to 30% PAR is 70% reduced	UC is compromised in scheduled case up to 60% in hybrid case and in individual testing to 50%
GWD	RTP	Reduces cost up to 17.87% and reduces the PAR to 26% in individual testing and hybrid case cost is minimized up to 30% PAR is 17% reduced	UC is compromised in scheduled case up to 60%
BPSO	RTP	Reduces cost up to 70% and reduces the PAR to 25%	UC is compromised up to 50%

5.4. Statistical Validation of GWD and Counter Part Algorithms Using ANOVA

In order to prove the metaheuristic algorithms' stochastic nature, we have done the statistical analysis for checking their correctness and efficiency. Two algorithms are taken for comparison with our proposed algorithm in terms of the variance. The ANOVA is based on three assumptions [44]: (i) all samples of the populations are normally distributed; (ii) all samples of the populations have equal variance; and (iii) all observations are mutually independent. In the table below, the analysis is described in detail for each sample population generated by the each individual algorithms.

Table 17. ANOVA results for the proposed algorithm with the existing algorithms.

Technique	Source of Variation	Sum of Squares	df	MS	F	Prob > F
WDO	Between Groups	1.4383	11	0.13075	0.48	0.9134
	Within Groups	29.5488	108	0.2736		
	Total	30.9871	119			
GA	Between Groups	3.058	11	0.27803	1.18	0.2956
	Within Groups	562.86	2388	0.2357		
	Total	565.918	2399			
GWD	Between Groups	0.6647	11	0.06043	0.61	0.813
	Within Groups	10.6203	108	0.09834		
	Total	11.285	119			

Here, df indicates the degrees of freedom; MS represents the mean square test; and F represents the F test (taken by dividing the sum of squares and MS); and these are calculated using the equations from [44]. We have done the ANOVA of three algorithms including our proposed algorithm. In this way, we have finally estimated that our proposed algorithm varies from them by a significant rate as shown in Table 17 above.

6. Conclusions

In this work, a DSM controller is designed in which five heuristic algorithms (GA, BPSO, WDO, BFOA and our proposed hybrid GWD) are implemented. The hybrid GWD scheme reduced the electricity cost by approximately 10% in comparison to GA and 33% to WDO. On the other hand, GA provided the global optimal solution in scheduling and faster convergence, even when the population size is large. The GA outperformed BPSO, WDO and BFOA in terms of electricity cost and energy consumption. In contrast to the BPSO, BFOA is suitable for a small population, because it converges at a faster rate when the population size is small. Explicit particle pressure values make WDO the slowest to converge among all of the compared algorithms. The stochastic behavior of these algorithms is analyzed by statistical analysis. Assigning priority to appliances helped with efficient scheduling. Statistical analysis is performed by the ANOVA test, which is used to measure the variation in the algorithms' performance metrics. In the future, we will focus on enhancing other heuristic algorithms to achieve the desired objectives.

Acknowledgments: This project was full financially supported by the King Saud University, through Vice Deanship of Research Chairs.

Author Contributions: Nadeem Javaid and Sakeena Javaid proposed and implemented the main idea. Wadood Abdul and Imran Ahmed performed the mathematical modeling and wrote the simulation section. Ahmad Almogren, Atif Alamri and Iftikhar Azim Niaz organized and refined the manuscript.

Conflicts of Interest: The authors declare no conflict of interest.

References

1. Logenthiran, T.; Srinivasan, D.; Shun, T.Z. Demand side management in smart grid using heuristic optimization. *IEEE Trans. Smart Grid* **2012**, *3*, 1244–1252.

2.   Palensky, P.; Dietrich, D. Demand side management: Demand response, intelligent energy systems, and smart loads. *IEEE Trans. Ind. Inform.* **2011**, *7*, 381–388.

3.   Shahidehpour, M.; Yamin, H.; Li, Z. Market overview in electric power systems. In *Market Operations in Electric Power Systems: Forecasting, Scheduling, and Risk Management*; Wiley-IEEE Press: Hoboken, NJ, USA, 2002; pp. 1–20.

4.   Popovic, Z.N.; Popovic, D.S. Direct load control as a market-based program in deregulated power industries. In Proceedings of the 2003 IEEE Bologna Power Tech Conference, Bologna, Italy, 23–26 June 2003; Volume 3, pp. 1–4.

5.   Maharjan, I.K. *Demand Side Management: Load Management, Load Profiling, Load Shifting, Residential and Industrial Consumer, Energy Audit, Reliability, Urban, Semi-Urban and Rural Setting*; LAP Lambert Academic: Saarbrücken, Germany, 2010.

6.   Gellings, C.W.; Chamberlin, J.H. *Demand Side Management: Concepts and Methods*; Fairmont: Liburn, GA, USA, 1988.

7.   Kothari, D.P.; Nagrath, I.J. *Modern Power System Analysis*; Tata McGraw-Hill Education: New Delhi, India, 2003.

8.   Awais, M.; Javaid, N.; Shaheen, N.; Iqbal, Z.; Rehman, G.; Muhammad, K.; Ahmad, I. An Efficient Genetic Algorithm Based Demand Side Management Scheme for Smart Grid. In Proceedings of the 2015 18th International Conference on Network-Based Information Systems (NBiS), Taipei, Taiwan, 2–4 September 2015; pp. 351–356.

9.   Ullah, I.; Javaid, N.; Khan, Z.A.; Qasim, U.; Khan, Z.A.; Mehmood, S.A. An Incentive-based Optimal Energy Consumption Scheduling Algorithm for Residential User. *Procedia Comput. Sci.* **2015**, *52*, 851–857.

10.   Rasheed, M.B.; Javaid, N.; Ahmad, A.; Khan, Z.A.; Qasim, U.; Alrajeh, N. An Efficient Power Scheduling Scheme for Residential Load Management in Smart Homes. *Appl. Sci.* **2015**, *5*, 1134–1163.

11.   Sousa, T.; Morais, H.; Vale, Z.; Faria, P.; Soares, J. Intelligent energy resource management considering vehicle-to-grid: A simulated annealing approach. *IEEE Trans. Smart Grid* **2012**, *3*, 535–542.

12.   Arabali, A.; Ghofrani, M.; Etezadi-Amoli, M.; Fadali, M.S.; Baghzouz, Y. Genetic algorithm based optimization approach for energy management. *IEEE Trans. Power Deliv.* **2013**, *28*, 162–170.

13.   Khan, M.A.; Javaid, N.; Mahmood, A.; Khan, Z.A.; Alrajeh, N. A generic demand side management model for smart grid. *Int. J. Energy Res.* **2015**, *39*, 954–964.

14.   Zhou, Y.; Chen, Y.; Xu, G.; Zhang, Q.; Krundel, L. Home energy management with PSO in smart grid. In Proceedings of the 2014 IEEE 23rd International Symposium on Industrial Electronics (ISIE), Istanbul, Turkey, 1–4 June 2014; pp. 1666–1670.

15.   Lugo-Cordero, H.M.; Fuentes-Rivera, A.; Guha, R.K.; Ortiz-Rivera, E.I. Particle swarm optimization for load balancing in green smart homes. In Proceedings of the 2011 IEEE Congress of Evolutionary Computation (CEC), New Orleans, LA, USA, 5–8 June 2011; pp. 715–720.

16.   Narendhar, S.; Amudha, T. A Hybrid Bacterial Foraging Algorithm for Solving Job Shop Scheduling Problems. *Int. J. Program. Lang. Appl. (IJPLA)* **2012**, *2*, 1–11.

17.   Chana, I. Bacterial foraging based hyper-heuristic for resource scheduling in grid computing. *Future Gener. Comput. Syst.* **2013**, *29*, 751–762.

18.   Molderink, A.; Bakker, V.; Bosman, M.G.; Hurink, J.L.; Smit, G.J. Domestic energy management methodology for optimizing efficiency in smart grids. In Proceedings of the 2009 IEEE Bucharest PowerTech, Bucharest, Romania, 28 June–2 July 2009; pp. 1–7.

19.   Soares, J.; Sousa, T.; Morais, H.; Vale, Z.; Faria, P. An optimal scheduling problem in distribution networks considering V2G. In Proceedings of the 2011 IEEE Symposium on Computational Intelligence Applications In Smart Grid (CIASG), Paris, France, 11–15 April 2011; pp. 1–8.

20.   Zhu, Z.; Tang, J.; Lambotharan, S.; Chin, W.H.; Fan, Z. An integer linear programming based optimization for home demand side management in smart grid. In Proceedings of the 2012 IEEE PES Innovative Smart Grid Technologies (ISGT), Washington, DC, USA, 16–20 January, 2012; pp. 1–5.

21.   Kriett, P.O.; Salani, M. Optimal control of a residential microgrid. *Energy* **2012**, *42*, 321–330.

22.   Wang, J.; Sun, Z.; Zhou, Y.; Dai, J. Optimal dispatching model of smart home energy management system. In Proceedings of the IEEE PES Innovative Smart Grid Technologies, Tianjin, China, 21–24 May 2012; pp. 1–5.

23.   Maringer, D.G. *Portfolio Management With Heuristic Optimization*; Springer Science and Business Media; Springer: New York, NY, USA, 2006.

24. Ullah, I.; Javaid, N.; Imran, M.; Khan, Z.A.; Qasim, U.; Alnuem, M.; Bashir, M. A Survey of Home Energy Management for Residential Customers. In Proceedings of the 2015 IEEE 29th International Conference on Advanced Information Networking and Applications, Guwangiu, Korea, 24–27 March 2015; pp. 666–673.
25. Wen, Z.; O'Neill, D.; Maei, H. Optimal demand response using device-based reinforcement learning. *IEEE Trans. Smart Grid* **2015**, *6*, 2312–2324.
26. Adika, C.O.; Wang, L. Smart charging and appliance scheduling approaches to demand side management. *Int. J. Electr. Power Energy Syst.* **2014**, *57*, 232–240.
27. Shirazi, E.; Jadid, S. Optimal residential appliance scheduling under dynamic pricing scheme via HEMDAS. *Energy Build.* **2015**, *93*, 40–49.
28. Mahmood, D.; Javaid, N.; Alrajeh, N.; Khan, Z.A.; Qasim, U.; Ahmed, I.; Ilahi, M. Realistic Scheduling Mechanism for Smart Homes. *Energies* **2016**, *9*, 202.
29. Hernández-Ocana, B.; Mezura-Montes, E.; Pozos-Parra, P. A review of the bacterial foraging algorithm in constrained numerical optimization. In Proceedings of the 2013 IEEE Congress on Evolutionary Computation, Cancun, Mexico, 20–23 Junuary 2013; pp. 2695–2702.
30. Jiménez, J.B. Electricity demand modeling for rural residential housing: A case study in Colombia. In Proceedings of the 2015 IEEE PES Innovative Smart Grid Technologies Latin America (ISGT LATAM), Montevideo, Uruguay, 5–7 October 2015; pp. 614–618.
31. Rottondi, C.; Barbato, A.; Chen, L.; Verticale, G. Enabling Privacy in a Distributed Game-Theoretical Scheduling System for Domestic Appliances. *IEEE Trans. Smart Grid* **2016**, *PP*, 1–11, doi:10.1109/TSG.2015.2511038.
32. Erol-Kantarci, M.; Mouftah, H.T. Energy-efficient information and communication infrastructures in the smart grid: A survey on interactions and open issues. *IEEE Commun. Surv. Tutor.* **2015**, *17*, 179–197.
33. Safdarian, A.; Fotuhi-Firuzabad, M.; Lehtonen, M. Optimal Residential Load Management in Smart Grids: A Decentralized Framework. *IEEE Trans. Smart Grid* **2016**, *7*, 1836–1845.
34. Liu, Y.; Yuen, C.; Yu, R.; Zhang, Y.; Xie, S. Queuing-based energy consumption management for heterogeneous residential demands in smart grid. *IEEE Trans. Smart Grid* **2016**, *7*, 1650–1659.
35. Ma, J.; Chen, H.H.; Song, L.; Li, Y. Residential load scheduling in smart grid: A cost efficiency perspective. *IEEE Trans. Smart Grid* **2016**, *7*, 771–784.
36. Barbato, A.; Capone, A. Optimization models and methods for demand-side management of residential users: A survey. *Energies* **2014**, *7*, 5787–5824.
37. Fernandes, F.; Morais, H.; Vale, Z.; Ramos, C. Dynamic load management in a smart home to participate in demand response events. *Energy Build.* **2014**, *82*, 592–606.
38. Soares, J.; Ghazvini, M.A.F.; Borges, N.; Vale, Z. A stochastic model for energy resources management considering demand response in smart grids. *Electr. Power Syst. Res.* **2017**, *143*, 599–610.
39. Rahman, A.; Liu, X.; Kong, F. A survey on geographic load balancing based data center power management in the smart grid environment. *IEEE Commun. Surv. Tutor.* **2014**, *16*, 214–233.
40. Chiu, W.Y.; Sun, H.; Poor, H.V. Energy imbalance management using a robust pricing scheme. *IEEE Trans. Smart Grid* **2013** *4*, 896–904.
41. Chiu, W.Y.; Sun, H.; Poor, H.V. A multi-objective approach to multimicrogrid system design. *IEEE Trans. Smart Grid* **2015**, *6*, 2263–2272.
42. Wang, L.; Wang, Z.; Yang, R. Intelligent multiagent control system for energy and comfort management in smart and sustainable buildings. *IEEE Trans. Smart Grid* **2012**, *3*, 605–617.
43. Electricity Tariff. Available online: http://www.lesco.gov.pk/3000063 (accessed on 2 April 2016).
44. One-Way Analysis of Variance (ANOVA) Example Problem. Available online: http://cba.ualr.edu/smartstat/topics/anova/example.pdf (accessed on 20 December 2016).

MDPI

St. Alban-Anlage 66

4052 Basel

Switzerland

Tel. +41 61 683 77 34

Fax +41 61 302 89 18

www.mdpi.com

Energies Editorial Office

E-mail: energies@mdpi.com

www.mdpi.com/journal/energies

www.ingramcontent.com/pod-product-compliance
Lightning Source LLC
Chambersburg PA
CBHW051714210326
41597CB00032B/5473